Fire/Explosion Risk Assessment and Loss Prevention of Hazardous Materials, Mines and Natural Gas

Fire/Explosion Risk Assessment and Loss Prevention of Hazardous Materials, Mines and Natural Gas

Guest Editors

Chuyuan Huang
Haipeng Jiang
Lijuan Liu

MDPI

Basel • Beijing • Wuhan • Barcelona • Belgrade • Novi Sad • Cluj • Manchester

Guest Editors
Chuyuan Huang
School of Safety Science and Emergency Management
Wuhan University of Technology
Wuhan
China

Haipeng Jiang
Department of Chemical Machinery and Safety Engineering
Dalian University of Technology
Dalian
China

Lijuan Liu
School of Safety Science and Emergency Management
Wuhan University of Technology
Wuhan
China

Editorial Office
MDPI AG
Grosspeteranlage 5
4052 Basel, Switzerland

This is a reprint of the Special Issue, published open access by the journal *Fire* (ISSN 2571-6255), freely accessible at: https://www.mdpi.com/journal/fire/special_issues/5MF7YK71ZF.

For citation purposes, cite each article independently as indicated on the article page online and as indicated below:

Lastname, A.A.; Lastname, B.B. Article Title. *Journal Name* **Year**, *Volume Number*, Page Range.

ISBN 978-3-7258-3557-7 (Hbk)
ISBN 978-3-7258-3558-4 (PDF)
https://doi.org/10.3390/books978-3-7258-3558-4

Cover image courtesy of Chuyuan Huang

Contents

About the Editors

Chuyuan Huang

Chuyuan Huang graduated from the Wuhan University of Technology with a Ph.D. in Safety Science and Engineering. He is currently an Associate Professor and Master's Supervisor at the School of Safety Science and Emergency Management, Wuhan University of Technology. He has presided over several research projects, including sub-projects of the National Key Research and Development Program, the National Natural Science Foundation of China and the Natural Science Foundation of Guangdong Province. As the first author or corresponding author, he has published more than 20 papers in high-level journals, one of which has been selected as an ESI Highly Cited Paper. He has been awarded the Outstanding Doctoral Dissertation Prize by the China Fire Protection Association, the First Prize for Science and Technology by the China Occupational Safety and Health Association, and the Second Prize for Safety Science and Technology Progress by the China Association of Work Safety.

Haipeng Jiang

Haipeng Jiang graduated from Dalian University of Technology with a Ph.D. in Chemical Process Machinery. He is currently an Associate Professor and Ph.D. supervisor at the School of Chemical Engineering, Dalian University of Technology. He has led various research projects, including sub-projects of the National Key Research and Development Program, as well as projects funded by the National Natural Science Foundation of China. As the first author or corresponding author, he has published 47 papers in high-impact journals, one of which has been selected as an ESI Highly Cited Paper and an ESI Hot Paper. He has received the Outstanding Doctoral Dissertation Award from the China Fire Protection Association and the First Prize in the Liaoning Provincial Science and Technology Progress Award.

Lijuan Liu

Lijuan Liu graduated with a bachelor's and Ph.D. degree from the Beijing Institute of Technology and is currently working as an Associate Professor and Master's Supervisor at the School of Safety Science and Emergency Management, Wuhan University of Technology. She has extensive experience as a researcher in sub-projects of the National Key R&D Program and the National Natural Science Foundation of China, the Natural Science Foundation of Guangdong Province, etc. She has been nominated for the Outstanding Doctoral Dissertation Prize in Armament Science and Technology by the China Ordnance Society and has has been awarded First Prize for Science and Technology by the China Occupational Safety and Health Association.

Preface

We are delighted to present this reprint of the Special Issue "Fire/Explosion Risk Assessment and Loss Prevention of Hazardous Materials, Mines and Natural Gas". This collection of research articles addresses the critical challenges and advancements in understanding, assessing, and mitigating risks associated with fire, explosions, and hazardous materials in industrial and natural environments. The scope of this reprint encompasses a wide range of topics, including risk assessment methodologies, safety management strategies, technological innovations, and case studies from mining operations, natural gas infrastructure, and hazardous material handling.

The motivation behind this Special Issue stems from the urgent need to enhance safety standards and reduce the potential for catastrophic incidents in industries that handle hazardous materials or operate in high-risk environments. With the increasing complexity of industrial processes and the growing demand for energy resources, the importance of proactive risk management cannot be overstated. This reprint aims to provide a comprehensive resource for researchers, industry professionals, and policymakers, offering insights that can inform safer practices and policies.

We extend our heartfelt gratitude to the contributing authors, whose expertise and dedication have made this reprint possible. Their work reflects the latest advancements in the field and underscores the importance of interdisciplinary collaboration. We also acknowledge the invaluable support of the reviewers, whose constructive feedback ensured the high quality of the published articles. Finally, we thank the editorial team for their guidance and assistance throughout the process.

This reprint is intended for a diverse audience, including academics, safety engineers, risk assessment professionals, and industry stakeholders. We hope that the findings and recommendations presented here will inspire further research and contribute to the development of safer, more sustainable industrial practices.

Chuyuan Huang, Haipeng Jiang, and Lijuan Liu
Guest Editors

fire

Article

Study on the Characteristics and Influence Factor of Methane and Coal Dust Gas/Solid Two-Phase Mixture Explosions

Yue Wang [1], Zhi Wang [2], Xingyan Cao [2,*] and Haoyue Wei [2]

[1] Key Laboratory of Coal Resources and Green Mining in Xinjiang, Ministry of Education, Xinjiang Institute of Engineering, Urumqi 830023, China
[2] College of Safety Science and Engineering, Nanjing Tech University, Nanjing 210009, China
* Correspondence: caoxingyan_007@163.com

Abstract: This research aimed to the characteristics and influence factor of methane and coal dust gas/solid two-phase mixture explosions by experiment. Through comparative analysis of flame propagation characteristics, pressure, flame temperature and products, the characteristics of gas/solid explosions and its influence factor were analyzed. And the influence mechanism was also revealed. Results indicate that the coal dust parameter and methane concentration were the important influence factor on mixture explosions. Explosion intensity could be indirectly affected by influencing the flame propagation. Under the determined coal dust parameter, the explosion parameter showed a change trend of increase firstly and then decrease as the methane concentration increased. And it was the greatest at 6% methane concentration. However, the concentration of coal dust corresponding to the maximum pressure was variable and was decreased successively as the methane concentration increased. The corresponding dust concentrations were 500 g/m^3 and 200 g/m^3 under 2% and 10% methane concentrations, respectively. Meanwhile, the pressure all presented an increasing trend with the reduction of coal dust diameter under five coal dust concentrations, and the explosion intensity was the greatest at 300 g/m^3 coal dust concentration. For 2% methane concentration, the explosion would not occur as the dust concentration was less than 400 g/m^3. And the same phenomena also appeared as the methane concentration exceeded 10%. The explosion parameter presented the same change trend with the changes of methane concentration and coal dust parameters. Besides, the thermal stability and decomposition oxidation characteristics of burned coal dust were evidently changed compared with unburned coal dust. The weight loss rate and oxidation reaction rate were decreased, and the corresponding temperature was increased. It indicates that coal dust participated in gas/dust two-phase explosion reactions, and the pyrolysis reaction of volatile matter led to an obvious reduction in the weight loss and oxidation reaction rate. And the precipitation of volatile matter also resulted in an obvious pore structure on its surface. The physical parameters and internal components of coal dust were important factors affecting the reaction rates of gas/dust mixture explosions.

Keywords: gas/dust two phases; explosion characteristics; influence factor; flame temperature; product analysis

Citation: Wang, Y.; Wang, Z.; Cao, X.; Wei, H. Study on the Characteristics and Influence Factor of Methane and Coal Dust Gas/Solid Two-Phase Mixture Explosions. *Fire* **2023**, *6*, 359. https://doi.org/10.3390/fire6090359

Academic Editors: Chuyuan Huang, Haipeng Jiang and Lijuan Liu

Received: 22 August 2023
Revised: 8 September 2023
Accepted: 11 September 2023
Published: 15 September 2023

1. Induction

The explosion accidents have become more frequent and severe with the expeditious development of economy and ceaseless progress of social, and it is more prone to a compound explosion consisting of two combustible substances, gas and dust [1–3]. The mechanism of gas/dust two-phase mixture explosions become more complex and the explosion power is greater, seriously causing the enormous economic loss and greatly restricting the national economy development [4,5]. In view of the industrial disaster, the explosion characteristics of gas/dust two-phase mixture should be researched in-depth [6–9]. Further, the effective protective method can be taken to avoid derivative accidents and disasters [10–13].

Relevant scholars have conducted the study on the characteristics of gas/dust two-phase mixture explosions. Kundu [14] proposed that the two-phase mixtures were easier to ignite than any single component as methane and coal dust coexisted and resulted in the more serious consequences. Benedetto [15] pointed out that the lower explosive limit of coal dust could be obviously decreased after adding methane and its flame propagation velocity was increased. Ji [16] also studied the influence of methane content on the lower explosive limit and found that the value was evidently reduced and the explosion intensity was enhanced after adding methane. Zhao [17] also pointed out that the addition of methane could increase the explosion pressure of coal dust and its reaction rate was also enhanced with the increase of methane content. Chen [18] researched the effect of coal dust concentration on gas/dust mixture explosions and proposed that the temperature and propagation velocity of explosion flame showed a change trend of increase firstly and then decrease with the rise of coal dust content. Jing [19] also proposed that the flame propagation velocity showed the same variation law with the increase of coal dust content. And it was also pointed out that the higher volatile content would result in the enhancement of mixture explosions. Cloney [20] found that the particle size of coal dust had an important impact on the velocity history of mixture explosions. And it was predicted by the model that the velocity was significantly reduced as the coal dust particle size exceeded a critical value. Gao [21] found that the lower explosive limit could be reduced as the coal dust particle size decreased, and the explosion risk caused by coal dust was enhanced.

Meanwhile, scholars also studied the impact of working condition on gas/dust mixture explosions. Wang [22] pointed out that the ignition delay time had an evidently effect on mixture explosions and the explosion intensity showed a trend of increase firstly and then decrease with the rise of ignition delay time under 9.5% methane concentration. Deng [23] found that ignition energy had also an evidently influence on the mixture explosion pressures, and the influence of ignition energy was more obvious under the smaller energy. Besides, Zhou [24] studied the impact of obstacles on the flame propagation process of methane and coal dust mixture explosions, and found that obstacles could significantly accelerate the flame propagation and increase its propagation velocity inside a closed pipeline. And the obstacle with the sharper corner would cause more violent explosion.

Totally, the research on gas/dust mixture explosion characteristics had been conducted, and mainly focused on the explosion pressure and its influence factor [25,26]. Nevertheless, it is rare to research the change law and influence mechanism under multi-parameter conditions based on the combination of gas/dust mixture explosion parameter and flame propagation characteristics. Especially for the changes in the flame temperature and products [27]. Comprehensive analysis of flame propagation and explosion parameters could deeply reveal the gas/dust two-phase mixture explosion characteristics and its influence mechanism. Hence, the main purpose of this experiment was to study the parameter of methane and coal dust mixture explosions (the pressure peak (ΔP_{max}), pressure rising rate (dP/dt), products and flame temperature (T)) and flame propagation characteristics (especially for the changes in flame morphology and its propagation velocity (v)) under multi-parameter operating conditions. According to the comprehensive analysis of explosion parameters, the characteristics of gas/dust mixture explosions and its influence mechanism can be analyzed in depth.

2. Experimental Setup

Figure 1 shows a schematic diagram of experiment apparatus for studying the characteristics and influence mechanism of gas/dust mixture explosions [28,29]. It mainly included eight subsystems. The main structure had been introduced in the previous work [30], and only part of the structure was slightly modified. The powder injection system consisted of a Hartmann powder spraying device, a globe valve and a solenoid valve. The Hartmann powder spraying device was installed at the center of the bottom flange through a threaded connection [31]. The premix tank was sequentially connected with a globe valve and a solenoid valve by the bottom end tube bundle. The coal dust was placed in the groove of the Hartmann powder spraying device. A bottom-up vortex was formed as the high-speed premixed gas flow passed through the Hartmann powder spraying device from the bottom tube bundle due to the obstruction of the hemispherical "mushroom cap", lifting the dust in the groove and forming a dust cloud. The powder spraying method could achieve the relatively uniform dust cloud. The mixture gas of methane (Purity: 99.99%) and air was prepared inside the premixed tank by the partial pressure method and the determined vacuum degree inside the vessel was achieved by a vacuum pump. The powder injection pressure and time were determined by the calculation and ensured the methane concentration inside the vessel after power spraying. And the initial pressure was normal. The ignition system was composed of a high voltage converter, an AC contactor, a solid-state relay and an ignition electrode. The ignition electrode was placed at a central position with 110 mm height distance from the bottom flange (ignition voltage (U) = 8 kV). A high-speed camera (Photron V1212 (Japan)) was adopted to collect the flame propagation process of gas/dust mixture explosions. And flame propagation velocity history can be computed by the derivation of flame front position over time. To ensure the accuracy of flame propagation velocity, the acquisition frequency was set to 2000 fps. Namely, the change in flame front for 0.5 ms. A high-frequency pressure transmitter (Accuracy: 0.5% FS; Acquisition frequency: 200 Hz) and a high-speed thermocouple (S-type platinum rhodium; Range: 0–1800 °C) were installed at the center of vessel sidewall to record the pressure and flame temperature of mixture explosions [32]. The order of powder injection and ignition, as well as data collection, were realized through a program control and data collection system with a single channel of 2 M/s [33].

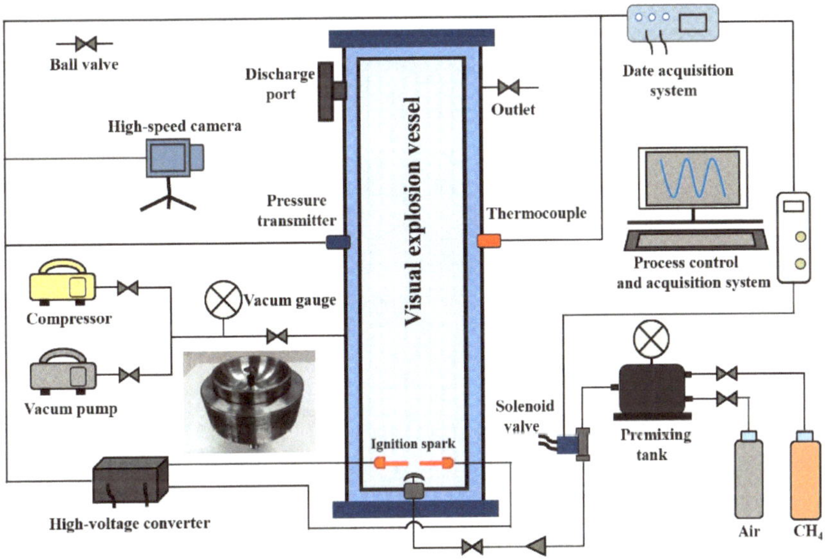

Figure 1. Schematic diagram of gas/dust mixture explosion experiment apparatus.

Before the experiment, the large coal block was pulverized by a grinder. And the coal dust with different particle size range was obtained by changing the screen mesh of different pore diameter. Three kinds of coal dust diameters (d_{50} = 35.6 μm, 76.6 μm and 141.4 μm), five kinds of coal dust concentrations (q = 100 g/m^3, 200 g/m^3, 300 g/m^3,

400 g/m^3 and 500 g/m^3) and five kinds of methane concentrations (c = 2%, 4%, 6%, 8% and 10%) were selected for studying mixture explosion experiments. Figure 2 shows the columnar distribution and microstructure of three coal dust diameters. The morphology of the three types of coal dusts presented a uniform state and their diameter histories were approximately normal distribution [34]. Meanwhile, the component content of coal dust was measured by industrial analysis (M_{ad} = 8.8%, A_{ad} = 6.3%, V_{ad} = 42.8%, FCad = 42.1%). Simultaneously, the pyrolysis characteristics of coal dust under the high temperature with a heating rate of 10 K/min was tested. It underwent three stages: the heat absorption, weight loss and stability. According to the comparison of explosion experiments under different powder injection pressures (P_{st}), the pressure was maximum as the P_{st} was 0.40 MPa. Hence, 0.40 MPa was selected as the powder injection pressure during experiments. Each condition was repeated 3–4 times to guarantee the experimental accuracy. The data deviation was an important parameter characterizing experimental repeatability and it was calculated by the ΔP_{max} and the average value of repetitive experiments. During this experiment, the deviation value of ΔP_{max} was less than 5.5% through the data analysis.

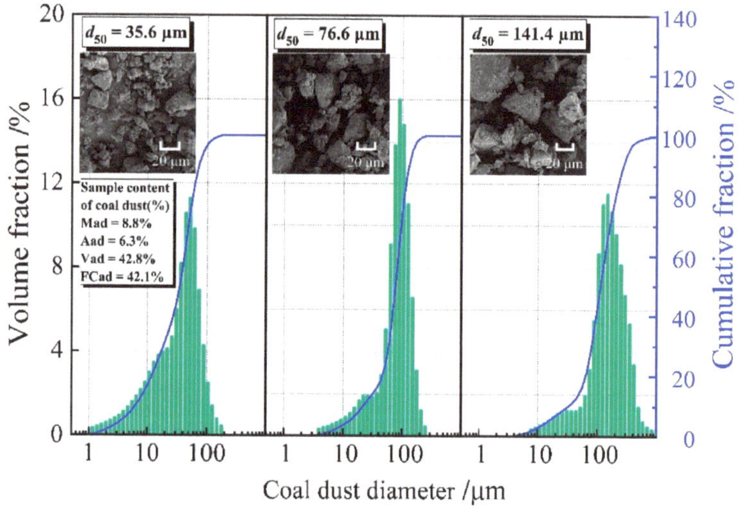

Figure 2. Columnar distribution of coal dust diameter.

3. Results and Discussion

3.1. Flame Propagation Characteristics

Figure 3 shows the comparison of mixture explosion flame propagation processes under five coal dust concentrations (d_{50} = 76.6 μm; c = 8%). The height of flame front at the same moment showed a change trend of increase firstly and then decrease as the coal dust concentration increased. As q = 300 g/m^3, the flame front was the highest and brightest. The initial flame exhibited an irregular structure [35]. Meanwhile, the flame got dark and the irregular flame became more significant with the increase and decrease of coal dust concentration. As 100 g/m^3 < q < 300 g/m^3, the quantity of coal dust per unit space was gradually increased. The heat released by methane explosion could be effectively absorbed by coal dust. Subsequently, the explosion intensity was increased due to the heat release from coal dust combustion. As the flame propagated, the flame front became the smooth and developed upward in an "elliptical" structure [36,37]. As q = 300 g/m^3, the heat release of gas/dust two phase reactions after heat absorption resulted in that the explosion reaction was the most severe and the flame was the brightest. As 400 g/m^3 < q < 500 g/m^3, the brightness was successively darkened and the flame began to become irregular. Especially the change was more significant as q = 500 g/m^3. It was because that the reaction heat of methane explosion could be evidently aborted by the coal dust of higher concentration per unit space. And the coal dust could not sufficiently occur the explosion reaction due to the limited oxygen concentration inside the confined space. Meanwhile, the combustion products could greatly consume the explosion reaction heat, leading to an evident reduction in flame brightness. And the flame front exhibited an evident wrinkle and instability. This explains that the dust concentration had a significant impact on the flame structure and its propagation characteristics of gas/dust mixture explosions [38].

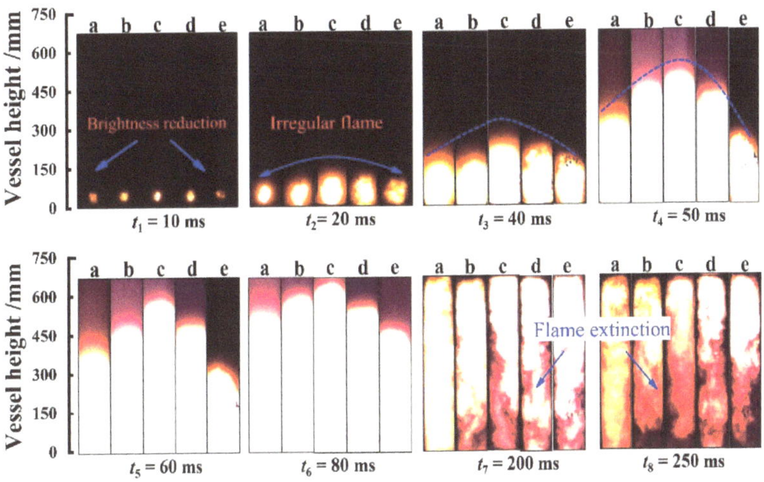

Figure 3. Comparison of flame propagation processes under five coal dust concentrations (a = 100 g/m^3, b = 200 g/m^3, c = 300 g/m^3, d = 400 g/m^3 and e = 500 g/m^3).

Figure 4 shows the comparison of velocity histories under different coal dust concentrations (d_{50} = 76.6 μm; c = 8%). The velocity history and its rising rate appeared a change trend of increase firstly and then decrease with the increase of coal dust concentration. As q = 300 g/m^3, the corresponding velocity value was largest and the required time for reaching the top end was the shortest. The $v_{\max}$ was evidently reduced from 17.7 to 12.9 m/s and 14.6 m/s with the change of coal dust concentration. Especially $v_{\max}$ was the smallest as q = 500 g/m^3. Figure 5 presents the effect of coal dust diameter on the velocity history of mixture explosions (c = 8%; q = 300 g/m^3). The velocity history was gradually reduced with the increase of coal dust diameter. In particular, $v_{\max}$ was reduced from 20.1 to 17.1 m/s and its corresponding moment was also obviously delayed. It was because that the larger coal dust possessed a smaller specific surface area, further resulting in a relatively smaller contact area with oxygen and exhibited a slower explosion reaction rate and a darker flame structure [39].

Figure 6 illustrates the influence of methane concentration on the velocity history of mixture explosions (d_{50} = 76.6 μm; q = 300 g/m^3). As the methane concentration increased, the velocity history was also appeared an alteration trend of increase firstly and then decrease. Under 6% methane concentration, the velocity value was the maximal ($v_{\max}$ = 21.5 m/s) and the corresponding moment of $v_{\max}$ was the smallest. With the increase and decrease of methane concentration, the velocity history was evidently reduced. In particular, the velocity history was smallest ($v_{\max}$ = 12.4 m/s) and the corresponding moment of $v_{\max}$ was increased from 44 to 82 ms as c = 10%. This was because that the methane and coal dust could not obtain sufficient oxygen during the explosion reaction as the methane concentration was increased to 10%, resulting in the inability of the heat released by the mixture system to maintain a high flame propagation velocity. Based on the Figures 4–6, the methane concentration and coal dust parameters could evidently affect the flame propagation characteristics of gas/dust two-phase mixture explosions.

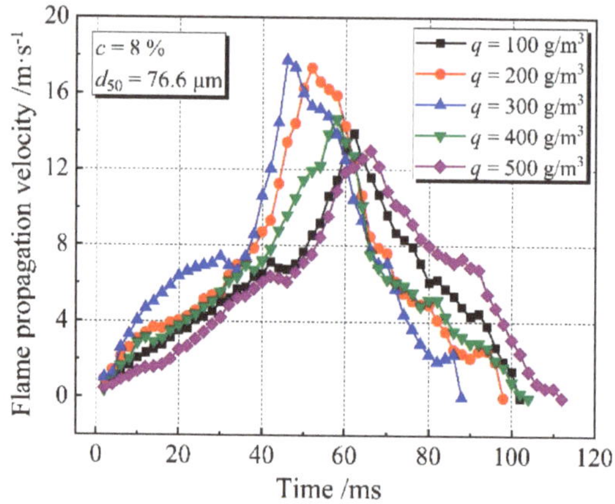

Figure 4. Comparison of velocity histories under five coal dust concentrations (d_{50} = 76.6 μm; c = 8%).

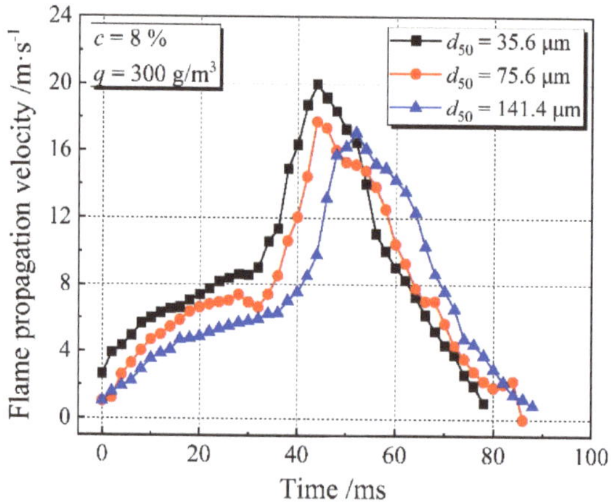

Figure 5. Comparison of velocity histories under three coal dust diameters (c = 8%; q = 300 g/m^3).

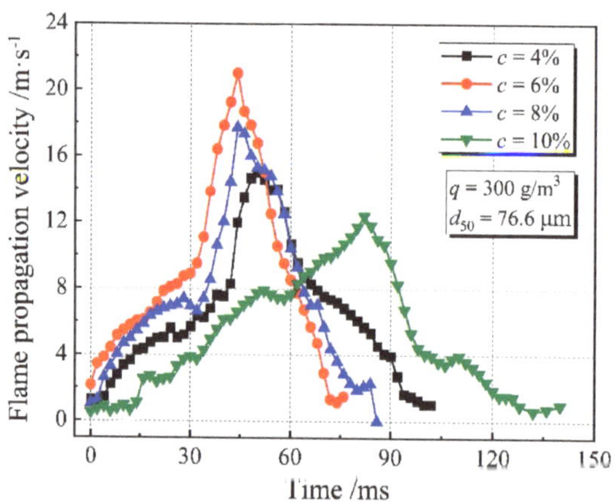

Figure 6. Comparison of velocity histories under five methane concentrations (d_{50} = 76.6 μm; q = 300 g/m^3).

3.2. Explosion Pressure

Figure 7 shows the comparison of mixture explosion pressure histories under five coal dust concentrations (d_{50} = 76.6 μm; c = 8%). The ΔP_{max} appeared a change trend of increase firstly and then decrease with the increase of coal dust concentration. The pressure history was maximal as q = 300 g/m^3 (ΔP_{max} = 0.61 MPa). With the increase and decrease of coal dust concentration, the pressure was significantly decreased. In particular, the ΔP_{max} was respectively reduced to 0.57 MPa and 0.52 MPa as q = 100 g/m^3 and 500 g/m^3. Meanwhile, the concentration of coal dust corresponding to the maximum pressure was decreased in turn as the methane concentration was increased. As c = 2% and 10%, the concentrations of coal dust corresponding to the maximum pressure were 500 g/m^3 and 200 g/m^3, respectively. Especially for 2% methane concentration, the mixture explosions would not occur as q < 400 g/m^3. This indicates that the coal dust content in the premixed gas of methane/air had a significant influence on the intensity of gas/dust two-phase mixture explosions.

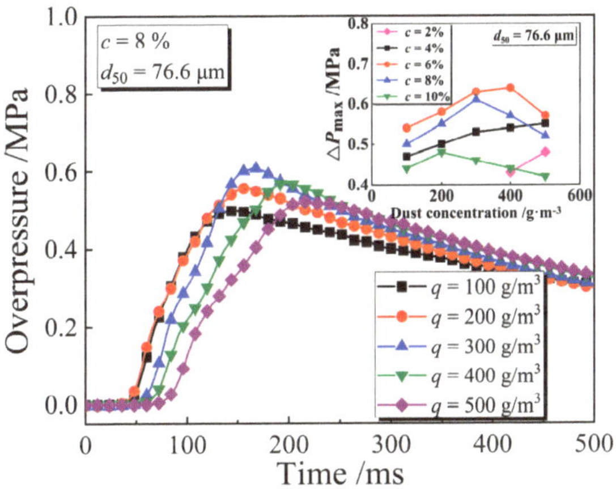

Figure 7. Comparison of pressure histories under different coal dust concentrations (d_{50} = 76.6 μm; c = 8%).

Figure 8 presents the comparison of mixture explosion pressure histories under three coal dust diameters (q = 300 g/m^3; c = 8%). The pressure was evidently reduced as the coal dust diameter increased. In particular, ΔP_{max} was decreased by 3.17% and 11.11%, respectively. As d_{50} = 141.4 μm, the pressure value was the smallest (ΔP_{max} = 0.56 MPa). Meanwhile, ΔP_{max} all appeared a decrease change trend as the coal dust diameter increased. Besides, the intensity of mixture explosions was maximal as q = 300 g/m^3. It also indicates that not only the concentration of coal dust could affect the explosion intensity, but also its specific surface area also could evidently affect the gas/dust two-phase mixture explosions. Figure 9 presents the comparison of mixture explosion pressure histories under different methane concentrations (q = 300 g/m^3; d_{50} = 76.6 μm). The pressure history also showed a trend of increase firstly and then decrease as the methane concentration increased. As c = 6%, ΔP_{max} was the maximal (ΔP_{max} = 0.63 MPa). With the increase and decrease of methane concentration, ΔP_{max} presented an obvious decrease trend. And the pressure all showed the same change trend under five coal dust concentrations. Especially for 2% methane concentration, the mixture explosions would not occur as q < 400 g/m^3. And the same phenomenon also occurred as c > 10%. It explains that the concentrations of combustible gas and dust could evidently affect the explosive limit of gas/dust two phases.

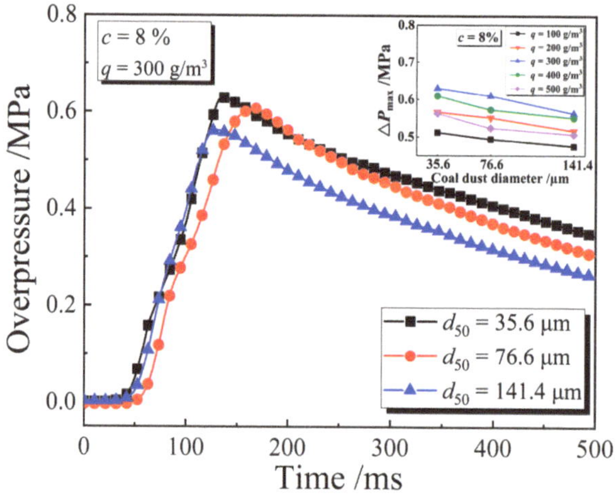

Figure 8. Comparison of pressure histories under different coal dust diameters (c = 8%; q = 300 g/m^3).

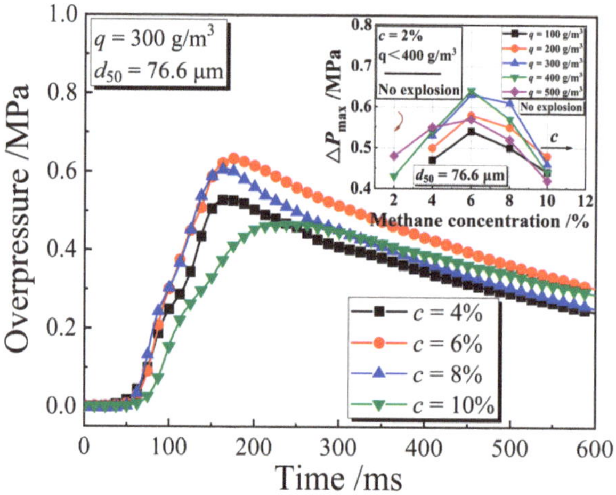

Figure 9. Comparison of pressure histories under different methane concentrations (d_{50} = 76.6 μm; q = 300 g/m^3).

3.3. Explosion Pressure Rising Rate

Figure 10 illustrates the comparison of (dP/dt) histories under five coal dust concentrations (c = 8%; d_{50} = 76.6 μm). It clear that the (dP/dt) history exhibited the same change trend as the pressure history. As the coal dust concentration increased, (dP/dt) history of mixture explosions also showed a change trend of increase firstly and then decrease. Meanwhile, the (dP/dt) history appeared two peaks (I and II peaks) and the I peak value was larger (($dP/dt)_{max}$). The two peak values were maximal as q = 300 g/m^3. With the decrease and increase of coal dust concentration, the two peak values showed a decrease trend and the decrease degree of II peak value was more significant. Besides, the coal dust concentration corresponding to $(dP/dt)_{max}$ of mixture explosions was decreased sequentially as the methane concentration increased under the determined coal dust diameter. As c = 4% and 10%, the corresponding coal dust concentrations were 500 g/m^3 and 200 g/m^3 (($dP/dt)_{max}$ = 8.4 MPa/s^{-1} and 6.1 MPa/s^{-1}), respectively.

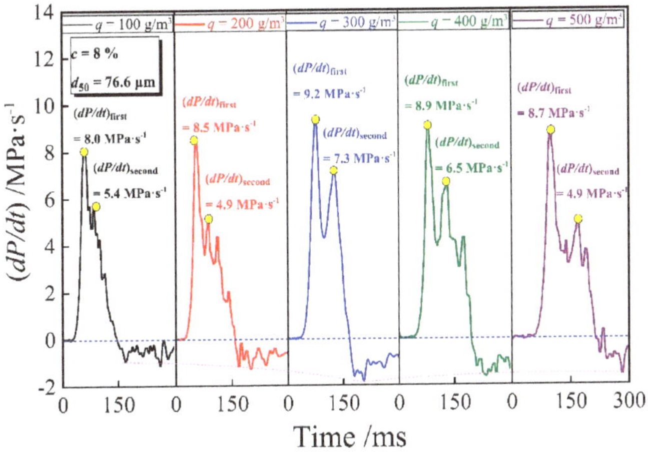

Figure 10. Comparison of (dP/dt) histories under different coal dust concentrations (d_{50} = 76.6 μm; c = 8%).

Figure 11 presents the comparison of mixture explosion (dP/dt) histories under three coal dust diameters (c = 8%; q = 300 g/m^3). As the coal dust diameter increased, the I and II peak values of (dP/dt) history presented an obvious decrease trend. In particular, the decrease extent of the II peak value was more significant and it was reduced by 30.39%. Meanwhile, $(dP/dt)_{max}$ all appeared a decrease trend and the maximum reduction extent reached 46.74% with the increase of coal dust diameter as c = 8%. Figure 12 shows the comparison of mixture explosion (dP/dt) histories under different methane concentrations (q = 300 g/m^3; d_{50} = 76.6 μm). It clear that the two peak values of (dP/dt) history showed a trend of increase firstly and then decrease as the methane concentration was raised. As c = 6%, the pressure rising rate was maximal. In addition, the $(dP/dt)_{max}$ were decreased by 25.49% and 33.33% under 4% and 6% methane concentrations. As d_{50} = 76.6 μm, the pressure rising rates corresponding to 6% and 10% methane concentrations were the maximal and minimal values, respectively. The above explains that the reaction rate of gas/dust two-phase mixture explosions could be evidently affected by the methane content, coal dust diameter and its concentration, thereby affecting the explosion intensity.

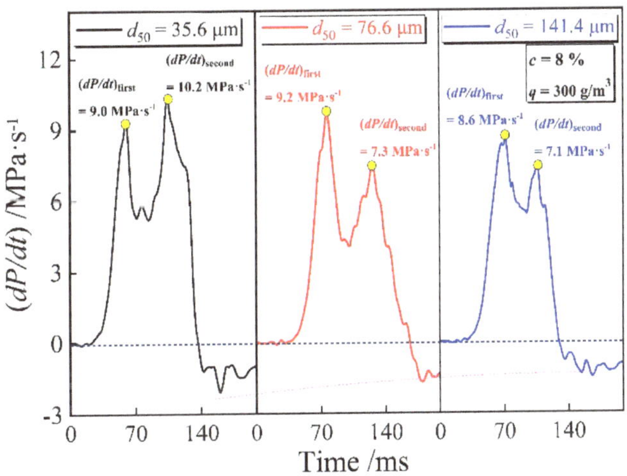

Figure 11. Comparison of (dP/dt) histories under different coal dust diameters (c = 8%; q = 300 g/m^3).

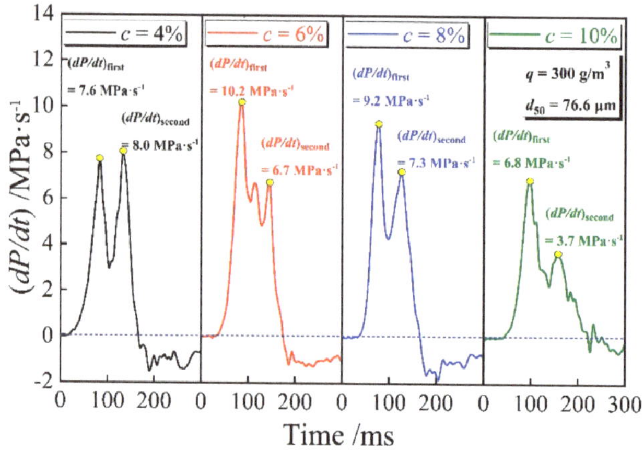

Figure 12. Comparison of (*dP/dt*) histories under different methane concentrations (d_{50} = 76.6 μm; q = 300 g/m^3).

3.4. Flame Temperature

Figure 13 is the comparison of flame temperature histories under five coal dust concentrations (c = 8%; d_{50} = 76.6 μm). The temperature also showed the same change trend as the pressure history with the increase of coal dust concentration. Especially the flame temperature was the highest and T_{max} reached 1687 °C as q = 300 g/m^3. For 100 g/m^3 and 500 g/m^3 methane concentrations, T_{max} were respectively reduced to 1315 °C and 1438 °C and their decrease extent were 22.05% and 14.76%. As c = 8%, T_{max} of mixture explosions all occurred at q = 300 g/m^3 under three dust diameters. Meanwhile, T_{max} presented an obvious decrease trend with the decrease and increase of coal dust concentration. Figure 14 presents the comparison of mixture explosion flame temperature histories under three coal dust diameters (c = 8%; q = 300 g/m^3). T_{max} presented an obvious decrease trend (from 1781 to 1569 °C) and the slope of temperature history was also decreased as the coal dust diameter was increased. In addition, the influence of coal dust diameter on T_{max} all showed a same change trend under five coal dust concentrations.

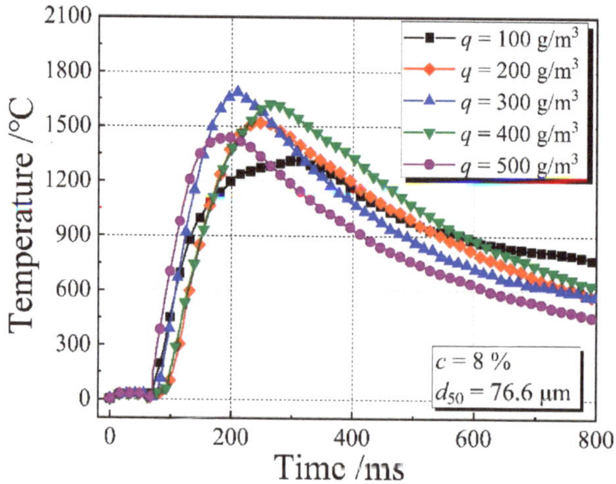

Figure 13. Comparison of temperature histories under different coal dust concentrations (d_{50} = 76.6 μm; c = 8%).

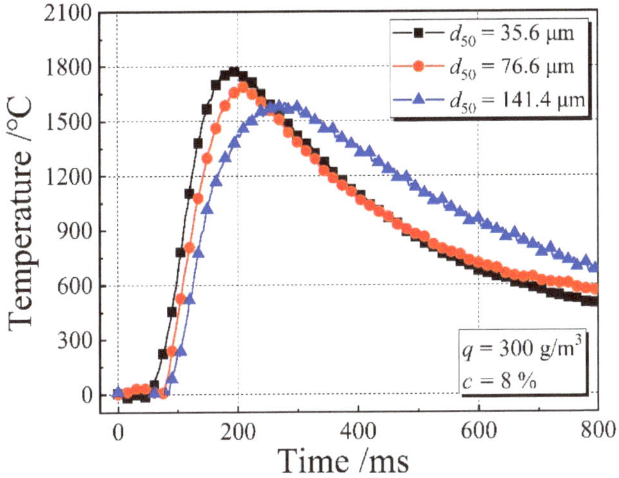

Figure 14. Comparison of temperature histories under different coal dust diameters (c = 8%; q = 300 g/m^3).

Figure 15 presents the comparison of mixture explosion flame temperature histories under different methane concentrations (q = 300 g/m^3; d_{50} = 76.6 μm). $T_{\max}$ of 6% methane concentration was the highest. With the change of methane concentration, the temperature history and its slope were decreased successively. Especially for 4% and 10% methane concentrations, $T_{\max}$ was decreased by 22.19% and 44.85%, respectively. Besides, the coal dust concentration corresponding to the highest flame temperature was also successively reduced with the increase of methane concentration under 76.6 μm coal dust diameter. Especially, the coal dust concentration corresponding to the highest flame temperature was 200 g/m^3 as c = 10%. This indicates that the combustion degree and heat release of gas/dust two-phase could be evidently affected by the combustion gas concentration, the dust diameter and its concentration.

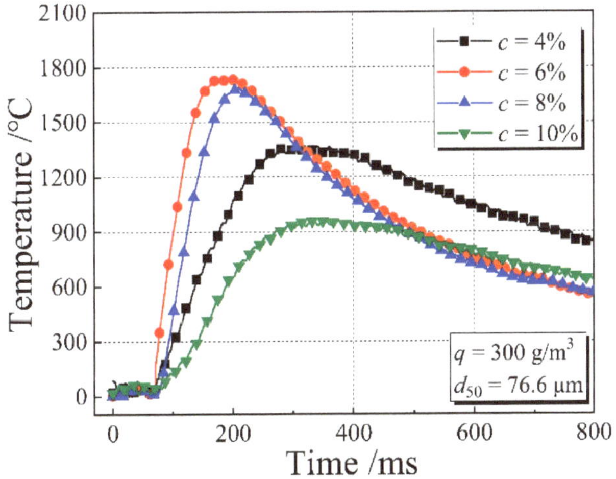

Figure 15. Comparison of temperature histories under different methane concentrations (d_{50} = 76.6 μm; q = 300 g/m^3).

4. Product Analysis

The DSC and TG histories of coal dust could intuitively reflect its thermal stability and decomposition oxidation characteristics after being heated. Figures 16 and 17 illustrate the TG and DSC histories of burned and unburned coal dusts in an oxygen atmosphere with the increase of temperature at a heating rate of 10 K/min, respectively [40,41]. The coal dust all presented three stages of the temperature rise, weight loss and stability after being heated, but there were the significant differences between the burned and unburned coal dust. It clear that from Figure 16, TG history of unburned coal dust occurred a slight change with the increase of temperature within the range of 0~263 °C. This was caused by the evaporation of moisture in coal dust after heating. At 263 °C, the coal dust occurred the pyrolysis. And TG history started to decline and the weight loss rate was gradually increased. The weight loss rate was the fastest and the corresponding DTG history reached a peak value (1.25%/°C) at 446 °C. This process was considered the acceleration weightlessness stage of coal dust. Meanwhile, the slope of coal dust DSC history was rapidly increased at this stage, indicating that the coal dust experienced a violent oxidation reaction and the heat release was continuously increased. During the whole thermal decomposition process (263 °C < T < 612 °C), there were two exothermic peaks on the DSC history at 446 °C and 484 °C, and the first peak (34.91 mW/mg) was significantly greater than the second peak (13.25 mW/mg). The appearance of the second exothermic peak resulted from the pyrolysis of stable aromatic rings inside the coal dust [42]. After the first peak value, the weight loss rate was rapidly decreased. As the temperature increased to 612 °C, the weight loss did not occur anymore and reached the thermal stability stage.

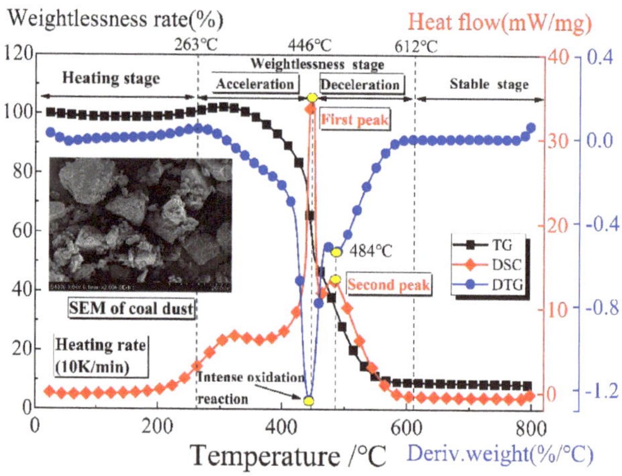

Figure 16. TG and DSC histories of unburned coal dust.

Compared with unburned coal dust, the thermal stability and decomposition oxidation of burned coal dust occurred an obvious change, as shown in Figure 17. At 274 °C, the burned coal dust appeared the pyrolysis and the corresponding temperature was slightly increased. After that, the coal dust entered the accelerated weight loss stage, but the weight loss rate was obviously decreased. Especially the peak value of weight loss rate (0.78%/°C) was evidently reduced and the corresponding temperature was increased (T = 479 °C). Meanwhile, the slope of DSC history during the accelerated weightlessness stage was also significantly decreased. Especially the exothermic peak value (18.18 mW/mg) of coal dust oxidation reaction was also significantly decreased and the corresponding temperature was evidently increased. Although the DSC history also appeared a small second peak, it was obviously reduced and the corresponding temperature was prolonged compared with the unburned coal dust. Besides, the temperature of coal dust reaching the stable stage was also increased. It indicates that the content of stable aromatic rings in coal dust was significantly decreased after the combustion, but there was still a small pyrolysis reaction of remaining stable aromatic rings at the high temperature.

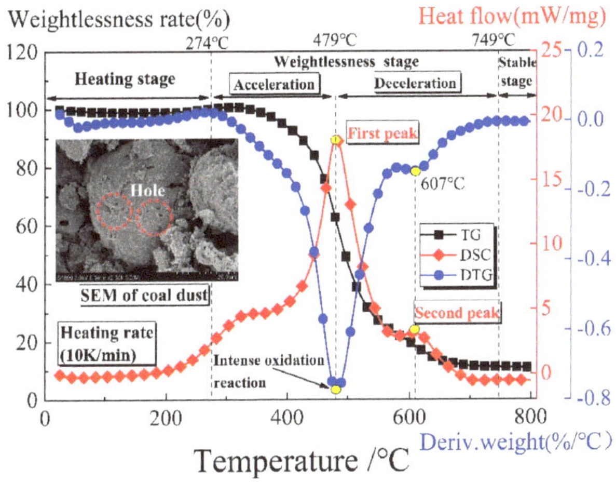

Figure 17. TG and DSC histories of burned coal dust.

The above indicates that coal dust participated in the gas/dust two-phase explosion reactions and the pyrolysis reaction of volatile matter resulted in an evident reduction in the weight loss and oxidation reaction rate of coal dust. It can be seen by the SEM that the unburned coal dust presented a blocky structure with sharp edges and corners [43]. Under the action of a high temperature flame, the volatile matter in coal dust continuously occurred the pyrolysis reaction. The macromolecular chemical bonds were decomposed into gaseous state (CH_4, CO, CO_2, N_2 and O_2, etc.) and small molecular liquids (hydrocarbons) [44], resulting in a continuous decrease in the functional groups involved in the oxidation reaction. Besides, the volatile matter gradually accumulated and expanded inside the coal dust, resulting in an obvious increase of coal dust volume [45]. Subsequently, the volatile substances were released from its surface after reaching a certain pressure, resulting in the numerous pore structure on the coal dust surface [46,47], as can be seen from Figure 17. The volatile gas generated by coal dust would promote the gas phase explosion reaction and heat transfer to the coal dust, further promoting the pyrolysis reaction of coal dust [48].

5. Conclusions

The characteristics and influence factor of gas/dust two-phase mixture explosions were studied experimentally. Based on the qualitative analysis of flame propagation, explosion intensity, flame temperature and products, the effects of methane and coal dust physical parameters on mixture explosion parameter were obtained. And the influence factor and its mechanism were analyzed deeply. Conclusions are summarized as follows:

First, the flame propagation of mixture explosions could be evidently affected by the coal dust parameters and methane concentration. The height of flame front at the same moment showed a change trend of increase firstly and then decrease with the rise of coal dust concentration. The v_{max} also presented the same change trend. In particular, the flame front was the highest and the brightness was the strongest as q = 300 g/m^3. With the increase and decrease of coal dust concentration, the flame brightness was gradually darkened and the irregular flame front was more significant. The velocity history also appeared the same change trend as the methane concentration increased. And the maximal velocity value appeared at 6% methane concentration. However, the time for reaching the vessel end was increased and v_{max} was gradually decreased as the coal dust dimeter increased.

Second, ΔP_{max} all appeared a change trend of increase firstly and then decrease with the increases of methane and coal dust concentrations. The coal dust concentration corresponding to the maximum pressure existed difference and was decreased in turn with the rise of methane concentration. As c = 2% and 10%, the corresponding coal dust concentrations were 500 g/m^3 and 200 g/m^3, respectively. However, ΔP_{max} was obviously reduced with the rise of coal dust diameter. The (dP/dt) history also showed the similar changing trend to the pressure with the changes of methane and coal dust concentrations and appeared two peak values. With the rise of coal dust diameter, the two peak values all showed a decrease trend. And the reduction extent of second peak value was more significant.

Third, T_{max} presented the similar changing trend to the pressure with the increases of methane and coal dust concentrations (increase firstly and then decrease). However, T_{max} showed a decreasing change trend as the coal dust diameter raised. Besides, the coal dust participated in the gas/dust two-phase explosion reaction at high temperature. The thermal stability and decomposition oxidation characteristics of burned coal dust appeared an obvious change compared with the unburned coal dust. The pyrolysis reaction of volatile matter led to an obvious reduction in the weight loss and oxidation reaction rate. And the precipitation of volatile matter also resulted in an obvious pore structure on its surface.

For better prevention of gas/dust two-phase mixture explosions, three suggestions are proposed: (1) The coal dust concentration corresponding to the maximum pressure was different with the change of methane concentration. In practice, the mixture concentration should be kept away from the concentration corresponding to the maximum mixture explosion pressures. (2) Enterprises involved in powder production should avoid the presence of small size dust as much as possible and the mixture of the two should exceed its explosion limit. (3) The space where dust existed should be sprayed with fine water mist to decrease the precipitation of volatile matter inside the dust and its combustion rate.

Author Contributions: Conceptualization, Y.W. and H.W.; methodology, Z.W. and X.C.; investigation, X.C.; writing—original draft preparation, Z.W. and X.C.; writing—review and editing, Z.W. and X.C.; funding acquisition, Y.W. and X.C.; All authors have read and agreed to the published version of the manuscript.

Funding: This research was funded by National project funding for National Natural Science Foundation of China (Project No. 52274210; 21865036; 51806056), the China Postdoctoral Science Foundation (Projects No. 2020M681573), the Natural Science Research Projects of Jiangsu Province Universities (Projects No. 20KJA620001), the Postdoctoral Research Funding Program of Jiangsu Province, State Key Laboratory of Coal Resources and Safe Mining-Xinjiang Institute Engineering Joint Open Research Fund Project (No. SKLCRSM-XJIEKF007).

Institutional Review Board Statement: Not applicable.

Informed Consent Statement: Not applicable.

Data Availability Statement: Not applicable.

Acknowledgments: We thank the National project funding for National Natural Science Foundation of China (Project No. 52274210; 21865036; 51806056), the China Postdoctoral Science Foundation (Projects No. 2020M681573), the Natural Science Research Projects of Jiangsu Province Universities (Projects No. 20KJA620001), the Postdoctoral Research Funding Program of Jiangsu Province, State Key Laboratory of Coal Resources and Safe Mining-Xinjiang Institute Engineering Joint Open Research Fund Project (No. SKLCRSM-XJIEKF007).

Conflicts of Interest: The authors declare no conflict of interest.

References

1. Dufaud, O.; Perrin, L.; Traore, M. Dust/vapour explosions: Mixture behaviors. *J. Loss Prev. Process Ind.* **2008**, *21*, 481–484. [CrossRef]
2. Rockwell, S.R.; Rangwala, A.S. Influence of coal dust on premixed turbulent methane air flames. *Combust. Flame* **2013**, *160*, 635–640. [CrossRef]
3. Guo, C.W.; Shao, H.; Jiang, S.G.; Wang, Y.J.; Wang, K.; Wu, Z.Y. Effect of low-concentration coal dust on gas explosion propagation law. *Powder Technol.* **2020**, *367*, 243–252. [CrossRef]
4. Lin, S.; Liu, Z.T.; Qian, J.F.; Li, X.L. Comparison on the explosivity of coal dust and of its explosion solid residues to assess the severity of re-explosion. *Fuel* **2019**, *251*, 438–446. [CrossRef]
5. Li, Q.Z.; Lin, B.Q.; Wang, K.; Zhao, M.Z.; Ruan, M.L. Surface properties of pulverized coal and its effects on coal mine methane adsorption behaviors under ambient conditions. *Powder Technol.* **2015**, *270*, 278–286. [CrossRef]
6. Sazal, K.K.; Jafar, Z.; Daniel, E.; Behdad, M. Explosion severity of methane-coal dust mixture mixtures in a ducted spherical vessel. *Powder Technol.* **2018**, *323*, 95–105.
7. Li, H.T.; Zhai, F.; Li, S.S.; Lou, R.Y.; Wang, F.C.; Chen, X.K.; Chu, C.M.; Yu, M.G. Macromorphological features and formation mechanism of particulate residues from methane/air/coal dust gas-solid two-phase mixture explosions: An approach for material evidence analysis in accident investigation. *Fuel* **2022**, *315*, 123209. [CrossRef]
8. Jiang, H.P.; Bi, M.S.; Gao, Z.H.; Zhang, Z.L.; Gao, W. Effect of turbulence intensity on flame propagation and extinction limits of methane/coal dust explosions. *Energy* **2022**, *239*, 122246. [CrossRef]
9. Li, L.; Qin, B.T.; Liu, J.S.; Leong, Y.K. Integrated experimentation and modeling of the formation processes underlying coal combustion-triggered methane explosions in a mined-out area. *Energy* **2020**, *203*, 117855. [CrossRef]
10. Xiao, F.; Ren, J.J.; Pu, M.Z.; Chen, B.; Bi, M.S. Suppression effect of ultrafine water mist on methane-coal dust mixture explosion. *Powder Technol.* **2022**, *406*, 117590.
11. Da, H.M.; Yin, H.P.; Liang, G.Q. Explosion inhibition of coal dust clouds under coal gasification atmosphere by talc powder. *Process Saf. Environ Prot.* **2022**, *165*, 286–294. [CrossRef]
12. Zhao, Q.; Chen, X.F.; Yang, M.J.; Zhang, H.M.; Huang, C.Y.; Dai, H.M.; Li, Y.; Liu, J.; Zhu, H.W. Suppression characteristics and mechanisms of ABC powder on methane/coal dust compound deflagration. *Fuel* **2021**, *298*, 120831. [CrossRef]
13. Wang, X.T.; Dai, H.M.; Liang, G.Q.; Zhu, H.W.; Zhang, B.Q.; He, S.; Zhao, Q.; Yin, H.P.; Chen, X.F. Flame propagation characteristics of mixed pulverized coal at the atmosphere of gasification. *Fuel* **2021**, *300*, 120954. [CrossRef]
14. Sazal, K.K.; Jafar, Z.; Daniel, E.; Nader, M.; Behdad, M. Explosion characteristics of methane-air mixtures in a spherical vessel connected with a duct. *Process Saf. Environ. Prot.* **2017**, *111*, 85–93.
15. Benedetto, A.D.; Garcia-agreda, A.; Russo, P.; Sanchirico, R. Combined effect of ignition energy and initial turbulence on the explosion behavior of lean gas/dust-air mixtures. *Ind. Eng. Chem. Res.* **2012**, *51*, 7663–7670. [CrossRef]
16. Ji, W.T.; Wang, Y.; Yang, J.J.; He, J.; Wen, X.P.; Wang, Y. Methods to predict variations of lower explosion limit associated with mixture mixtures of flammable gas and dust. *Fuel* **2022**, *310*, 122138. [CrossRef]
17. Zhao, P.; Tan, X.; Schmidt, M.; Wei, A.Z.; Huang, W.X.; Qian, X.M.; Wu, D.J. Minimum explosion concentration of coal dusts in air with small amount of $CH_4/H_2/CO$ under 10-kJ ignition energy conditions. *Fuel* **2020**, *260*, 116401. [CrossRef]
18. Chen, D.L.; Sun, J.H.; Liu, Y. Temperature profiles of the methane-coal dust mixture flame. *J. Saf. Environ.* **2008**, *8*, 123–125.
19. Jing, G.X.; Shao, H.Y.; Wu, Y.L.; Guo, S.S.; Liu, C.; Zhang, S.Q. Experimental study on the influence of different coal species on gas and coal dust explosion. *Saf. Coal Mines* **2020**, *51*, 1–5.
20. Cloney, C.T.; Ripley, R.C.; Pegg, M.J.; Amyotte, P.R. Role of particle diameter in laminar combustion regimes for mixture mixtures of coal dust and methane gas. *Powder Technol.* **2020**, *362*, 399–408. [CrossRef]
21. Gao, C.; Li, H.; Su, D.; Huang, W.X. Explosion characteristics of coal dust in a sealed vessel. *Expl Shock wave* **2010**, *30*, 164.
22. Wang, X.B. Effect of ignition delay time on explosion characteristics of methane coal dust mixture mixtures. *Saf. Coal Mines* **2020**, *51*, 23–27.
23. Addai, E.K.; Gabel, D.; Kamal, M.; Krause, U. Minimum ignition energy of mixture mixtures of combustible dusts and gases. *Process Saf. Environ. Prot.* **2016**, *102*, 503–512. [CrossRef]
24. Zhou, Y.H.; Bi, M.S.; Qi, F. Experimental research into effects of obstacle on methane-coal dust mixture explosion. *J. Loss Prev. Process Ind.* **2012**, *25*, 127–130. [CrossRef]
25. Yu, X.Z.; Zhang, Z.H.; Yan, X.Q.; Wang, Z.; Yu, J.L.; Gao, W. Explosion characteristics and combustion mechanism of hydrogen/tungsten dust hybrid mixtures. *Fuel* **2023**, 332, 126017. [CrossRef]
26. Ji, W.T.; Gan, X.Y.; Li, L.; Li, Z.; Wen, X.P.; Wang, Y. Prediction of the explosion severity of hybrid mixtures. *Powder Technol.* **2022**, *400*, 117273. [CrossRef]
27. Xiong, X.X.; Gao, K.; Mu, J.; Li, B.; Zhang, D.; Xie, L.F. Study on explosion characteristic parameters and induction mechanism of magnesium powder/hydrogen hybrids. *Fuel* **2022**, *326*, 125077. [CrossRef]
28. Cao, X.Y.; Lu, Y.W.; Jiang, J.C.; Wang, Z.R.; Wei, H.Y.; Li, Y.M. Experimental study on explosion inhibition by heptafluoropropane and its synergy with inert gas. *J. Loss Prev. Process Ind.* **2021**, *71*, 104440. [CrossRef]
29. Cao, X.Y.; Bi, M.S.; Ren, J.J.; Chen, B. Experimental research on explosion suppression affected by ultrafine water mist containing different additives. *J. Hazard. Mater.* **2019**, *368*, 613–620. [CrossRef]
30. Cao, X.Y.; Ren, J.J.; Zhou, Y.H.; Wang, Q.J.; Gao, X.L.; Bi, M.S. Suppression of methane/air explosion by ultrafine water mist containing sodium chloride additive. *J. Hazard. Mater.* **2015**, *285*, 311–318. [CrossRef]

31. Cao, X.Y.; Wei, H.Y.; Wang, Z.R.; Fan, L.T.; Zhou, Y.Q.; Wang, Z. Experimental research on the inhibition of methane/coal dust mixture explosions by the ultrafine water mist. *Fuel* **2023**, *331*, 125937. [CrossRef]
32. Cao, X.Y.; Wang, Z.; Wang, Z.R.; Zhou, Y.Q. Effect of wire mesh structure parameter on the flame propagation characteristics of syngas explosion. *Fuel* **2023**, *334*, 126658. [CrossRef]
33. Cao, X.Y.; Zhou, Y.Q.; Wang, Z.R.; Fan, L.T.; Wang, Z. Experimental research on hydrogen/air explosion inhibition by the ultrafine water mist. *Int. J. Hydrogen Energy* **2022**, *47*, 23898–23908. [CrossRef]
34. Castellanos, D.; Carreto-Vazquez, V.H.; Mashuga, C.V.; Trottier, R.; Mejia, A.F.; Mannan, M.S. The effect of particle size polydispersity on the explosibility characteristics of aluminum dust. *Powder Technol.* **2014**, *254*, 331–337. [CrossRef]
35. Zhao, Q.; Liu, J.; Huang, C.Y.; Zhang, H.M.; Li, Y.; Chen, X.F. Characteristics of coal dust deflagration under the atmosphere of methane and their inhibition by coal ash. *Fuel* **2021**, *291*, 120121. [CrossRef]
36. Chen, X.F.; Hou, Z.X.; Zhao, Q.; Li, Q.; Li, Y.; Huang, C.Y.; Dai, H.M. Suppression of methane/coal dust deflagration by $Al(OH)_3$ based on flame propagation characteristics and thermal decomposition. *Fuel* **2022**, *311*, 122530. [CrossRef]
37. Zhang, H.M.; Chen, X.F.; Zhang, Y.; Niu, Y.; Yuan, B.H.; Dai, H.M.; He, S. Effects of particle size on flame structures through corn starch dust explosions. *J. Loss Prev. Process Ind.* **2017**, *50*, 7–14. [CrossRef]
38. Zhang, L.; Wang, H.Y.; Chen, C.; Wang, P.P.; Xu, L.W. Experimental study to assess the explosion hazard of CH_4/coal dust mixtures induced by high-temperature source surface. *Process Saf. Environ. Prot.* **2021**, *154*, 60–71. [CrossRef]
39. Li, H.T.; Deng, J.; Chen, X.K.; Shu, C.M.; Kuo, C.H.; Zhai, X.W.; Wang, Q.H.; Hu, X.Y. Qualitative and quantitative characterisation for explosion severity and gaseous-solid residues during methane-coal particle mixture explosions: An approach to estimating the safety degree for underground coal mines. *Process Saf. Environ. Prot.* **2020**, *141*, 150–166. [CrossRef]
40. Du, Y.Y.; Jiang, X.G.; Lv, G.J.; Ma, X.J.; Jin, Y.Q.; Wang, F.; Chi, Y.; Yan, J.H. Thermal behavior and kinetics of bio-ferment residue/coal blends during co-pyrolysis. *Energ. Convers. Manag.* **2014**, *88*, 459–463. [CrossRef]
41. Wang, G.W.; Zhang, J.L.; Shao, J.G.; Liu, Z.J.; Zhang, G.H.; Xu, T.J.; Guo, H.Y.; Xu, R.S.; Lin, H. Thermal behavior and kinetic analysis of co-combustion of waste biomass/low rank coal blends. *Energ. Convers. Manag.* **2016**, *124*, 414–426. [CrossRef]
42. Wang, J.H.; Du, J.; Chang, L.P.; Xie, K.C. Study on the structure and pyrolysis characteristics of Chinese western coals. *Fuel Process Technol.* **2010**, *91*, 430–433. [CrossRef]
43. Meng, X.L.; Gao, M.Q.; Chu, R.Z.; Wu, G.G.; Fang, Q. Multiple linear equation of pore structure and coal-oxygen diffusion on low temperature oxidation process of lignite. *Chin. J. Chem. Eng.* **2016**, *24*, 818–823. [CrossRef]
44. Liu, Z.T.; Zhang, S.S.; Xi, R.Z.; Guo, R.L.; Liu, B.B.; Lin, S. Analysis on residual gas characteristics of coal dust explosion in confined space. *J. Coal Sci.* **2015**, *40*, 1574–1579.
45. Fu, W.B. *Coal Combustion Theory and Its Macro General Law*; Tsinghua University Press: Beijing, China, 2003.
46. Nie, B.S.; Liu, X.F.; Yang, L.L.; Meng, J.Q.; Li, X.C. Pore structure characterization of different rank coals using gas adsorption and scanning electron microscopy. *Fuel* **2015**, *158*, 908–917. [CrossRef]
47. Li, H.Y.; Ogawa, Y.; Shimada, S. Mechanism of methane flow through sheared coals and its role on methane recovery. *Fuel* **2003**, *82*, 1271–1279. [CrossRef]
48. Zhou, Y.H.; Gan, B.; Jiang, H.P.; Huang, L.; Gao, W. Investigations on the flame propagation characteristics in methane and coal dust mixture explosions. *Shock Waves* **2022**, *42*, 167–175.

fire

Article

The Application of Nitrogen Curtain Technology to Longwall Goaf to Prevent the Spontaneous Combustion of Coal: A Case Study in Shajihai Coalmine, China

Hai-Jiang Geng [1,2], Ya-Ming Zhao [2,3], Xiang-Lan Liu [2] and Fu-Chao Tian [1,2,3,*]

1 School of Mechanics and Civil Engineering, China University of Mining and Technology (Beijing), Beijing 100083, China; genghaijiang@ccteg.com
2 China Coal Research Institute, Beijing 100013, China; zhaoyaming@syccri.com (Y.-M.Z.); tsg636@126.com (X.-L.L.)
3 State Key Laboratory of Coal Mine Safety Technology, China Coal Technology and Engineering Group Shenyang Research Institute, Fushun 113122, China
* Correspondence: tianfuchao@cumt.edu.cn

Abstract: To enhance the inerting effect of nitrogen on large longwall goaf, a novel goaf-inerting method of using a full cross-section nitrogen curtain for a U-shaped ventilation working face is developed. The working principle of the full cross-section nitrogen curtain is elucidated. The design principle and key parameters of the nitrogen curtain needed to achieve an optimum nitrogen injection effect are established. The nitrogen curtain technology is successfully applied to the scenario of fire prevention in the underground goaf. The field study shows that the full cross-section curtain injection of nitrogen exhibits many advantages such as simple operation and homogeneous diffusion. The implementation of the nitrogen curtain reduced the maximum width of the goaf oxidation zone from 70 m to 20 m. And the CO concentration in the upper corner decreased from 21.8 ppm to 11.2 ppm after 18 h of nitrogen injection. After 48 h of injection, the CO concentration in the upper corner remained unchanged and the concentration had reduced to 0 in the lower corner. It was demonstrated that the inerting efficiency of the full cross-section nitrogen curtain reached 86% in the upper corner and 100% in the lower, which is significantly superior to the traditional buried-pipe nitrogen injection method. The effect verifies the effectiveness of the curtain injection method of nitrogen, which can ensure a safe working face production.

Keywords: spontaneous combustion; full cross-section; curtain nitrogen injection; fire prevention

Citation: Geng, H.-J.; Zhao, Y.-M.; Liu, X.-L.; Tian, F.-C. The Application of Nitrogen Curtain Technology to Longwall Goaf to Prevent the Spontaneous Combustion of Coal: A Case Study in Shajihai Coalmine, China. *Fire* **2023**, *6*, 363. https://doi.org/10.3390/fire6090363

Academic Editors: Chuyuan Huang, Haipeng Jiang and Lijuan Liu

Received: 28 August 2023
Revised: 10 September 2023
Accepted: 12 September 2023
Published: 18 September 2023

1. Introduction

The safety of coal production has been greatly improved with the development of longwall mining in China, while this also brings several safety problems, like the frequent occurrence of spontaneous combustion of coal in the goaf due to severe air leakage and a large amount of coal leaving [1–5]. To control spontaneous combustion of residual coal in longwall goaf, many techniques like grouting, inert gases injection and gel and foam injection have been developed and applied [6–9]. However, nitrogen injection has been the most widely used practice due to its easy operation and good adaptability [10,11]. When taking the technical means of nitrogen injection and fire prevention in goaf into account, the selection of a reasonable ventilation mode is one of the effective means of preventing the spontaneous combustion of the coal seam and gas overlimit, and also an important way to mitigate accidents [12]. The U-type ventilation mode is the most common application in the coal face. The characteristics of nitrogen in U-shaped ventilation are as follows: U-shaped ventilation is conducive to nitrogen moving along the potential leakage flow, and the inerting effect near the nitrogen release outlet is very obvious. The cooling range gradually decreases from near to far from the nitrogen release outlet. The time by which nitrogen injection delays the inerting of the goaf is influenced by many factors, such as the area of the goaf, the amount of air leakage, the temperature of the bedrock and the degree of spontaneous combustion of the remaining coal [13]. After the release of nitrogen, the migration trend of the nitrogen in the goaf can be divided into three categories: part of the nitrogen develops at a certain distance and forms a complex upwards vortex after kinetic energy loss or continues to flow upward slowly; the second part of nitrogen flows directly deeply into the goaf; the third part is carried by the leaked air and flows to low pressure areas such as the upper corner or drainage lane [14]. A large-flow nitrogen injection in goaf can achieve good a inerting effect, which decreases gradually with the advance of the working face position. While the low-flow nitrogen injection moves forward with the position of the working face, the position of the fireproof line retracts to the working face, and then expands to the depth of the goaf, that is, the fire prevention effect has an inflection point at the position of nitrogen injection. With U-shaped ventilation, the oxidation zone of goaf presents an asymmetric distribution on the inlet and return air sides, and the spontaneous combustion zone has the widest range on the inlet air side. The closer it is to the return air side, the smaller the range of the spontaneous combustion zone is and the more forward the location is. The influence of nitrogen injection on oxygen distribution can be summarized by two points: the local influence is that the oxygen concentration near the nitrogen injection port changes obviously, while the macro influence is that the oxidation zone retracts and moves forward, and the influence degree decreases from the inlet side to the return side.

Currently, nitrogen injection is mainly accomplished through pre-buried pipe, drag pipe, borehole and intubation pipe [15–21]. For the pre-buried pipe method, seamless steel pipe is placed along the main gate of the working face first. And then, when the working face advances to a certain position, the pipeline valve is opened, and nitrogen is released at the end of the pipe. Additionally, for the drag pipe nitrogen injection method, the injection pipe is placed in the intake roadway along the goaf, and the pipe moves forward under the screw motor traction in the transportation traction head or mechanical winch in the roadway. For this method, the stepping distance for the drag pipe is definite, which ensures the exit position is always in the oxidation zone. The borehole nitrogen injection method has a significant advantage compared with other methods, which is directionality, which can infuse the large-flow nitrogen into the spontaneous combustion area. The intubation pipe nitrogen injection method is appropriate to use when the mine fire originates in the open-off cut or stopping line of the working face.

The air leakage greatly impacts the inerting effect of nitrogen injection. The air leakage in the goaf is mainly distributed in three characteristic areas, namely, the upper corner air leakage area, the rear frame air leakage area and the lower corner air leakage area. There is a nonlinear relationship between the air leakage intensity and the distance from a certain point in the goaf to the working face. The air leakage rate in the goaf gradually decreases along the working face tendency and flows out of the goaf in the reverse direction, to a position in the middle. U-shaped ventilation is conducive to nitrogen moving along the potential leakage flow, and the cooling range decreases from points near to the nitrogen injection outlet to points far from it. There are three main migration laws of nitrogen in the goaf after release.

In addition, the nitrogen injection methods mentioned above also have three shortcomings. First, the nitrogen diffusion radius is small and the effective scope is limited. Second, the nitrogen diffusion is very likely to be affected by air leakage which results in an inhomogeneous distribution in the goaf. Third, the nitrogen injection quantity is small, which restricts the nitrogen diffusion in the goaf. However, the more serious challenge for nitrogen injection is the successive emergence of large longwall panels, where the mining height of the top-coal caving face can reach up to 20 m and the working face is generally longer than 200 m. In addition, a weakened air leakage and a poor nitrogen diffusion in the goaf, because of working face enlarges the distance between the return and intake roadway, is a challenge [22–25]. The inferior applicability of traditional nitrogen injection techniques greatly restricts their inert effectiveness in a huge goaf. In this paper, a new full cross-section curtain method of injecting nitrogen is developed to improve the nitrogen inerting effect, by creating multi-row nitrogen injection ports to replace the single horizontal nitrogen injection ports. The issues around conventional nitrogen injection are solved through forming a universal coverage in the goaf. Moreover, it exhibits a more obvious superiority in the tailgate and middle goaf due to its special design in the three-dimensional space. To investigate the performance of the curtain nitrogen injection method, the basic principles and the key design concepts are introduced in this paper. Additionally, in accordance with the above results, the industrial practice is carried out to evaluate the inert effect of full cross-section nitrogen injection.

2. Principle and Methods

2.1. The Principle of Full Cross-Section Curtain Injection of Nitrogen

The conventional nitrogen injection layout of pre-buried pipe is shown in Figure 1. As can be seen from Figure 1, the pre-buried pipe is placed at the bottom of the main gate and the pipe port is positioned in the goaf with a distance of 20–40 m to the working face. And the nitrogen discharge valve is open when the working face advances to an appropriate position. Under the driving effects of air leakage along the working face, the nitrogen is introduced to the entire goaf, diluting the oxygen concentration and restraining the gas explosion in the spontaneous combustion zone [26].

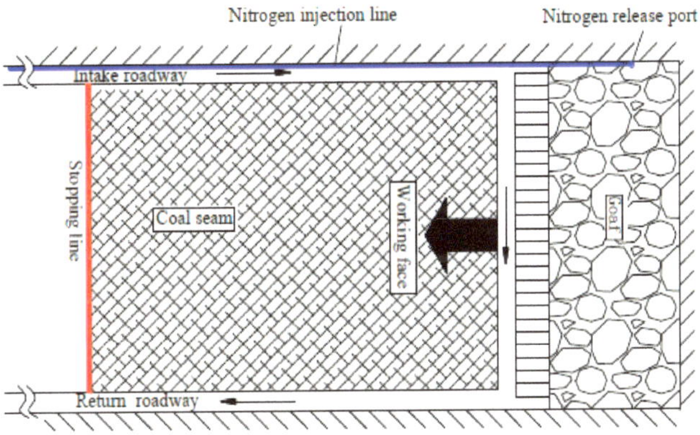

Figure 1. Open buried pipe nitrogen injection process.

The full cross-section curtain injection of nitrogen is established based on the existing pre-buried pipe, as shown in Figure 2. And the key technical innovation is the nitrogen injection port layout which is a circumferential collocation around the nitrogen pipe. The nitrogen pipes are arranged along the working face, and are uniformly spaced and parallel mounted in the goaf. To achieve a uniform inerting effect and reduce the nitrogen injection amount, the cell hole sizes are specially set in the nitrogen pipe. With the nitrogen pipe laid successively in the scraper-trough conveyer, the nitrogen is continuously released, and a nitrogen curtain is formed in the goaf. The three-dimensional nitrogen diffusion created by the curtain is preferable for dealing with the large-scale hidden fire zone. And the diffusion effect of the full cross-section curtain injection of nitrogen is shown in Figure 3.

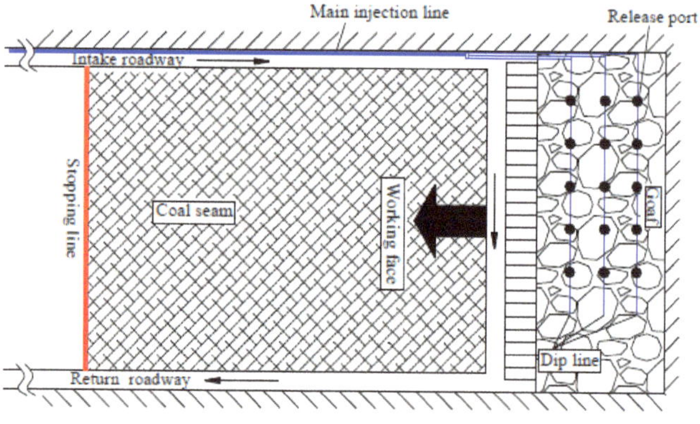

Figure 2. Full cross-section nitrogen curtain injection process.

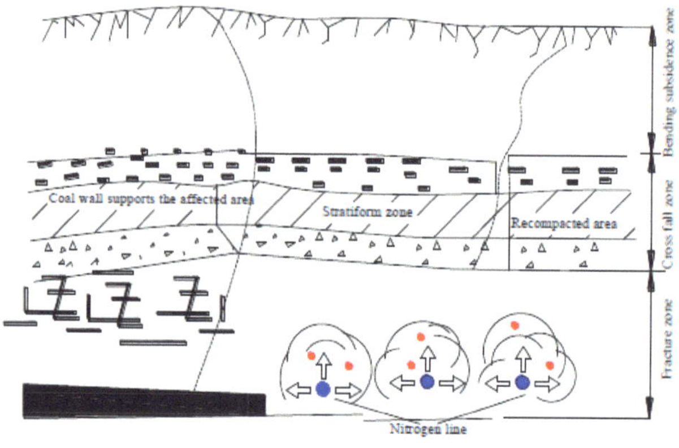

Figure 3. Nitrogen-injecting curtain stereo effect.

2.2. The Design of Full Cross-Section Curtain Injection of Nitrogen

The basic design principle for full cross-section curtain injection of nitrogen is to obtain a rapid efficient inerting in the goaf with a low cost. The oxidation zone advances in the goaf as the working face moves forward. Thus, it is a challenge to realize a homogeneous nitrogen diffusion in the goaf, especially in the oxidation zone. The key technique is to design the hole group and cell hole of the nitrogen curtain pipe, as shown in Figure 4. To obtain a better diffusion effect, one hole group is designed for six cell holes, which are evenly distributed around the pipe with an angle of 60° and a radius of 1 cm. The nitrogen pipe array pitch is two times the nitrogen diffusion radius to avoid the overlap of the coverage of two groups of nitrogen release ports.

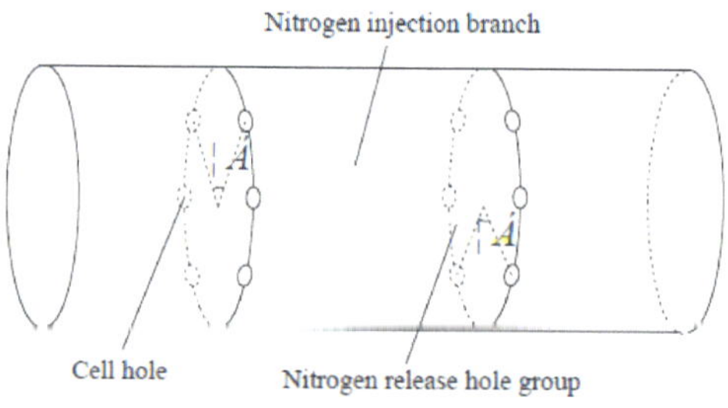

Figure 4. Nitrogen release hole group distribution.

Since nitrogen release is a coupling process of gas injection and expansion, related to the exchange of momentum, mass and heat between nitrogen and the gas mixture in the goaf, a turbulent free jet is always produced at the nitrogen outlet and the momentum remains constant along the jet axis. The expansion process originates from the hole port and throughout the whole goaf. The nitrogen generates a sudden expansion with a pressure decline, forming a semicircular ball around the hole port [27–30]. The movement of nitrogen after its release from the pipeline can be divided into two stages, both of which are closely related to time. At the initial stage, nitrogen migration is mainly affected by jet flow and expansion, and the effects of gravity and drag can be ignored. As gas diffuses between coal voids in the goaf, resistance, gravity and buoyancy gradually become the main influencing factors, and the migration velocity decreases gradually. When the velocity attenuates to a certain extent, the resultant force of nitrogen reaches an equilibrium state. Accordingly, there are three migration paths for nitrogen in the goaf: first, it diffuses directly to the deep part of the goaf under the influence of the jet flow; second, it slowly flows upward after reaching the equilibrium state; and third, it flows downward to the corner and other areas under the influence of negative pressure.

The nitrogen injection number n_0 and hole group number n_1 are calculated according to Equations (1) and (2).

$$n_0 = \frac{L_\mathrm{y}}{L_\mathrm{b}} \tag{1}$$

$$n_1 = \frac{L_\mathrm{z}}{L_\mathrm{b}} + 1 \tag{2}$$

where L_y is the maximum width of the goaf oxidation zone, m; L_z is the length of the working face, m; L_b is the nitrogen injection pipe array pitch, m; and L_b is calculated using Equation (3).

$$L_b = 2(S_c + L_c) \tag{3}$$

where S_c is the displacement caused by jet action when nitrogen in the goaf reaches critical state, m; L_c is displacement caused by expansion action when nitrogen in the goaf reaches a critical state, m; S_c is calculated [31] using Equation (4); and L_c is calculated using Equation (5).

$$S_c = \sqrt{24.8 V_0 r t_c} \tag{4}$$

$$L_c = \sqrt[3]{3 V_0 r^2 \frac{\rho_1}{\rho_0} t_c} \tag{5}$$

where V_0 is jet velocity when nitrogen is released from the hole, m/s; r is nitrogen release hole radius, m; ρ_1 is nitrogen density in nitrogen injection tube, kg/m^3; ρ_0 is nitrogen density under ambient conditions, kg/m^3; and t_c is time for nitrogen in goaf to reach critical state, s. V_0 is calculated using Equation (6) and t_c is calculated using Equation (7).

$$V_0 = k\lambda \sqrt{\frac{2k_j}{k_j - 1} R T_1 [1 - (\frac{p_0}{p_1})^{\frac{k_j - 1}{k_j}}]} \tag{6}$$

where λ is the orifice velocity coefficient, generally considered 0.98; k_j is the nitrogen adiabatic coefficient, taken to be 1.4; k is goaf porosity; R is molar gas constant, J/(mol·K); T_1 is the temperature prior to nitrogen release, K; p_1 is the pressure prior to nitrogen release, Pa; and p_1 is standard atmospheric pressure, Pa.

$$t_c = \frac{6.2 r V_0}{V_C{}^2} \tag{7}$$

where V_c is the velocity when the nitrogen in the goaf reaches a critical state, m/s, and V_c is calculated by Equation (8).

$$V_C = \sqrt{\frac{8}{3C_D}(\frac{\rho_2 - \rho_0}{\rho_2})gr} \tag{8}$$

where ρ_2 is gas density in goaf, kg/m^3; C_D is the gas release coefficient; and g is acceleration of gravity, m/s^2.

3. Field Application

To study the effect of the process of full cross-section curtain injection of nitrogen, the field demonstration is carried out in the B1003W01 working face of the Shajihai coalmine of Shenhua Guoneng Group.

The discharge pressure of nitrogen injection in the B1003W01 working face was 250,000 Pa, the temperature was 293.15 K, the porosity of the goaf was 0.2 and the radius of the single hole of nitrogen injection was 1 cm. The nitrogen density in the nitrogen injection pipeline was 2.30 kg/m^3, the nitrogen density in the ambient state was 1.16 kg/m^3, the gas density in the goaf was assumed to be 1.19 kg/m^3 and the gas release coefficient was 1.00. The length of the working face is 219.4 m, and the maximum width of the oxidation zone in the goaf is 70 m. By adopting the relevant formulae from Section 3, the spacing of the full cross-section nitrogen injection curtain pipelines in the goaf is 39 m, the number of nitrogen injection pipelines in the goaf is three and the groups of nitrogen release ports is 22.

3.1. Layout and Operation

Figure 5 illustrates the field layout of the full cross-section curtain injection of nitrogen in the goaf. In the field application, the first nitrogen injection pipe is laid along the direction of the cut-off line in the working face. The nitrogen discharges from the pipe port after flowing through the nitrogen generator, main nitrogen pipe, bypass valve and pressure hose. The second nitrogen injection pipe is immediately established when the working face advances at the distance of the array pitch. According to this operation, the goaf will always be effectively covered by nitrogen.

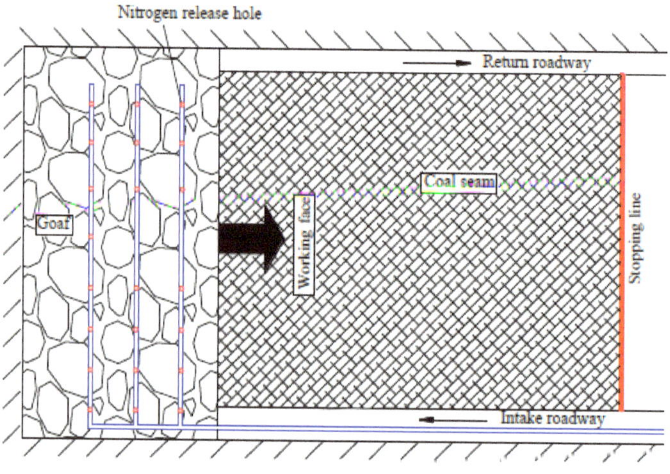

Figure 5. Sketch map of full cross-section nitrogen curtain layout.

Five monitors are placed behind the longwall support of the working face to investigate the inerting effect of curtain injection of nitrogen. The monitors are marked as 1#, 2#, 3#, 4#, and 5# from the longwall tailgate, respectively. The 1# monitoring point is set 28 m from the intake roadway and the 5# monitoring point is set 13 m from the return roadway. The distance between the 1# monitoring point and the 2# monitoring point is 30 m. The other monitoring point distances are all 48 m, as shown in Figure 6.

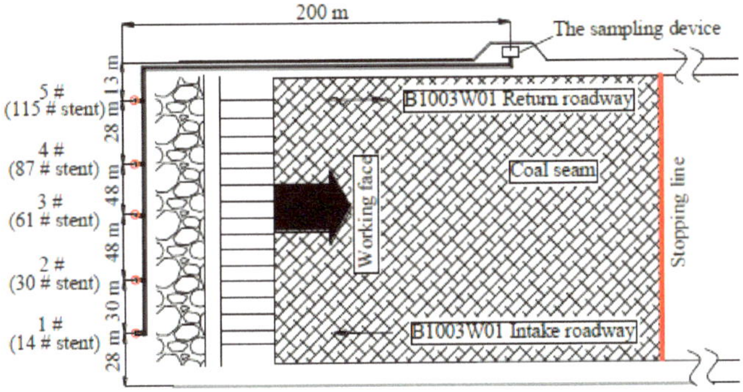

Figure 6. Layout of the monitoring points at the rear of the bracket.

3.2. *Effect Investigation*

3.2.1. Change in Oxidation Zone Width

Figure 7 presents the evolution of oxygen concentration in the goaf after being affected by the full cross-section curtain method of injecting nitrogen. It can be seen that the variation trend for all monitoring points is consistent, indicating that the inerting effect of curtain injection of nitrogen is uniform. The oxygen concentration reduces with the increase in buried distance, to about 18% when the monitoring point is placed about 7 m in to the goaf. And when the oxygen concentration presents a sharp reduction at the buried distance of 7–25 m, the oxygen concentration of the 5# monitoring point is the minimum while at the 1# monitoring point, it is the maximum. Additionally, since the oxygen concentration remains lower than 7% when the buried distance is larger than 25 m, this remarkable oxygen concentration reduction indicates that the nitrogen curtain exhibits a superior inerting performance in the goaf. The oxygen distribution is measured all across the goaf based on the monitoring point date, as shown in Figure 8.

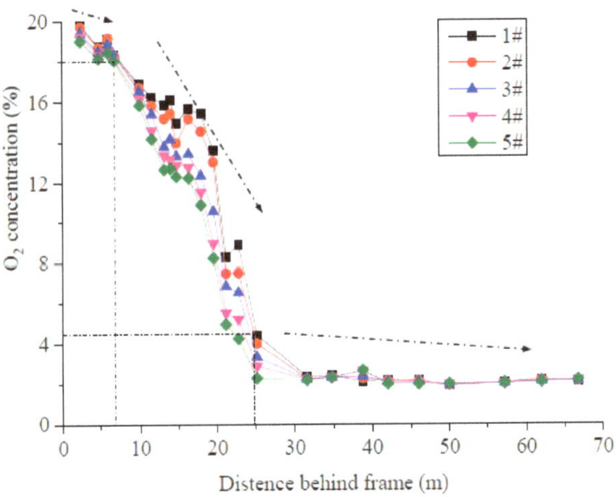

Figure 7. Oxygen concentration change in the goaf.

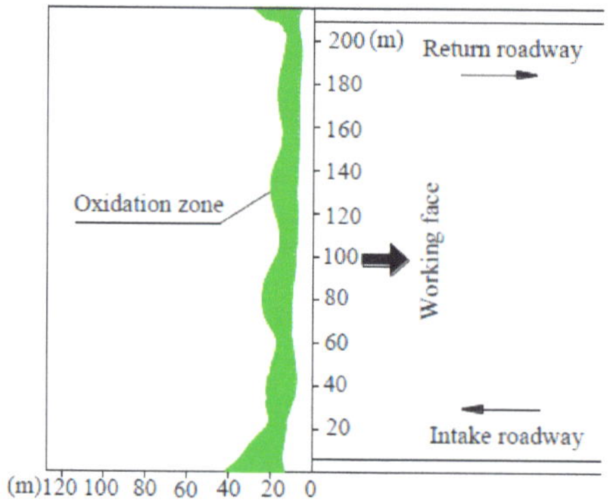

Figure 8. Distribution of oxidation zone in the goaf.

The oxygen concentration range in which coal may spontaneously combust is 7–18% in the goaf [32]. Thus, the width of the oxidation zone is compared between the conventional pre-buried pipe and the full cross-section curtain injection of nitrogen. The oxidation region width for the conventional pre-buried pipe is 27–100 m on the intake side, 20–90 m in the middle and 18–80 m on the return side. The oxidation region width has been significantly reduced with curtain injection of nitrogen, to 10–40 m on the intake side, 10–30 m in the middle and 7–30 m on the return side. Meanwhile, the oxidation zone's position clearly advances, which also verifies the excellent inerting effect of the full cross-section curtain injection of nitrogen. The inerting effect is significantly better than the existing nitrogen injection tested [33], where eight nitrogen injection trials were conducted and the inerting effect was confirmed to be limited for the working face at a flow ratio of 900 m^3/h. In addition, Zhu et al. 2013 also carried out research and discovered that the conventional pre-buried pipe may adversely enlarge the oxidation zone as the injection port position embeds in the goaf.

3.2.2. CO Concentration in the Working Face Corner

The CO concentration change is a key indicator of gas produced by coal oxidation, which can be used to evaluate the inerting effect of the full cross-section curtain injection of nitrogen and the buried pipe injection of nitrogen. The evolution of CO concentration with injection time in the upper corner is plotted in Figure 9. The nitrogen injection amounts are both 900 m^3/h, and the CO concentrations in the upper and lower corner were measured every 6 h for a total of 72 h.

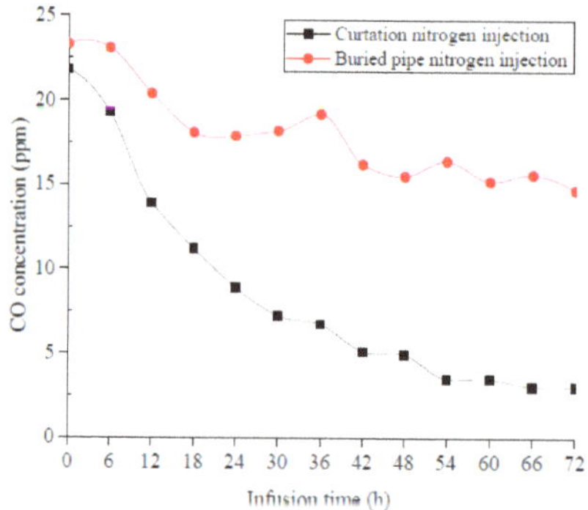

Figure 9. The change in upper corner CO concentration.

It is found that the CO concentration reduces in both scenarios after nitrogen injection. The upper corner CO concentration decreases from 21.8 ppm to 11.2 ppm after curtain injection of nitrogen for 18 h and the CO concentration tends to stabilize at 3 ppm after 48 h. However, the CO concentration is still as high as 18.1 ppm after 18 h for buried pipe injection of nitrogen. Figure 10 displays the CO concentration change in the lower corner, which decreases from 11.4 ppm to 0 ppm after 48 h for the curtain injection of nitrogen, while the CO concentration is still 6.5 ppm for the buried pipe injection of nitrogen.

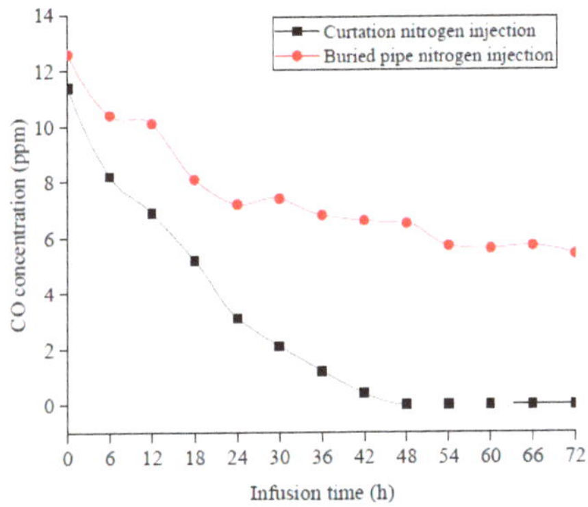

Figure 10. The change in lower corner CO concentration.

It is a significant result that the effect of buried pipe nitrogen injection is only obvious in the early stage, while it is difficult to maintain its effectiveness over most of the relevant time period and the region of the spontaneous combustion zone due to its limited inerting ability. Therefore, the CO concentration remains at a high level and even rebounded because of the coal re-oxidation. In comparison, the inerting effect of curtain injection of nitrogen is distinctly superior to the conventional buried pipe method because the new nitrogen injection method results in a lower CO concentration.

3.2.3. The Nitrogen Inerting Efficiency

The nitrogen release process in the goaf is the coupling of a jet process and an expansion process. In essence, the jet process during nitrogen release is the exchange of momentum, mass and heat between nitrogen emitted from the orifice and the mixed gas in the goaf. A turbulent free jet is usually formed at the nitrogen release orifice, and the momentum remains unchanged along the jet axis. The expansion process of nitrogen release is carried out on the circular orifice, and nitrogen is rapidly expanded and depressurized after being released from the orifice. Its expansion shape is assumed to be semicircular.

The nitrogen inerting efficiency refers to the reduction amplitude of the upper (lower) corner CO concentration under different nitrogen injection conditions, which is expressed by Equation (9).

$$\eta = \frac{\Delta C_{co}}{C_{initialco}} - \frac{C_{initialco} - C_{terminalco}}{C_{initialco}} \quad (9)$$

where $C_{initialco}$ is the original CO concentration without nitrogen injection, ppm, and $C_{terminalco}$ is the CO concentration after nitrogen injection, ppm.

A total of seven experiment groups are conducted in an underground field with nitrogen injection amounts of 300 m^3/h, 400 m^3/h, 500 m^3/h, 600 m^3/h, 700 m^3/h, 800 m^3/h and 900 m^3/h. The reductions in the upper and lower corner CO concentration are tested successively. Figures 11 and 12 present the nitrogen inerting efficiency at the upper and lower corner under different nitrogen injection amounts, respectively. It is found that the nitrogen inerting efficiency is improved significantly with the application of a curtain injection of nitrogen. As the nitrogen injection amount rises from 300 m^3/h to 900 m^3/h, the upper corner inerting efficiency increases from 11% to 86% and the lower corner inerting efficiency grows from 28% to 100%. Nevertheless, the upper corner inerting efficiency only remains at 10–37% and the lower corner at 27–57% when the traditional buried pipe nitrogen injection is used, which also reflects that the curtain injection of nitrogen works in terms of reducing the nitrogen injection amount and enhancing the inerting efficiency.

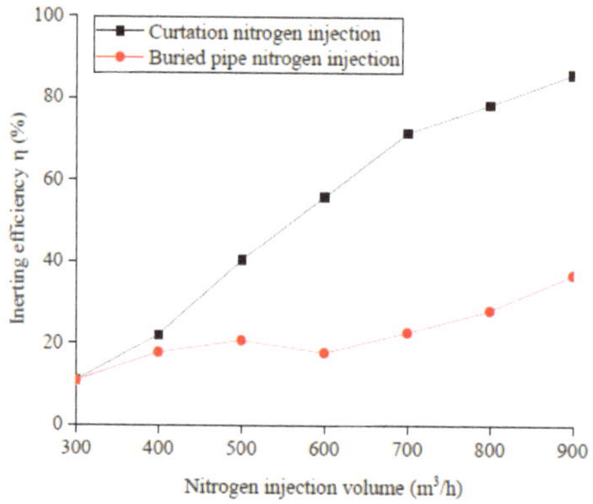

Figure 11. Inerting efficiency in upper corner of the working face.

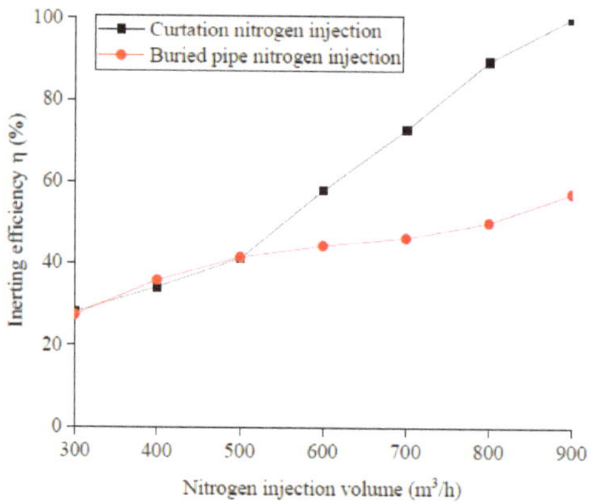

Figure 12. Inerting efficiency in lower corner of the working face.

4. Conclusions

In this paper, the nitrogen curtain technology applied on longwall goaf to prevent the spontaneous combustion of coal was studied in Shajihai Mine, Ta Cheng China. The conclusions can be summarized as follows:

(1) The working principle of full cross-section curtain injection of nitrogen is elucidated and the design principles and key parameters of the nitrogen curtain are established, such as the nitrogen injection port layout, the nitrogen injection pipe array pitch, the number of hole groups and the cell hole radius. And from the above results, we know that the nitrogen injection hole groups should be evenly spaced, and that the closer the cell hole is to the inlet side, the smaller the radius is, which ensures the consistency of the amount of nitrogen released.

(2) The field application was carried out in the B1003W01 working face to investigate the inerting effect of the nitrogen curtain technology. It is proved that the full cross-section curtain injection of nitrogen exhibits a superior extinguishment performance of goaf fires compared with the conventional buried pipe nitrogen injection method. After implementing curtain injection of nitrogen, the maximum width of the goaf oxidation zone is reduced from 70 m to 20 m and the upper corner CO concentration

decreases from 21.8 ppm to 11.2 ppm after 18 h of nitrogen injection. Additionally, the upper corner CO concentration tends to stabilize at 3 ppm and the lower corner CO concentration drops to 0 ppm after 48 h.

Author Contributions: Conceptualization, H.-J.G.; methodology, H.-J.G., X.-L.L. and F.-C.T.; formal analysis, H.-J.G. and Y.-M.Z.; investigation, H.-J.G. and F.-C.T.; resources, Y.-M.Z.; data curation, H.-J.G. and X.-L.L.; writing—original draft preparation, H.-J.G.; writing—review and editing, Y.-M.Z., X.-L.L. and F.-C.T.; project administration, F.-C.T.; funding acquisition, Y.-M.Z. and F.-C.T. All authors have read and agreed to the published version of the manuscript.

Funding: This research was funded by National Natural Science Foundation of China (Grant No. 52174230, 52174229), the Natural Science Foundation of Liaoning Province (2023-MS-355).

Data Availability Statement: Not applicable.

Conflicts of Interest: The authors declare no conflict of interest.

References

1. Wang, D.M. *Mine Fire Science*; China University of Mining and Technology Press: Xuzhou, China, 2008.
2. Wang, L.K.; Ren, T.X.; Nie, B.S.; Chen, Y.; Lv, C.Q.; Tang, H.Y.; Zhang, J.F. Development of a spontaneous combustion TARPs system based on BP neural network. *Int. J. Min. Sci. Technol.* **2015**, *25*, 803–810. [CrossRef]
3. Liang, Y.T.; Hou, X.J.; Luo, H.Z. The current situation and countermeasures of coal mine fire control in China. *Coal Sci. Technol.* **2016**, *44*, 1–6. [CrossRef]
4. Mohalik, K.; Ray, M.; Varma, S. Application of CFD techniques to assess spontaneous heating/fire during extraction of thick coal seam using blasting gallery (BG) method. *Combust. Sci. Technol.* **2020**, *192*, 1772–1784. [CrossRef]
5. Tian, L.; Liu, S.; Li, Y. Application of inorganic solidified foam to control the coexistence of unusual methane emission and spontaneous combustion of coal in the Luwa coal mine, China. *Combust. Sci. Technol.* **2020**, *192*, 638–656. [CrossRef]
6. Zhou, F.B.; Ren, W.X.; Wang, D.M. Application of the three-phase foam to fight an extraordinarily serious coal mine fire. *Int. J. Coal Geol.* **2006**, *67*, 95–100. [CrossRef]
7. Ray, S.K.; Singh, R.P. Recent developments and practices to control fire in underground coal mines. *Fire Technol.* **2007**, *43*, 285–300. [CrossRef]
8. Lu, X.X.; Wang, D.M.; Qin, B.T.; Tian, F.C.; Shi, G.Y.; Dong, S.J. Novel approach for extinguishing large-scale coal fires using gas-liquid foams in open pit mines. *Environ. Sci. Pollut. Res.* **2015**, *22*, 18363–18371. [CrossRef]
9. Guo, Q.; Ren, W.X.; Shi, J.T. Experimental investigations on foaming device used for coal spontaneous combustion prevention. *Combust. Sci. Technol.* **2020**, *192*, 1850–1862. [CrossRef]
10. Adamus, A. Technical note: Review of nitrogen as an inert gas in underground mines. *J. Mine Vent. Soc. S. Afr.* **2001**, *54*, 60–61.
11. Zhou, F.B.; Shi, B.B.; Cheng, J.W. A new approach to control a serious mine fire with using liquid nitrogen as extinguishing media. *Fire Technol.* **2015**, *51*, 325–334. [CrossRef]
12. Yu, Q.X.; Wang, K.; Yang, S.Q. Study on Pattern and Control of Gas Emission at Coal Face in China. *J. China Univ. Min. Technol.* **2000**, *1*, 9–14. [CrossRef]
13. Zhu, H.Q.; Liu, P.F.; Liu, X.K.; Yao, M.Q. Numerical simulation of temperature change and spontaneous combustion area in goaf during nitrogen injection process. *J. Hunan Univ. Sci. Technol. (Nat. Sci. Ed.)* **2012**, *27*, 1–6. [CrossRef]
14. Liu, X.K.; Wang, G.Z.; Zhou, A.T. Nitrogen injection parameters design for goaf and Its 3D flow field analysis. *J. Taiyuan Univ. Technol.* **2015**, *46*, 312–317. [CrossRef]
15. Li, Z.X.; Shan, L.B.; Zhang, W.J. Numerical simulation study of nitrogen injection process for fire prevention and extinguishment in goaf. *J. Hunan Univ. Sci. Technol.* **2004**, *19*, 5–9. [CrossRef]
16. Yang, Z.G.; Mu, S.L. Practice of nitrogen injection and fire extinguishing technology in goaf of fully mechanized top coal caving face. *Coal Mine Saf.* **2008**, *39*, 30–34.
17. Zhu, H.Q.; Li, F.; Zhang, Y.; Wang, W.H.; Liang, S.K. Design and analysis of the effect of non-interval nitrogen injection on fire extinguishing technology. *Saf. Coal Mines* **2013**, *44*, 175–178. [CrossRef]
18. Liu, H.F. Optimization on calculation method of nitrogen injection and effect analysis of nitrogen injection. *Coal Technol.* **2016**, *35*, 202–203. [CrossRef]
19. Wang, Y.H.; Wu, Y.; Zhang, J.L.; Guan, Y.J. Study on technological parameters of nitrogen injection for fire prevention and extinguishing in goaf. *Min. Res. Dev.* **2019**, *39*, 96–101. [CrossRef]
20. Cui, C.F. Research on gas drainage drilling parameters and nitrogen injection for fire prevention in goaf. *J. Mine Autom.* **2020**, *46*, 12–20. [CrossRef]
21. Yang, W.; Shi, J.L.; Chen, M.Q.; Zhang, L.K. Application of step type full section buried pipe nitrogen injection technology in preventing and extinguishing fire in goaf of fully mechanized top coal caving face. *Coal Technol.* **2023**, *2*, 154–157. [CrossRef]
22. Hong, L.; Feng, C.C.; Zhou, X.H. Hongmiao mine fully mechanized caving face reasonable parameter of nitrogen injection fire prevention. *J. Liaoning Tech. Univ.* **2012**, *31*, 729–732. [CrossRef]

23. He, Z.L. Optimization of nitrogen injection parameter for U-shaped ventilation goaf. *Saf. Sci. Technol. China* **2013**, *9*, 32–37. [CrossRef]
24. Zhao, H.B.; Wang, Z.W.; Zhang, H.; Li, W. Effects of dynamic loads on development of internal microstructure and distribution of new surface fractures of coal. *Chin. J. Rock Mech. Eng.* **2016**, *5*, 971–979. [CrossRef]
25. Zhang, L.; Chen, S.; Xue, J.H.; Ren, T.X.; Zhang, C. Permeability of bituminous coal and its stress sensitivity under liquid nitrogen fracturing. *J. Min. Saf. Eng.* **2020**, *37*, 401–408. [CrossRef]
26. Zhou, J.; Hou, B.Y. Tongxin mine goaf numerical simulation of nitrogen injection and preparation technology. *Comput. Appl. Chem.* **2016**, *33*, 343–347. [CrossRef]
27. Cai, Z.J.; Yu, T.L. *Fluid Mechanics Pumps and Fans*, 4th ed.; China Building Industry Press: Beijing, China, 1999.
28. Li, Y.L.; Yao, A.L.; Li, Y.J. Study on Diffusion Model of Gas Pipeline Leaking. *Nat. Gas Ind.* **2004**, *24*, 102–104.
29. Yan, S.Y.; Pan, K. Development and application of the software on flammable, explosive & toxic cloud diffusion based on JAVA technique. *China Saf. Sci. J.* **2009**, *19*, 152–158.
30. Wang, G.D.; Ren, T.X.; Qi, Q.X.; Zhang, J.; Zhang, L. Mathematical model of coal seam methane displacement by injecting CO_2/nitrogen. *Chin. J. Rock Mech. Eng.* **2016**, *A02*, 3930–3936. [CrossRef]
31. Yu, C.Z. *Turbulent Jet*; Higher Education Press: Beijing, China, 1993.
32. Lv, Z.J.; Ou, Y.H. Distribution of three zones in goaf of the first mining face with shallow buried depth. *Saf. Coal Mines* **2018**, *49*, 238–241. [CrossRef]
33. Luo, X.R.; Jiang, S.G.; Li, Z.H. Experiment study on preventing goaf fire by nitrogen injection in fully-mechanized working faces. *J. China Univ. Min. Technol.* **1997**, *26*, 42–45.

fire

MDPI

Article

Experimental Study on the Isolation Effect of an Active Flame-Proof Device on a Gas Explosion in an Underground Coal Mine

Zichao Huang [1,2,3], Rongjun Si [2,3,*], Guangcai Wen [1,2,3], Songling Jin [4,*] and Shaoqian Xue [2,3]

1 School of Resources and Safety Engineering, Chongqing University, Chongqing 400044, China; passion_hzc@163.com (Z.H.); wgc139@126.com (G.W.)
2 Industrial Safety Research Branch, China Coal Technology and Engineering Group Chongqing Research Institute, Chongqing 400037, China; 2012109@cqccteg.com
3 State Key Laboratory of Coal Mine Disaster Prevention and Control, Chongqing 400037, China
4 State Key Laboratory of Fine Chemicals, Department of Chemical Machinery and Safety Engineering, Dalian University of Technology, Dalian 116024, China
* Correspondence: sirj73@163.com (R.S.); jinsongling@mail.dlut.edu.cn (S.J.); Tel.: +86-13-7094-25791 (R.S.)

Abstract: Passive explosion-isolation facilities in underground coal mines, such as explosion-proof water troughs and bags, face challenges aligned with current trends in intelligent and unmanned technologies, due to restricted applicability and structural features. Grounded in the propagation laws and disaster mechanisms of gas explosions, the device in this paper enables accurate identification of explosion flames and pressure information. Utilizing a high-speed processor for rapid logical processing enables judgments within 1 ms. Graded activation of the operating mechanism is enabled by the device. The tunnel flame-proof device's flame-extinguishing agent has a continuous action time of 6075 ms. Experiments on the active flame-proof effect of a 100 m^3 gas explosion were conducted using a cross-sectional 7.2 m^2 large-tunnel test system. With a dosage of 5.6 kg/m^2, the powder flame-extinguishing agent completely extinguished the explosion flame within a 20 m range behind the explosion isolator. Numerical calculations unveiled the gas-phase chemical suppression mechanism of the powder flame-extinguishing agent $NH_4H_2PO_4$ in suppressing methane explosions. Building upon these findings, application technology for active flame-proofing was developed, offering technical support for intelligent prevention and control of gas explosions in underground coal mines.

Keywords: gas explosion; active flame-proofing; flame velocity; explosion pressure; powder flame suppressor; application technology

Citation: Huang, Z.; Si, R.; Wen, G.; Jin, S.; Xue, S. Experimental Study on the Isolation Effect of an Active Flame-Proof Device on a Gas Explosion in an Underground Coal Mine. *Fire* **2023**, *6*, 468. https://doi.org/10.3390/fire6120468

Academic Editor: Thomas H. Fletcher

Received: 23 November 2023
Revised: 9 December 2023
Accepted: 12 December 2023
Published: 13 December 2023

1. Introduction

Coal stands as a cornerstone of China's energy landscape, contributing coal production of 4.56 billion tons in 2022, according to the China Coal Industry Economic Operation Report 2022. Despite advancements in alternative energy sources, coal consumption continues to hold a significant share [1,2]. The intricate nature of coal endowment conditions, coupled with the potential for disasters [3,4], means that gas explosions in coal mines can unleash high temperatures, intense pressures, shockwaves, and toxic gases. The repercussions of such explosions are profound, leading to substantial casualties and colossal economic losses [5,6]. Between 2007 and 2019, Chinese coal mines experienced a staggering total of 456 gas-related accidents, resulting in 2112 fatalities. Of these incidents, 223 were classified as gas explosion accidents. This alarming data underscore the persistently severe challenges facing coal mine safety in relation to production [7,8].

Presently, passive explosion-isolation technologies are widely employed in underground coal mines. These technologies leverage the temporal discrepancy between the explosion shockwave and the advancing flame front during the explosion propagation process. Utilizing the shockwave as a triggering force, passive explosion-isolation facilities, including rock powder sheds, explosion-isolation water troughs, and explosion-isolation water bags, create a suppression zone ahead of the flame, halting the further spread of the explosion. The effectiveness of passive explosion-isolation technology facilities depends on various factors, such as installation location and explosion energy. However, their activation often lacks synchronization with the arrival of the explosion flame, and the short-lived explosion-isolation water curtain limits their performance. Additionally, these facilities are typically disposable, rendering them ineffective in the face of secondary or multiple explosions within a coal mine [9,10]. The operational range of passive explosion-isolation facilities typically spans 60 to 140 m from the explosion source. Furthermore, facilities like explosion-isolation water sheds necessitate regular watering for dust removal and frequent relocation, resulting in a substantial daily maintenance workload. Given their structural limitations and on-site applicability, these facilities struggle to meet the evolving demands of intelligent and unmanned coal mining practices.

Due to inherent limitations in technical performance, daily management, and maintenance of passive explosion-isolation facilities, there are constraints on their effectiveness in isolating explosions. In recent years, active explosion-isolation technologies, exemplified by automatic fine water mist explosion-isolation systems [11,12] and automatic powder spray explosion-isolation systems [13–15], have garnered significant attention. Active flame-proof technology proves instrumental in effectively isolating secondary or multiple gas explosions. Shao et al. [16,17] devised a vacuum chamber installed on one side of an experimental pipeline. In the event of a gas explosion, the negative pressure inside the vacuum chamber draws in the flame and shockwave, thereby mitigating the explosion shockwave and suppressing the flame's propagation. Jiang et al. [18] developed an automatic spraying system employing nitrogen and ABC powder to suppress gas explosions. Lu et al. [19] engineered integrated equipment that automatically detects flames and dispenses suppressants, evaluating the suppressive effects of active flame-proof devices positioned at various ventilation points during gas explosions. Cui et al. [20] created equipment involving a vacuum chamber coupled with CO_2. The injection of CO_2 into the vacuum chamber enhances the suppressive effect of an active flame-proof device on gas explosions, reducing the cost of maintaining high vacuum levels.

The previous studies predominantly focused on a flame-proof device that suppresses the spread of gas explosion flames, with a limited exploration into flame detection. Active flame-proof technology should comprehensively employ sensors to collect explosion information. It utilizes a control system to guide the injection system in creating explosion suppression zones, demonstrating the capability to suppress many explosions, demanding high system stability. Therefore, a comprehensive active flame-proof device that combines flame detection and explosion suppression is crucial for coal mine safety. To effectively advance the intelligent prevention and control of gas and coal-dust explosions in coal mines, and to promote practical engineering applications, this study applies the large underground tunnel test system of the Chongqing Research Institute of China Coal Technology and Engineering Group Corporation. Combining the requirements of intelligent coal mining construction, an intelligent active flame-proof device was developed. The study investigates the dosage and application technology of powder flame-extinguishing agents under different explosion conditions, providing important theoretical and technical support for the intelligent control of gas and coal-dust explosions in underground coal mines. Additionally, through numerical calculations, the study reveals the gas-phase chemical suppression mechanism of the flame suppressant $NH_4H_2PO_4$ powder in suppressing methane explosions. The proposed application technology of active flame-proof devices contributes significantly to the intelligent management of gas and coal-dust explosions in underground coal mines.

2. Principles of Active Flame-Proof Technology

An active flame-proof device typically encompasses flame sensors, pressure sensors, controllers, tunnel isolators, and power sources, as depicted in Figure 1. The operational principle involves advanced detection of explosion flames and pressure information, following an explosion incident. The controller promptly triggers the isolator's activation, initiating the release of powder flame-extinguishing agents to create an explosion-isolation barrier, covering a specific area. This action extinguishes the explosion flames, mitigates the explosion pressure waves, and hinders the further spread of the explosion.

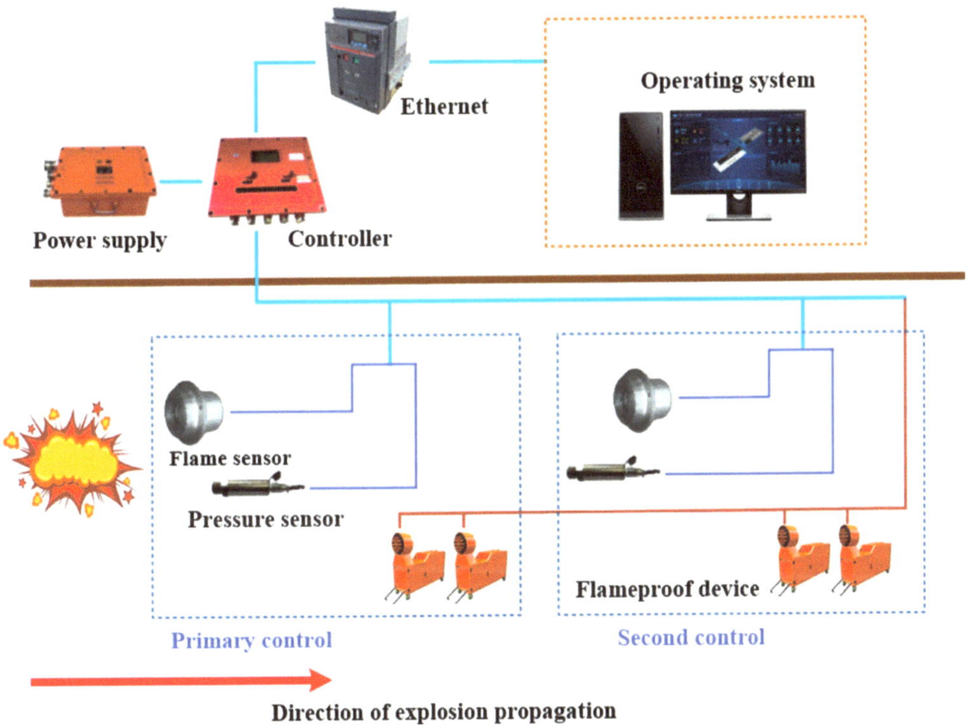

Figure 1. Diagram of the active flameproof device.

By considering the positional relationship and characteristic changes in flames and pressure waves during various stages of explosion propagation, a comprehensive detection and analysis of the explosion flame and pressure information is carried out. This leads to the graded activation of different quantities of isolators. The explosion-isolation effectiveness of active-isolation devices is influenced by the cross-sectional dimensions of the tunnel, the location of potential explosion sources, and the explosive equivalent. In light of these conditions, the technical indicators of the isolator, the powder mass concentration of the flame-extinguishing agent, and the installation position, emerge as direct factors influencing the explosion-isolation effectiveness. They are also pivotal elements in the application technology of active flame-proof devices.

3. Experimental Setup

3.1. Flame and Pressure Sensors

To mitigate false triggering caused by sunlight in the ultraviolet wavelength band and underground mining lights, the flame sensor utilizes a narrow-band spectrum with a spectral response range of 185 to 260 nm. The flame sensor circuit incorporates background light detection. Through an intelligent analysis of light intensity and filtering, it can effectively differentiate between explosion flames and ambient light sources. Additionally, the two ultraviolet photosensitive elements operate in an AND relationship, ensuring high reliability. The high frequency of ultraviolet light enables swift detection in the early stages of a flame, resulting in a rapid response time. The duration from the sensor receiving the flame signal to outputting the electrical signal is within 1 ms.

The pressure sensor employs a pressure-resistant silicon crystal fabricated through MEMS processing for detecting pressure signals. This crystal is utilized to measure the dynamic pressure resulting from gas explosions. The upper-limit frequency response of dynamic pressure measurement is dependent on the design form and placement position of the diaphragm. Given the small mass of the silicon diaphragm, the frequency response exceeds 10 kHz, with a rapid sensor response time of 2 ms. The sensor circuit encompasses measurement components, signal amplification circuitry, A/D conversion circuitry, processing unit, control output, and other integral parts. It has a pressure-sensor activation threshold of 25 kPa. Flame and pressure sensors were shown in Figure 2.

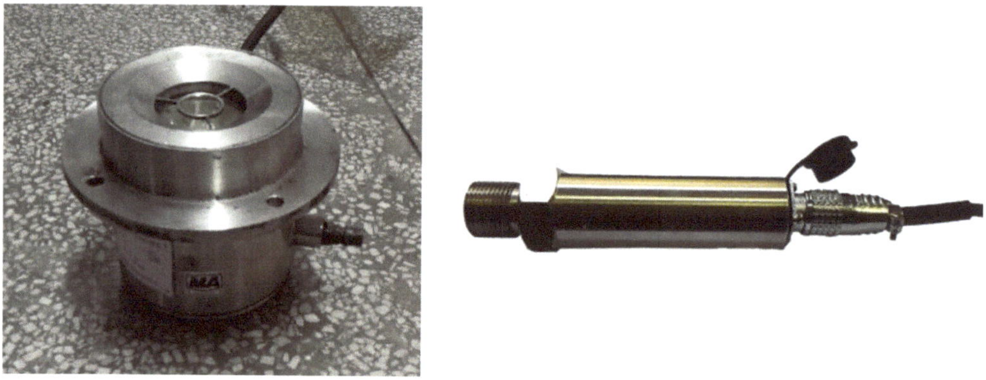

(a) (b)

Figure 2. (**a**) flame sensor (**b**) pressure sensor.

3.2. *Graded Start Controller for Active Flame Proofing*

The controller serves as the central component of the tunnel explosion-isolation device, exerting a pivotal influence on the execution and effectiveness of explosion-isolation actions. As illustrated in Figure 3, it receives detection signals from the sensors, scrutinizes and evaluates the information, and generates control signals to initiate the isolator's action. The controller can connect to, and power, two flame sensors and two pressure sensors. It monitors the operational status of each flame sensor and pressure sensor, while concurrently receiving signals from multiple flame and pressure sensors. Intelligently designed, the controller boasts features such as automatic fault diagnosis, automatic sensor and isolator detection, power supply monitoring, and prompt information-transmission functions. The controller adheres to communication protocols to both receive and provide feedback on relevant data, offering real-time insights into the system's operational status, to a remote terminal. Equipped with a high-speed processor, the controller conducts logical processing, intelligently and precisely identifying flame and pressure signals, and rendering logical judgments within 1 ms. Based on the laws governing flame and pressure propagation during explosions, the controller can establish distinct triggering thresholds for pressure sensors. By considering the detected location and pressure magnitude of the flame and pressure signals, it achieves "stepwise control, multi-level activation". The controller also stores triggering information and displays status indications on the screen.

Figure 3. The controller.

3.3. Tunnel Isolator

The tunnel isolator serves as the central element of the tunnel explosion-isolation device, predominantly comprised of a tank body, a rapid opening structure, spray pipelines, and a powder spraying mechanism, as illustrated in Figure 4. Operating on the principle of stored pressure, the isolator stores powder flame-extinguishing agents and high-pressure inert gas, functioning as the terminal execution mechanism to isolate gas explosions within the tunnel. The isolator's tank body adopts a standard high-pressure vessel structure, featuring a design with a large-capacity storage unit that can accommodate 30 kg of powder. It utilizes high-pressure nitrogen gas as the driving force, operating at a pressure of 8.0 ± 1.0 MPa, and holds an IP65 protection rating. The rapid-opening mechanism of the isolator employs an electronic trigger to rupture the diaphragm, enabling the isolator opening valve to swiftly open within 2 ms. This establishes a two-phase explosion-isolation barrier comprised of gas and powder, substantially amplifying the efficacy of explosion isolation.

Figure 4. The active flame-proof device.

The dispersion characteristics of the sprayed flame-extinguishing agent in space significantly influence the ultimate explosion-isolation effectiveness of the isolator. To assess the spraying effect of the tunnel isolator, high-speed camera technology was employed, with a capture frame rate of 1000 fps. The spraying experiment utilized a simulated signal to trigger the opening mechanism, acting as the starting point for action control. The final configuration of the powder spraying mechanism was determined by adjusting the nozzle angle, quantity, and distribution method. Through image analysis, the lag time, fogging time, and duration of action of the isolator were determined. According to the test results, the powder flame-extinguishing agent is rapidly sprayed under high-pressure driving, forming a cone-shaped cloud with an initial area of 8.04 m^2 within the first 120 ms. It propels forward rapidly, while simultaneously diffusing radially. The effective spraying distance can reach up to 30 m, and the effective action time of the extinguishing agent in space extends to 6075 ms. The effect of the powder cloud in the spraying experiment is depicted in Figure 5.

Figure 5. Distribution of cloud curtain of powders.

3.4. *Experimental System and Methodology*

The experimental tests were conducted to assess the effectiveness of flame-extinguishing agents in isolating explosion propagation under varying gas explosion intensity conditions. These experiments were performed using an engineering-scale large underground-explosion tunnel test system. The experimental system consists primarily of a 7.2 m^2 cross-sectional underground experimental tunnel, an explosion data acquisition system, a gas-distribution and circulation system, an ignition device, and a hydraulic system. By adjusting conditions such as the length of the gas chamber, the system can facilitate gas explosion propagation and isolation experiments on a realistic scale. The system arrangement is depicted in Figure 6.

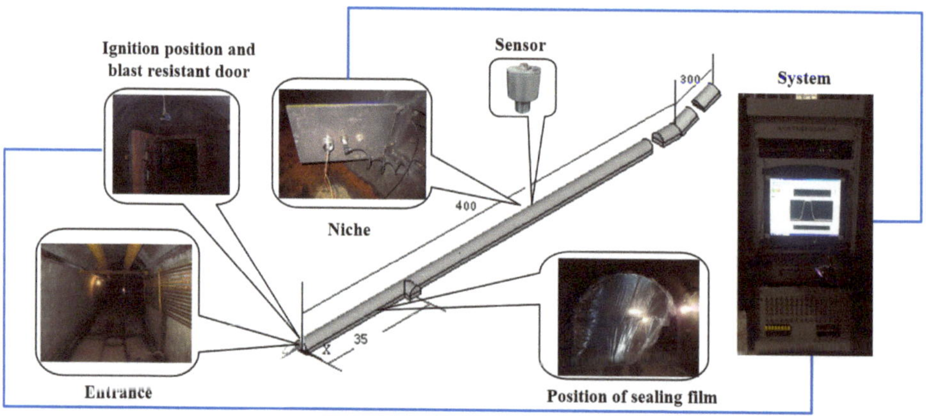

Figure 6. Layout of the large laneway test system.

The underground experimental tunnel spans a total length of 896 m, with the main section designated for explosion testing extending to 710 m. The cross-section exhibits a semi-circular arch shape, featuring both flat and inclined sections with a 24° slope. Double explosion-proof doors are positioned at the ends of two ignition chambers. The closure of these doors divides the tunnel into a closed section and an open section. Sealing rings for closed sections are strategically placed at intervals of 4.15 m, 7 m, 14 m, and 28 m, starting from the closed end of the tunnel. By sealing these points with plastic film, gas explosion chambers of four distinct capacities (30 m^3, 50 m^3, 100 m^3, and 200 m^3) can be configured. For the initial 40 m from the starting end, measurement points along the tunnel walls are spaced at 10 m intervals. Beyond the 40 m mark, measurement points are set at intervals of 20 m. The arrangement of sensors within the tunnel is illustrated in Figure 7.

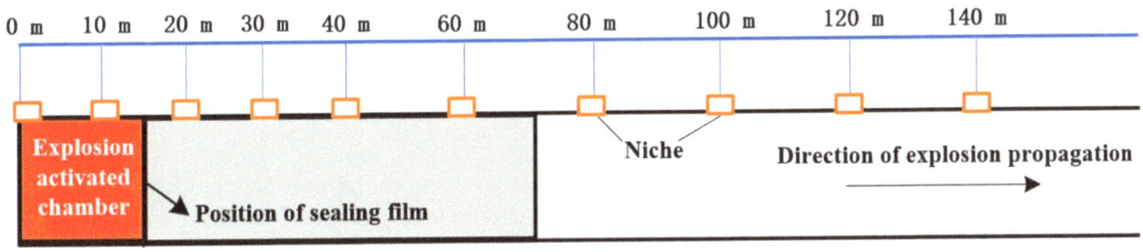

Figure 7. Schematic diagram showing the layout of instrument niches.

Utilizing the experimental system, simulations were executed to assess the explosion-isolation capabilities for both a 100 m^3 gas explosion and a 100 m^3 gas explosion, each subjected to varying mass concentrations of powder flame-extinguishing agents. The experiments emphasized the critical need to establish an effective explosion-isolation barrier before the explosion flame reached the specified point. Consequently, the controller was configured to trigger the tunnel isolator's spray mechanism, simultaneously with the initiation of the explosion experiment. The isolation experiments employed ABC ultrafine dry powder as the flame-extinguishing agent, with its primary component being $NH_4H_2PO_4$. Variations in explosion-isolation effectiveness were explored by adjusting the powder filling amount or the number of isolators to modify the extinguishing agent usage. Preceding the experiment, gas was introduced into the gas explosion chamber, via an inflation pipeline, generating a gas–air explosive mixture, through circulation. In both the explosion and isolation experiments, an igniter was employed to ignite the gas mixture. Synchronous data collection of the explosion flame's propagation speed and range during the experiment was conducted, using the PXIe signal processing module, featuring multiple channels and a sampling rate of 1 million samples per second. To evaluate the performance and effectiveness of the isolation device under different explosion intensity conditions, the tunnel isolator was positioned at the front end, where the maximum gas explosion pressure and maximum flame speed occurred. This setup allowed for a comparative analysis with the explosion characteristic values observed in experiments with equivalent propagation. In the 100 m^3 gas explosion-isolation experiment, the tunnel isolator was installed within the range of 12–15 m from the closed end inside the main tunnel, based on the results of explosion propagation experiments.

At a position 14 m from the closed end of the tunnel, a volume of 100 m^3 was isolated using a membrane as the ignition chamber. At a position 12–15 m from the closed end, an explosion-isolation device was installed. Three different quantities of ABC powder suppressors, specifically 20 kg, 30 kg, and 40 kg, were chosen for suppression tests. The corresponding mass densities of the suppressors are approximately 2.8 kg/m^2, 4.2 kg/m^2, and 5.6 kg/m^2, respectively.

3.5. Numerical Calculations

CHEMKIN-PRO was studied for the gas-phase chemical suppression mechanism of $NH_4H_2PO_4$ in suppressing methane explosions. The chemical reactions of CH_4/air were modeled using the GRI-Mech 3.0 model. The $NH_4H_2PO_4$ powder completely decomposes into gaseous substances NH_3 and H_3PO_4 in the flame preheating zone. The mechanism model for phosphorus-containing substances involving H_3PO_4 is based on the research data of Pitz et al. [21], Twarowski [22], and Korobeinichev et al. [23]. The mechanism model for nitrogen-containing substances involving NH_3 is based on the research data of Konnov [24] and Li et al. [25]. These mechanism models collectively encompass 86 species and 573 reactions. A Closed 0–D Homogeneous batch reactor was employed, with a constrained volume and energy-solution equation approach. The reactor volume was set at 100 m^3, the initial temperature at 1200 K, and the ambient pressure at 1 atm. Methane was used as a substitute for gas during numerical calculations, considering that methane is the primary component of the gas.

4. Results and Discussion

4.1. Gas-Explosion-Flame Velocity and Pressure

The progression of the explosion flame serves as a direct indicator of the explosion's intensity and hazard, offering a visual representation of the destructive energy unleashed by the explosion. Concurrently, the explosion pressure stands as a pivotal parameter for gauging the reaction process and the extent of damage caused by the explosion. To quantify the average speed of flame propagation between two adjacent measurement points, the calculated arrival times of the explosion flame at various measurement locations and the extent of flame coverage in explosion tests are instrumental. The determination can be expressed as follows:

$$v = L/(t_2 - t_1) \quad (1)$$

In a 7.2 m^2 large underground experimental tunnel, a gas explosion with a volume fraction of 9.5 vol% in a 100 m^3 chamber was simulated. Figure 8 depicts the changes in flame speed and explosion pressure in relation to propagation distance.

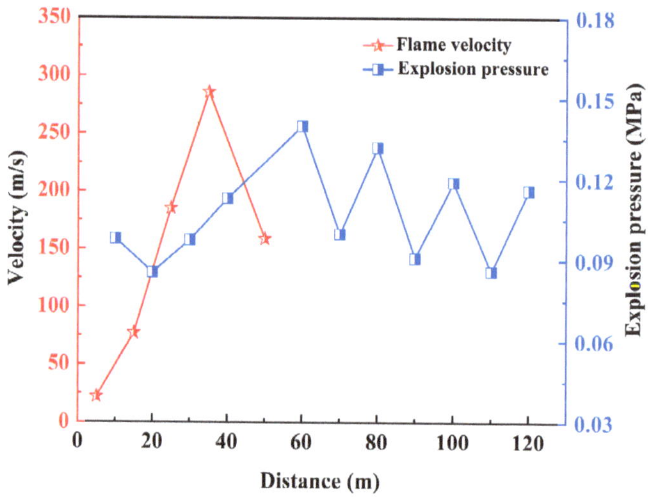

Figure 8. Flame velocity and explosion pressure for gas explosion.

Examining the graph, it is evident that the maximum explosion pressure at various measurement points does not consistently decrease with the increasing propagation distance of the gas explosion. Near the explosion source, combustible gas ignites, initiating the rapid outward expansion of the explosion wave from the point source. The gas moves forward under the influence of the compression wave, reaching a peak pressure before diminishing. Due to the spatial constraints of the tunnel and the forward progression of the explosion wave, more gas undergoes ignition, resulting in a rapid increase in explosion pressure. As the explosion reaction proceeds and no additional gas participates in the reaction, under unchanged conditions, the subsequent explosion pressures at various measurement points gradually decrease. During the 100 m^3 gas explosion process, the maximum explosion pressure occurs at the 60 m measurement point, reaching 0.14 MPa. In the initial stage of the explosion, the flame-propagation speed exceeds the expansion speed of the pressure wave. As the explosion unfolds, a plane pressure wave forms, surpassing the flame front, creating a typical "two waves, three zones" structure. In the gas explosion experiment, influenced by the volume of combustible gas, the range covered by the explosion flame is 60 m. This aligns with the understanding that the length of the explosion flame zone is 3–5 times the initial gas accumulation area. The maximum flame-propagation speed generally corresponds to the region with the maximum explosion pressure, occurring in the range of 30 m to 40 m, with a peak value of 285.7 m/s. Beyond 60 m, the explosion pressure continues to propagate, but no explosion flame is detected.

4.2. Impact of Flame Suppressants on Gas-Explosion-Flame Velocity

At a gas volume fraction of 9.5 vol%, Figure 9 illustrates the variations in explosion-flame arrival positions and propagation speeds of a 100 m^3 gas explosion, influenced by different quantities of ABC powder flame suppressant.

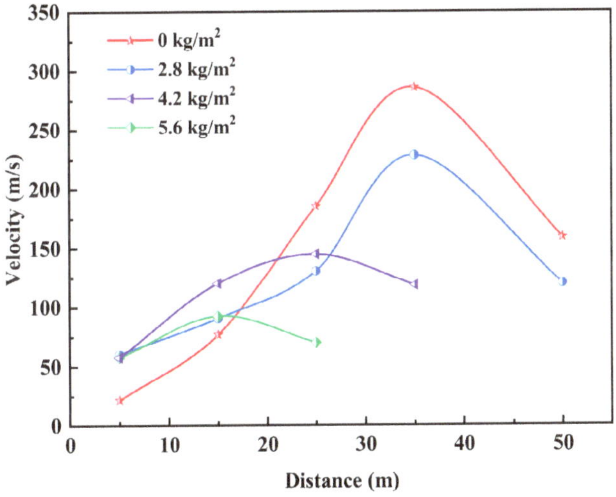

Figure 9. Variation of flame propagation in 100 m^3 gas flameproof experiments.

When the energy released by the combustion explosion surpasses the heat consumed and lost during the action of the suppressors, the explosion persists. In the explosion-isolation experiment, the tunnel explosion-isolation device sprayed ABC powder suppressors, to create an effective explosion-isolation barrier, extinguishing the flames. With the active flame-proof device in place, there exists a buffer zone for the explosion flame under the impact of the explosion pressure, leading to a notable reduction in both flame area and speed. However, at lower concentrations of ABC powder suppressors, the explosion-isolation barrier fails to effectively extinguish the flames. As the concentration of the flame suppressant increases, the explosion-isolation effect improves. In the explosion-propagation experiment, the flame range of a single 100 m^3 gas explosion, with a volume fraction of 9.5 vol%, is 50 m. In the 100 m^3 gas explosion-isolation experiment, when the powder mass is 40 kg (with a concentration of 5.6 kg/m^2), the maximum flame speed shifts from the 30–40 m range to the 10–20 m range, with the maximum value decreasing from

285.7 m/s to 92.6 m/s, with the maximum decrease in flame speed being 67.59%. The explosion flame is completely extinguished within 20 m behind the explosion-isolation device. Based on the analysis of the explosion-isolation experiment results, the active flame-proof device promptly detects explosion flame and pressure information, acts in the early stages of the explosion, and has a prolonged duration of action, effectively impeding the propagation of gas explosion flames. In the explosion-isolation experiment, when the mass concentration of ABC powder suppressors reaches a certain level, it reduces the thermal effect, leading to the extinguishing of the explosion flame.

4.3. Impact of Flame Suppressants on Gas Explosion Pressure

Figure 10 depicts the fluctuation in maximum explosion pressure in relation to propagation distance in gas explosion-isolation experiments, under different concentrations of powder media.

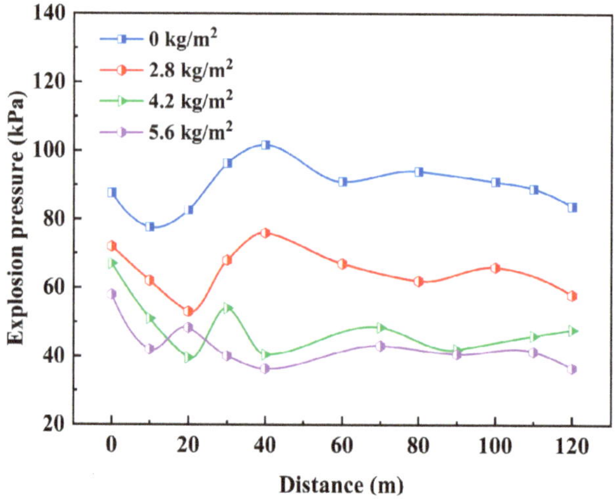

Figure 10. Effects on explosion pressure under different concentrations of powders.

From the graph, it is evident that the application of ABC powder suppressors effectively mitigates the peak explosion pressure. In all three sets of suppression experiments, the maximum explosion pressures at various distances exhibit varying degrees of reduction, indicating the suppression effect of ABC powder suppressors on gas explosions. Notably, in suppression experiments with different mass densities, there is a discernible difference in the attenuation of peak pressures, at various distances. Specifically, as the powder-mass density increases, the degree of attenuation of peak pressure continues to rise. Although the active flame-proof system cannot eliminate the explosion pressure wave, it significantly diminishes its intensity and destructiveness. In the gas explosion-isolation experiment, when the powder concentration is 2.8 kg/m^2, the maximum explosion pressure decreases, and the explosion overpressure at the same location is reduced by 25%, indicating a certain suppression effect on the explosion. As the powder concentration increases, the isolation system's effectiveness in reducing the shock wave gradually intensifies. When the powder concentration is 4.2 kg/m^2 and 5.6 kg/m^2, the maximum explosion pressure occurs near the explosion source, and the peak pressures at various distances behind the isolation device remain below 60 kPa, throughout the explosion process. Along the tunnel, they exhibit a gradual attenuation trend. The maximum explosion pressure at the rear of the isolation system decreases to 40.5 kPa and 36.4 kPa, respectively, representing a reduction of 60.2% and 64.2%, compared to the maximum pressure at the same location in the experiment without the isolation device.

4.4. Gas-Phase Chemical-Suppression Mechanism of Ammonium Dihydrogen Phosphate on Methane Explosions

In the process of methane explosions, chain reactions in chemical reactions are primarily driven by free radicals O, H, and OH. The chemical reaction rates of these free radicals determine the propagation speed and severity of the explosion flame [26,27]. Preventing the ignition process can help avoid certain accidents. Figure 11 illustrates the molar fraction changes of CH_4, O_2, CO_2, H_2O, H, O, OH, and CO after the explosion of 9.5 vol% methane. The concentrations of H, O, OH, and CO exhibit abrupt peaks, indicating the generation of a large number of free radicals in the initial reaction stage. Subsequent chain reactions consume some of these free radicals, leading to a decrease in their concentrations. Figure 12 shows the impact of 5.6 kg/m^2 $NH_4H_2PO_4$ powder on the molar fractions of various components in a 9.5 vol% methane explosion. The addition of the flame suppressant results in a reduction in the molar fractions of CO and CO_2.

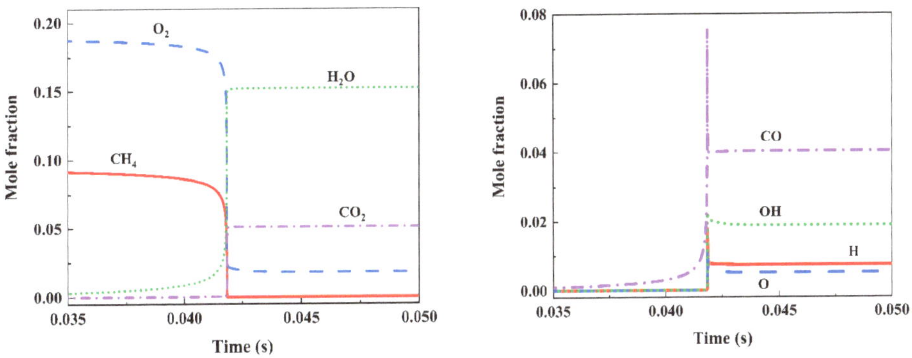

Figure 11. Variation of mole fraction of components in 9.5 vol% methane explosion.

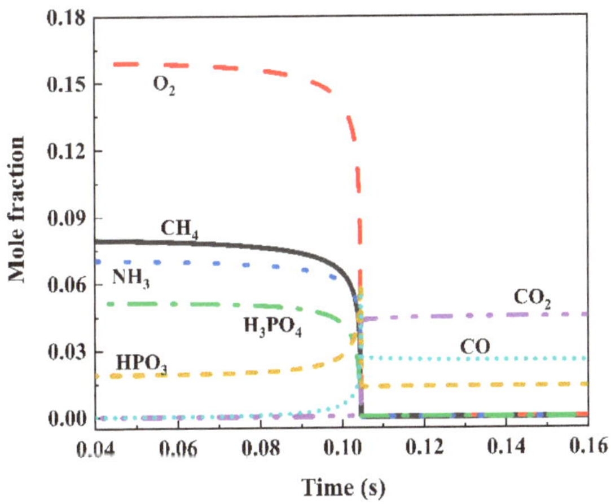

Figure 12. A total of 5.6 kg/m^2 $NH_4H_2PO_4$ inhibits the change in mole fraction of each component in 9.5 vol% methane explosion.

Known to play a crucial supporting role in flame development, H, O, and OH undergo significant changes in their molar fractions upon the addition of $NH_4H_2PO_4$ powder at different concentrations, as depicted in Figure 13. Following the introduction of the flame-suppressant $NH_4H_2PO_4$ powder, the concentrations of the three free radicals—H, O, and OH—markedly decreased. $NH_4H_2PO_4$ exhibits effective clearance effects on these three free radicals, contributing to the suppression and development of the flame.

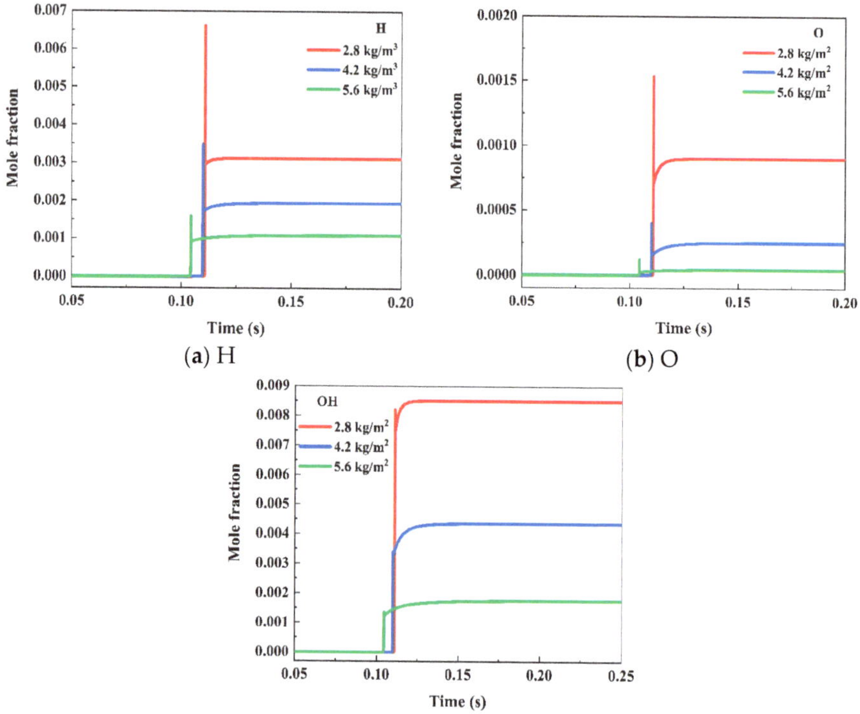

(**a**) H (**b**) O (**c**) OH

Figure 13. Effects of $NH_4H_2PO_4$ on H, O, and OH at different concentrations.

The key intermediate product yields in $NH_4H_2PO_4$ powder suppressing methane explosions are illustrated in Figure 14. The reactions depicted in the figure play a crucial role in eliminating the free radicals H, O, and OH. These reactions consist of three groups, forming a catalytic cycle:

$$PO_2 + H + M = HOPO + M$$

$$HOPO + OH = PO_2 + H_2O$$

$$HOPO + O = OH + PO_2$$

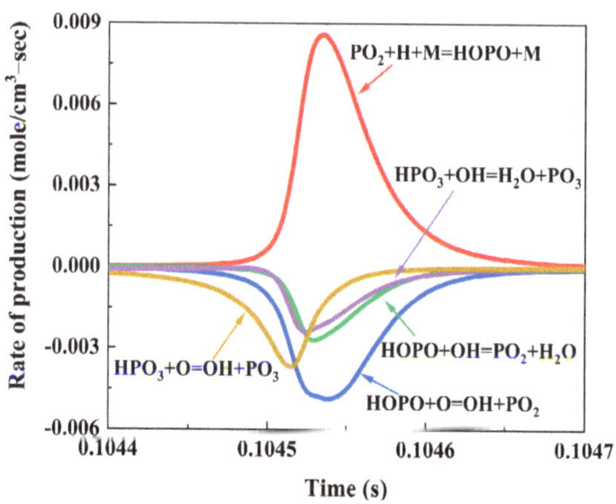

Figure 14. Rate of production of P due to reactions in $CH_4/NH_4H_2PO_4/air$ flame.

This catalytic cycle continuously catalyzes the consumption and generation of crucial intermediate products, PO_2 and HOPO, derived from the decomposition of the flame suppressant $NH_4H_2PO_4$, forming a reversible reaction $PO_2 \Leftrightarrow HOPO$. Additionally, the critical intermediate product HPO_3 generated from the decomposition of $NH_4H_2PO_4$ combines with free radicals O and OH to produce H_2O and PO_3:

$$HPO_3 + OH = H_2O + PO_3$$

$$HPO_3 + O = OH + PO_3$$

These two reactions convert highly reactive O and OH into stable combustion products, H_2O, and other free radicals, which are less relevant to the chain reactions of methane, playing a role in terminating the chain reactions. The catalytic recombination of these critical intermediate products and the removal of key free radicals result in a reduction in the quantities of free radicals H, O, and OH, leading to a decrease in chain branching and a reduction in heat release. The decomposition of the flame suppressant $NH_4H_2PO_4$, producing NH_3, consumes O_2, to generate H_2O and N_2, thereby reducing the concentration of O_2 within a unit volume and decreasing the collision probability between CH_4 and O_2 and O free radicals.

5. Application Technology of Active Flame-Proof Devices

In compliance with the stipulations outlined in the 'Coal Mine Safety Regulations' and based on a comprehensive analysis of the risks associated with gas explosions in underground coal mines, it is imperative to install explosion-proof devices in areas susceptible to explosions. These areas encompass coal galleries, semi-coal–rock galleries in the advancing working face, intake and return air-tunnels in the coal mining face, adjacent coal mining faces, adjacent mining areas, and connecting tunnels.

The active flame-proof device, designed for proactive detection and precise control of explosion incidents, finds applicability in locations within underground coal mines prone to the risk of gas or coal-dust explosions. It promptly extinguishes explosion flames following an eruption, thereby mitigating the severity of the accident. During on-site implementation, the design considerations should revolve around the shape and dimensions of the tunnel section, potential sources of explosion, and explosion equivalents. This determination is crucial for establishing the optimal number and placement of explosion-isolation devices. The quantity of powder flame suppressant must be computed based on the tunnel section's area. For effectively isolating gas explosions, the required quantity of flame suppressant should exceed 1.5 times the unit quantity derived from type testing of the active flame-proof device, multiplied by the tunnel section's area, as per Equation (2).

$$m \geq 1.5 \times c \times S \quad (2)$$

In the equation: m—the required quantity of flame suppressant for isolating gas explosions, kg; c—the unit-area quantity of flame suppressant determined in the inspection of the tunnel explosion-isolation device, kg/m^2; and S—the area of the tunnel section, m^2.

During on-site application, the flame and pressure sensors of the active flame-proof device must feature anti-pollution measures on their windows or probes, with these elements oriented toward potential hazard sources. The flame sensor should be positioned within 5 m of the potential explosion source. Installation of the tunnel explosion-isolation device should not disrupt normal tunnel operation. The isolator nozzles should be directed toward the hazard source, and the on-site configuration must guarantee that the emitted mist from the flame-suppressant spray effectively covers either the potential hazard source or the entire cross-section of the tunnel.

When deploying an active flame-proof device in a coal mine's explosive hazard tunnel, a system of active flame-proof devices should be installed in the straight tunnel, within a 200 m range behind the suppression device. Moreover, at intersections, changes in slope and turns in other tunnels susceptible to gas and coal-dust explosions, a system of active flame-proof devices should be installed in the straight tunnel within a 60 m range from the high-risk point.

6. Conclusions

This study conducted independent research into, and the development of, an active flame-proof device and gas explosion isolation. The key findings are as follows:

- We developed an active flame-proof device with flame and pressure sensors, a start controller, and tunnel explosion isolators. The device uses spectral recognition and dual ultraviolet sensors for flame detection, and pressure-resistant silicon crystal components for precise identification. The controller makes rapid decisions within 1 ms, allowing for graded activation. The tunnel explosion isolator has large-capacity storage and efficient spraying, for a continuous action time of up to 6075 ms.
- We conducted experiments on 100 m^3 gas explosions to evaluate the effectiveness of a powder flame-extinguishing agent. The explosion barrier significantly reduce the flame-propagation range, effectively stopping its spread within a 20 m range behind the isolation device.
- We investigated the gas-phase chemical-suppression mechanism of $NH_4H_2PO_4$ in suppressing methane explosions. The flame suppressant's intermediate products play a significant role in eliminating free radicals, reducing chain branching, and in heat release.
- We formulated an application technology for active flame-proof devices in underground coal mines. Guidelines for device installation, calculation of extinguishing agent dosage, and determination of protective zone range were established, supporting intelligent explosion prevention and control.

Author Contributions: Conceptualization, Z.H. and R.S.; methodology, G.W.; software, S.J.; validation, Z.H. and S.X.; formal analysis, G.W.; data curation, Z.H.; writing—original draft preparation, Z.H.; writing—review and editing, R.S.; project administration, Z.H., R.S. and G.W. All authors have read and agreed to the published version of the manuscript.

Funding: This research was funded by the National Nature Science Foundation of China (52104238, 52174226, and 52174227) the Natural Science Foundation of Chongqing, China (CSTB2022NSCQ-MSX0867), and the grants from the State Key Laboratory of Gas Disaster Detecting, Preventing and Emergency Controlling (2022SKLKF03).

Institutional Review Board Statement: Not applicable.

Informed Consent Statement: Not applicable.

Data Availability Statement: Data available on request, due to restrictions.

Conflicts of Interest: The authors declare no conflict of interest.

References

1. Teng, J.; Li, S.; Jia, M.; Lian, J.; Liu, H.; Liu, G.; Wang, W.; Volker, S.; Feng, L.; Yao, X.; et al. Research and application of in-seam seismic survey technology for disaster-causing potential geology anomalous body in coal seam. *Acta Geol. Sin. Engl. Ed.* **2020**, *94*, 10–26. [CrossRef]
2. Wang, Q.; Lu, X.; Ma, C.; Luo, Z.; Li, Q.; Deng, J.; Sheng, Y.; Peng, B. Comparative study of the kinetic characteristics of coal spontaneous combustion. *J. Therm. Anal. Calorim.* **2023**, *148*, 4463–4476. [CrossRef]
3. Meng, X.; Liu, Q.; Luo, X.; Zhou, X. Risk assessment of the unsafe behaviours of humans in fatal gas explosion accidents in China's underground coal mines. *J. Clean. Prod.* **2019**, *210*, 970–976. [CrossRef]
4. Cheng, C.; Si, R.; Wang, L.; Jia, Q.; Xin, C. Explosion and explosion suppression of gas/deposited coal dust in a realistic environment. *Fuel* **2024**, *357*, 129710. [CrossRef]
5. Lin, C.L.; Chien, C.F. Systems thinking in a gas explosion accident–lessons learned from Taiwan. *J. Loss Prev. Process Ind.* **2019**, *62*, 103987. [CrossRef]
6. Wang, K.; Jiang, S.; Ma, X.; Wu, Z.; Zhang, W.; Shao, H. Study of the destruction of ventilation systems in coal mines due to gas explosions. *Powder Technol.* **2015**, *286*, 401–411. [CrossRef]

7. Zhang, J.; Xu, K.; Reniers, G.; You, G. Statistical analysis the characteristics of extraordinarily severe coal mine accidents (ESCMAs) in China from 1950 to 2018. *Process Saf. Environ. Prot.* **2020**, *133*, 332–340. [CrossRef]
8. Zhang, J.; Fu, J.; Hao, H.; Fu, G.; Nie, F.; Zhang, W. Root causes of coal mine accidents: Characteristics of safety culture deficiencies based on accident statistics. *Process Saf. Environ. Prot.* **2020**, *136*, 78–91. [CrossRef]
9. Liu, W.; Mu, C.; Li, Z. Influence of cavity structure on gas explosion characteristics in coal mine. *Powder Technol.* **2022**, *398*, 117084. [CrossRef]
10. Cao, X.; Wang, C.; Wang, Y.; Wang, Z.; Wei, H.; Lu, Y. Research on the inhibition characteristics of ultrafine water mist on gas/dust two-phase mixture explosions. *Fuel* **2024**, *357*, 129967. [CrossRef]
11. Yang, K.; Zhang, P.; Yue, C.; Chen, K.; Ji, H.; Xing, Z.; Hao, Y.; Jiang, J. Experimental research on methane/air explosion inhibition using ultrafine water mist containing methane oxidizing bacteria. *J. Loss Prev. Process Ind.* **2020**, *67*, 104256. [CrossRef]
12. Song, Y.; Zhang, Q. Quantitative research on gas explosion inhibition by water mist. *J. Hazard. Mater.* **2019**, *363*, 16–25. [CrossRef] [PubMed]
13. Luo, Z.; Wang, T.; Tian, Z.; Cheng, F.; Deng, J.; Zhang, Y. Experimental study on the suppression of gas explosion using the gas–solid suppressant of CO_2/ABC powder. *J. Loss Prev. Process Ind.* **2014**, *30*, 17–23. [CrossRef]
14. Song, Y.; Zhang, Q. The quantitative studies on gas explosion suppression by an inert rock dust deposit. *J. Hazard. Mater.* **2018**, *353*, 62–69. [CrossRef]
15. Zhao, T.; Chen, X.; Luo, Z.; Cheng, F.; Lu, K.; Shi, X.; Yu, W. Effect of N_2 inerting on the inhibition of methane explosions by a multicomponent powder. *Fuel* **2023**, *337*, 127203. [CrossRef]
16. Shao, H.; Jiang, S.; Zhang, X.; Wu, Z.; Wang, K.; Zhang, W. Influence of vacuum degree on the effect of gas explosion suppression by vacuum chamber. *J. Loss Prev. Process Ind.* **2015**, *38*, 214–223. [CrossRef]
17. Shao, H.; Jiang, S.; Wu, Z.; Zhang, W.; Wang, K. Influence of diaphragm thickness on gas explosion suppression by vacuum chamber. *Powder Technol.* **2016**, *295*, 245–253. [CrossRef]
18. Jiang, B.; Liu, Z.; Tang, M.; Yang, K.; Lv, P.; Lin, B. Active suppression of premixed methane/air explosion propagation by non-premixed suppressant with nitrogen and ABC powder in a semi-confined duct. *J. Nat. Gas Sci. Eng.* **2016**, *29*, 141–149. [CrossRef]
19. Lu, C.; Zhang, Y.; Zhu, H.; Meng, Q.; Pan, R.; Yu, M. Spurting $NH_4H_2PO_4$ powder to prevent the propagation of gas explosion along the duct. *Combust. Sci. Technol.* **2021**, *193*, 2534–2552. [CrossRef]
20. Cui, C.; Shao, H.; Jiang, S.; Zhang, X. Experimental study on gas explosion suppression by coupling CO_2 to a vacuum chamber. *Powder Technol.* **2018**, *335*, 42–53. [CrossRef]
21. Dounia, O.; Vermorel, O.; Poinsot, T. Theoretical analysis and simulation of methane/air flame inhibition by sodium bicarbonate particles. *Combust. Flame* **2018**, *193*, 313–326. [CrossRef]
22. Twarowski, A. The temperature dependence of H + OH recombination in phosphorus oxide containing post-combustion gases. *Combust. Flame* **1996**, *105*, 407–413. [CrossRef]
23. Korobeinichev, O.P.; Shvartsberg, V.M.; Shmakov, A.G.; Knyazkov, D.A.; Rybitskaya, I.V. Inhibition of atmospheric lean and rich CH_4/O_2/Ar flames by phosphorus-containing compound. *Proc. Combust. Inst.* **2007**, *31*, 2741–2748. [CrossRef]
24. Konnov, A.A. Implementation of the NCN pathway of prompt-NO formation in the detailed reaction mechanism. *Combust. Flame* **2009**, *156*, 2093–2105. [CrossRef]
25. Li, B.; He, Y.; Li, Z.; Konnov, A.A. Measurements of NO concentration in NH_3-doped CH_4+air flames using saturated laser-induced fluorescence and probe sampling. *Combust. Flame* **2013**, *160*, 40–46. [CrossRef]
26. Zhang, Z.; Gao, W.; Jiang, H.; Bai, Q.; Zheng, G. Explosion mechanism of nano-sized dust cloud in interconnected vessels. *Combust. Flame* **2024**, *259*, 113135. [CrossRef]
27. Nie, Z.; Gao, W.; Jiang, H.; Zhao, F.; Zheng, G.; Zhang, Z. Flameless venting characteristics of hydrogen explosion under the coupling of carbon dioxide and metal foam. *Process Saf. Environ. Prot.* **2023**, *180*, 375–385. [CrossRef]

fire

MDPI

Article

Explosive Characteristics Analysis of Gasoline–Air Mixtures within Horizontal Oil Tanks

Xinsheng Jiang [1,2], Dongliang Zhou [1,*], Peili Zhang [1,2,*], Yunxiong Cai [1,2], Ri Chen [1], Donghai He [1,2], Xizhuo Qin [1], Keyu Lin [1,2] and Sai Wang [1]

1 Petroleum, Oil and Lubricants Department, Army Logistics Academy, Chongqing 401331, China; jxs_dy@163.com (X.J.); cyunxiong@163.com (Y.C.); crljqwxy@163.com (R.C.); 15998991983@163.com (D.H.); 15823599036@163.com (X.Q.); linky3939@163.com (K.L.); ws_1662023@163.com (S.W.)

2 Chongqing Key Laboratory of Fire and Explosion Safety Protection, Chongqing 401331, China

* Correspondence: zdl974641@163.com (D.Z.); zpl612323@163.com (P.Z.)

Abstract: Horizontal oil tanks, like other oil storage containers, carry the risk of explosion when gasoline–air mixtures are ignited. With the widespread application of horizontal oil tanks in the petrochemical industry, attention to safety risks is increasing. However, currently, a limited amount of experimental research on such tanks exists. To explore the characteristics of gasoline–air mixtures combustion within the confined space of horizontal oil tanks, this study constructed a medium-scale simulated horizontal oil tank (L/D = 3, V = 1.0 m^3) platform. By investigating the effects of different initial gasoline–air mixture volume fractions and ignition positions on explosion overpressure characteristic parameters, an analysis of the combustion characteristics was conducted. It was found that the most dangerous gasoline–air mixture volume fraction is 1.9% when ignited at the top position and 2.1% at the middle. It was also observed that the ignition position has a significant impact on the variation in explosion overpressure characteristic parameters, with ignition at the middle position resulting having a greater explosive force compared to ignition at the top position. Furthermore, using ignition at the middle position as an example, a study was conducted on the flame morphology characteristics at initial gasoline–air mixture volume fractions of 1.1%, 1.9%, and 2.7%. The conclusions from this research deepen our understanding of the explosion characteristics of different containers, providing theoretical insights for the safe storage and transportation of oil materials in horizontal oil tanks.

Keywords: horizontal oil tanks; gasoline–air mixtures; explosive characteristics analysis; flame morphology characteristics analysis

Citation: Jiang, X.; Zhou, D.; Zhang, P.; Cai, Y.; Chen, R.; He, D.; Qin, X.; Lin, K.; Wang, S. Explosive Characteristics Analysis of Gasoline–Air Mixtures within Horizontal Oil Tanks. *Fire* **2024**, 7, 24. https://doi.org/10.3390/fire7010024

Academic Editors: Chuyuan Huang, Haipeng Jiang and Lijuan Liu

Received: 21 November 2023
Revised: 5 January 2024
Accepted: 8 January 2024
Published: 11 January 2024

1. Introduction

Horizontal oil tanks, as common storage containers, have advantages over vertical oil tanks in terms of a lower center of gravity, higher stability, and ease of transportation and installation. Their structure and design provide good resistance to explosions, and the degree of attention to explosion risks is relatively small compared to other types of containers. As a result, there is relatively little research on explosion experiments specifically targeting this type of container. However, horizontal oil tanks are crucial containers for storing liquid fuels and chemical products. Widely utilized in the fields of petroleum, petrochemicals, and chemicals, their safety is of paramount importance.

Oil materials are hazardous, flammable, explosive, and volatile. The vapors generated from their evaporation can mix with air to form a combustible gasoline–air mixtures. When these mixtures reach a certain volume fraction, they can cause severe fires and explosions upon exposure to open flames, high temperatures, or static electricity. Throughout the processes of oil production, transportation, storage, and filling, incidents involving gasoline–air mixtures fires and explosions have consistently been ranked among the most destructive and expensive accidents [1,2]. For instance, on 13 June 2020, a major explosion involving a liquefied petroleum gas transport tank truck at the exit ramp of Wenzhou West on the Wenling section of the Shenhai Expressway in Wenling, Taizhou, resulted in 20 fatalities, 175 injuries, and direct economic losses of RMB 94.778 million. On 24 January 2004, at 20:45, a flash explosion suddenly occurred when a railroad tanker was unloaded at an oil depot in Jinan. This occurred when the railroad tanker, located at the far end of the railway, was nearly empty. When the crane valve was closed and the crane vertical air intake valve of the pipe was opened, a fireball emerged from the opening of the railroad tanker [3]. To prevent accidents, reduce casualties, and minimize economic losses, it is imperative to intensify research on the explosion characteristics of horizontal tank gasoline–air mixtures.

Numerous scholars have conducted research on confined spaces with different shapes and volumes. Wei and Du [4] conducted experimental studies on the explosion characteristics of a gasoline–air mixture in different–shaped confined spaces. Their research revealed that the pressure–time curves of pipelines (short and long pipes) are more complex than those of volume–type confined spaces (spherical and cubic containers). With an increase in the initial volume fractions of gasoline–air mixture, the explosion overpressure values and average pressure rise rates increased and later decreased. Zhang and Wang [5] conducted experiments on the explosion characteristics in sealed containers of different shapes and volumes (20 L spherical, 280 L rectangular, 1.15 m^3 cylindrical, and 5.19 m^3 cylindrical), and different shapes but the same volume (straight rectangular pipes and one–, two–, three–, and four–branched rectangular pipes). They found that under constant initial conditions, the maximum pressure rise rate and explosion index of spherical containers were greater than those of rectangular and cylindrical containers, respectively. When the container shapes were similar, an increase in the container volume resulted in a decrease in the maximum pressure increase rate and an increase in the explosion index. When the sizes of the test containers were similar but the shapes were different, the maximum overpressure and maximum pressure rise rate of the gasoline–air mixture explosions increased with the number of branches in the rectangular pipes. Cai [6] conducted experimental studies to simulate gasoline–air mixture explosions in vertical–domain oil tanks. They investigated the effects of the different initial volume fractions of the gasoline–air mixture, ignition positions, and liquid levels on explosion overpressure parameters and flame development. They found that the ignition positions and liquid levels considerably impacted the explosion overpressure parameters of the gasoline–air mixture. From the results of the aforementioned studies, current experimental research mainly focuses on vertical tanks, rectangular pipelines, cylinders, and spherical containers, while very few experimental studies have been carried out on horizontal tanks, and almost no experimental studies have been carried out on this type of mesoscale bench. Hence, it is essential to study the explosion characteristics of gasoline–air mixtures in horizontal tanks. Further, existing research has focused more on flammable gases, such as methane [7,8], acetylene [9], ethylene [10], coal dust [11], and methane/coal dust mixtures [12]. Limited experimental research has been conducted on gasoline–air mixture. Castellanos et al. [12] investigated the primary factors influencing the explosion process of methane/coal dust mixtures in a 20 L spherical standard container. Dong et al. [13] conducted experiments on methane–hydrogen–air mixture explosions in a 20 L spherical pressure vessel apparatus. All these studies found that flammable gases exhibited an increase in the explosion overpressure and pressure rise rate within a certain concentration range, followed by a decrease. Although gasoline–air mixtures explosions share some similarities with flammable gases, the complexity of their composition (possessing a greater variety of components) results in distinct explosion characteristics [14].

Previous studies explored the influence of different ignition positions on the explosion characteristics of premixed gases. Ferrara G. et al. [15] conducted experiments on pipeline releases and found that ignition in the middle position resulted in the greatest explosion overpressure destruction in containers with smaller spatial dimensions. Bauwens et al. [16], using the FM Global explosion test chamber, discovered that the ignition position, obstacles, and exhaust port size were all considerable factors affecting the maximum flame area, burning rate, and external explosion overpressure in the combustion chamber. Y. Cao et al. [17], using small cylindrical containers for experimentation, studied the effects of the ignition position and exhaust burst pressure on the pressure–time history of hydrogen–air mixed gas explosions and external flame propagation. The results showed that center ignition always resulted in the highest internal peak overpressure. Jin Guo et al. [18] investigated the venting behavior of hydrogen mixed gas with equivalence ratios ranging from 0.6 to 5.0 under three different ignition modes (rear, middle, and front ignition). They found that front ignition always resulted in the smallest internal and external overpressures, whereas center ignition generated the highest internal overpressure. The aforementioned studies demonstrated that the ignition position has a considerable impact on the explosion overpressure and flame behavior of different premixed flammable gases in different containers. When a horizontal oil tank is ignited at the middle position, the explosion pressure wave and turbulence exhibit a flattened lateral spread, distinct from other containers. Experimental research on this operating condition is equally indispensable.

Based on the aforementioned findings, the volume and shape of the experimental container and ignition source position considerably influence the explosion characteristics. In the industrial production, storage, and transportation of oil, oil leakage and volatilization are prone to hazardous situations. Any slight negligence could potentially lead to a fire or explosion, posing a safety hazard [19]. Based on this premise, we first investigated the influence of different initial gasoline–air mixture volume fractions and ignition positions on the explosion overpressure characteristic parameters of horizontal oil tanks and analyzed the detonation characteristics. Subsequently, the flame morphology was studied for low, medium, and high initial gasoline–air mixture volume fractions when ignited at the middle position. Finally, a comparative analysis of explosion overpressure characteristic parameters was conducted, revealing significant effects of different initial gasoline–air mixture volume fractions and ignition positions on these parameters and flame morphology. The purpose of the experimental research is to provide theoretical references for gasoline–air mixtures safety protection and equipment research to ensure the utmost safety in storage and transportation processes.

2. Experimental Configuration

2.1. Experimental System

The simulated horizontal oil tank had a weight, volume, total length, and diameter of 1010 kg, 1.0 m^3, 2467 mm, and 800 mm, respectively, as shown in Figure 1. The main body was made of Q345R steel, while other pressure–bearing components were made of 16MnII. The design pressure was 2.5 MPa, and the operating pressures were 2 MPa each. It was equipped with 11 M20 $\times$ 1.5 threaded holes used in installing pressure sensors, ignition devices, and valves. Additionally, there were three DN100 mm sight glasses on the side for observing flame morphology after an explosion. The experimental system comprised a gasoline–air mixture supply system, high–energy non–interference igniter, GXH–1050 infrared analyzer, high–speed camera, dynamic data acquisition system, synchronous trigger controller, and the simulated horizontal oil tank. Experimental data were collected using a Dong Hua Testing DH8301 dynamic data acquisition system and a ZXP series of high–frequency transient pressure sensors. The schematic is shown below.

2.2. Experimental Conditions

The experimental medium used was a 92–octane automotive gasoline–air mixture, with gasoline as the national VI. Although preliminary experimental research found that the most dangerous gasoline–air mixture explosion volume fractions vary slightly in different confined spaces, the majority are approximately 1.7%. Therefore, in this study, initial gasoline–air mixture volume fractions were set within the range of the upper and lower explosive limits. Two different typical ignition positions were selected for the experiment and placed at the top (A12 in Figure 2) and middle (A7 in Figure 2). Each ignition position was tested for gasoline–air mixture explosions under lean and rich conditions. Nine initial gasoline–air mixture volume fractions (1.1%, 1.3%, 1.5%, 1.7%, 1.9%, 2.1%, 2.3%, 2.5%, and 2.7%) were selected. Each test group was equipped with 3 pressure sensors in the range of 0–2 MPa, the non–linear error was not greater than 0.25% FS, the repeatability and hysteresis error were not greater than 0.1% FS, and the supply voltage was 5 V, allowing an overload of 150%. IPT1 (A3 in Figure 2), IPT2 (A4 in Figure 2), and IPT3 (A13 in Figure 2) were the pressure measurement points for the A12 ignition position, whereas IPT4 (A3 in Figure 2), IPT5 (A12 in Figure 2), and IPT6 (A13 in Figure 2) were the pressure measurement points for the A7 ignition position. A high–speed camera was placed in the middle position on the side of the container (directly facing A7) to record the flame behavior, with a synchronous trigger controller set to synchronize pressure with flame images. When igniting at the top position, the ignition source A12 was set as the origin with coordinates (0, 0, 0). As shown in Figure 2b, the coordinates of the three sensors are A3 (645, 400, 0), A4 (1245, 400, 0), A13 (2467, 0, 0). When igniting at the middle position, the ignition source A7 was set as the origin with coordinates (0, 0, 0). As illustrated in Figure 2d, the coordinates of the three sensors are A3 (0, 400, 0), A12 (−1233.5, 0, 400), A13 (1233.5, 0, 400).

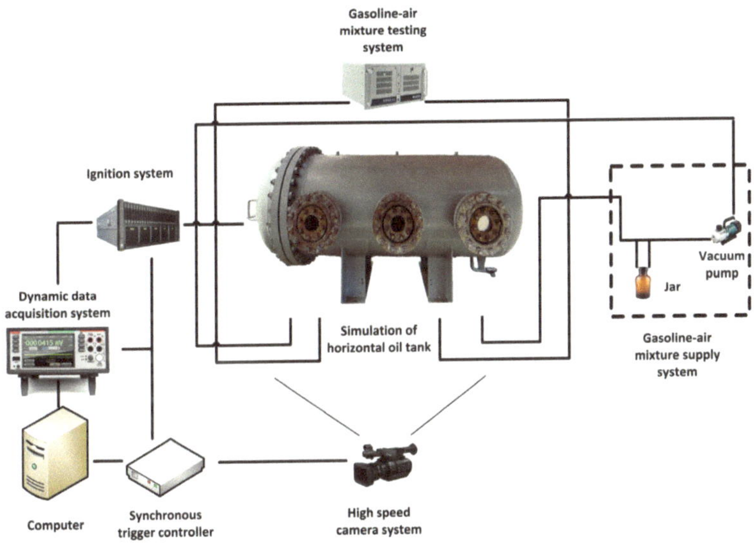

Figure 1. Schematic of the experimental system.

The GXH–1050 infrared analyzer was set at a flow rate of 0.5 L/min, ignition voltage of 1500 V, and ignition energy of 1.5 J. The pressure sensor used was the ZXP610 model, with a measurement range of 0–2 MPa. A high–speed camera was configured to capture images at a frame rate of 500 frames per second (fps). We set the data sampling frequency of the dynamic data acquisition system (DH8301) to 100 k, utilizing Origin 2021 software (version 9.8.0.200). The adjacent averaging method and Savitzky–Golay method were employed for noise reduction processing on the explosive overpressure data. The mixture was generated through the gasoline–air mixture supply system, allowing the mixture to circulate within the test stand for at least 3 min. Subsequently, the circulation was halted, and the gasoline–air mixture volume fractions were adjusted to the desired target value. This target value was maintained for at least 1 min, following which all valves on the experimental stand were closed, sealing the confined space. These operations aimed to achieve a uniform gas mixture. Upon completion of the explosion experiment, all valves were reopened, and exhaust gases were purged using a fan for no less than 5 min. Subsequently, a resting period of no less than 30 min ensued to ensure the complete exhaustion of exhaust gases. The stand was allowed to fully cool down before proceeding with the next set of experiments.

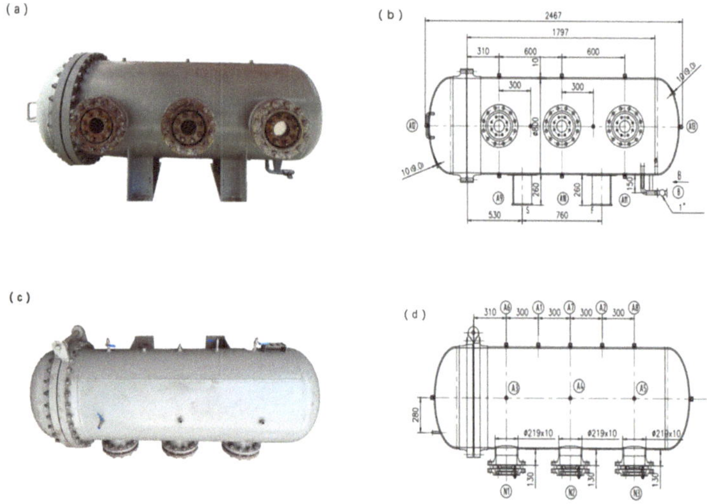

Figure 2. Front view (**a**,**b**) and top view (**c**,**d**) of the simulated horizontal oil tank.

To prevent unexpected incidents and ensure the accuracy of the experimental results, each experiment was repeated thrice, as described in the literature [20]. The experimental settings and initial conditions are presented in Table 1.

Table 1. Testing contents and initial conditions.

Initial Conditions	Value of Initial Parameters
Gas volume concentration	1.1%, 1.3%, 1.5%, 1.7%, 1.9%, 2.1%, 2.3%, 2.5%, 2.7%
Ignition voltage	1500 V
Ignition energy	1.5 J
Ambient temperature	294 k
Ambient pressure	1 atm
Relative humidity	65%
Experimental repetition times	3

3. Explosion Overpressure Analysis in a Horizontal Oil Tank When Ignition Occurs at the Top Position

The explosion of a gasoline–air mixture involves rapid physical and chemical changes, instantly releasing a significant amount of energy accompanied by a tremendous sound Through experimental research, the behavior of the explosion can be reflected through the characteristic parameters of explosion overpressure. This chapter primarily analyzes the developmental patterns of numerical values, such as the maximum overpressure peak (p_{max}), time to reach the maximum overpressure peak (t_{max}), average pressure rise rate (v_{ave}), and explosion power index (E_{max}) under different volume fractions.

3.1. Effect of Different Initial Gasoline–Air Mixture Volume Fractions of Hydrocarbons on the Overpressure Characteristics of Explosions

To some extent, the variation pattern of the explosion overpressure values reflects the development process of the explosion. Figure 3 shows the explosion overpressure time series curves for nine different initial gasoline air mixture volume fractions (YCH) when ignition occurs at the top position. The development of the explosion overpressure shows an initial increase followed by a decrease. In other words, as YCH continues to increase, the overpressure peak increases and later decreases.

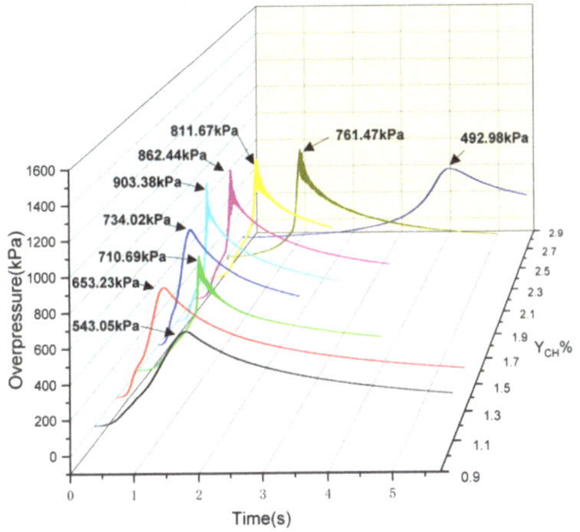

Figure 3. Explosion overpressure time series curves for different YCH when igniting at the top position.

As shown in Figure 4, we conducted a detailed analysis of the explosion overpressure time series curves for three typical initial gasoline–air mixture volume fractions (YCH) of 1.1%, 1.9%, and 2.7%. The overpressure time series curves for different YCH values exhibit similar trends. As YCH continued to increase, the time required to reach the peak explosion pressure initially decreased and later increased. The times required to reach the peak explosion pressure for the YCH values of 1.1%, 1.9%, and 2.7% were 1.482, 0.547, and 4.178 s, respectively. This indicates that the time required for YCH = 1.1% was 171% longer than that for YCH = 1.9%, while the time required for YCH = 2.7% was 664% longer than that for YCH = 1.9%. This indicates that the effects of the lean and rich mixtures on t_{max} were considerable, while the explosion developed rapidly at YCH = 1.9%. At YCH = 1.9%, when the explosion shockwave passed through the pressure sensor, the explosion overpressure rapidly increased and reached its maximum value of 903.38 kPa at 0.547 s. This was the maximum explosion overpressure obtained when igniting the top position of the experimental container with the most hazardous gasoline—air mixture volume fraction. At this point, the time required was the shortest and the peak value was the highest, thereby indicating that YCH = 1.9% was closer to the stoichiometric concentration, wherein the chemical reaction was most intense. After the explosion shockwave passed, the explosion overpressure gradually decreased. When YCH was low (high), the explosion inside the tank was characterized by lean (rich) combustion. At YCH = 1.1%, the maximum explosion overpressure was 1.482 s with a peak value of 543.05 kPa. At YCH = 2.7%, the gasoline—air mixture ratio increased, thereby intensifying the oxygen consumption and resulting in a longer time to reach the maximum explosion overpressure of 4.178 s, with a peak value of only 492.98 kPa. When compared to YCH = 1.9%, the peak value at YCH = 1.1% was 60% of that at 1.9%, and the peak value at YCH = 2.7% was only 55% of that at 1.9%. At this point, it took longer to reach the explosion overpressure value and the peak value was smaller [21], indicating that the chemical reaction was not as intense and that the severity of the explosion was somewhat reduced.

The maximum overpressure peak value is an important parameter that characterizes the explosive destructive capability of a gasoline–air mixture. This value can be obtained experimentally, while the average rate of the increase in pressure can be calculated by dividing the maximum overpressure peak value by the time required to reach that value. Experiments have found that both the flammability index [5,22,23] and explosive power index [24] can characterize the explosive destructive strength. Based on the characteristics of the experimental research, we selected the explosive power index as an important parameter for studying explosive strength. This can be obtained by multiplying the maximum overpressure peak value by the average rate of the rise in pressure.

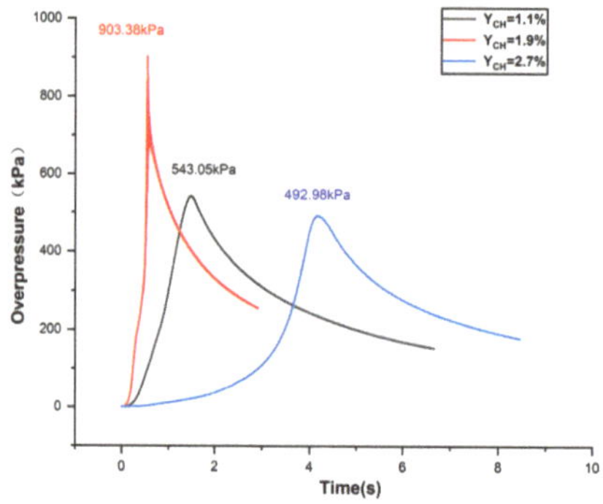

Figure 4. Explosion overpressure time series curves for YCH values of 1.1%, 1.9%, and 2.7%.

The average rate of the rise in pressure can be expressed using Equation (1) [25], which mainly reflects the intensity of the changes in the explosion overpressure:

$$v_{\text{ave}} = \left(\frac{dp}{dt}\right)_{ave} = \frac{P_{\text{max}}}{t_{\text{max}}} \tag{1}$$

The explosive power index can be used to characterize the magnitude of the explosive destructive strength and assess the consequences and safety of gas explosions, as expressed in Equation (2) [26]:

$$E_{\text{max}} = p_{\text{max}} \times v_{\text{ave}} \tag{2}$$

By performing statistical and data analyses on the experimental data, the characteristic parameters were obtained, as presented in Table 2.

Table 2. The overpressure parameters for different initial gasoline–air mixture volume fractions when igniting at the top position.

Y_{CH}/%	P_{max}/kPa	t_{max}/s	v_{ave}/kPa·s^{-1}	E_{max}/kPa^2·s^{-1}
1.1	543.05	1.482	366.43	198,989.81
1.3	653.23	0.794	822.71	537,418.85
1.5	710.69	0.637	1115.68	792,902.62
1.7	734.02	0.569	1290.02	946,900.48
1.9	903.38	0.547	1651.52	1,491,950.14
2.1	862.44	0.678	1272.04	1,097,058.18
2.3	811.67	0.873	929.75	754,650.18
2.5	761.47	1.435	530.64	404,066.44
2.7	492.98	4.178	117.99	58,166.71

To provide a more intuitive representation of the trends in terms of reaching the maximum overpressure time, t_{max}, maximum overpressure, p_{max}, average rate of pressure rise, v_{ave}, and explosive power index, E_{max}, of the initial gasoline–air mixture volume fractions, YCH, Figure 5 illustrates the variation curves of t_{max}, p_{max}, v_{ave}, *and* E_{max}.

As shown in Figure 5a, with an increase in YCH, t_{max} exhibits a trend of initially decreasing and then increasing. When the initial gasoline–air mixture volume fraction is 1.5–2.1%, the explosion overpressure reaches its peak value in 0.70 s. At YCH = 1.9%, the time to reach the maximum explosion overpressure is the shortest (0.547 s), while at YCH = 2.7%, owing to lean oxygen combustion, the limited oxygen for the reaction results in the longest time to reach the maximum explosion overpressure. As shown in Figure 5b, with an increase in YCH, p_{max} increases and later decreases. Between 1.1%

and 1.9% YCH, p_{max} increases with YCH, thereby attaining its maximum at YCH = 1.9%. However, between 1.9% and 2.7% YCH, p_{max} decreases as YCH increases, as shown in Figure 5c. With an increase in YCH, v_{ave} increases and later decreases, thereby attaining its maximum at YCH = 1.9%. Figure 5d shows the curve of the E_{max} variation with different YCH values. As YCH continues to increase, the trend in the E_{max} change increases and later decreases. p_{max}, v_{ave}, and E_{max} exhibit a similar trend of increasing and later decreasing with YCH. E_{max} is at its maximum at YCH = 1.9%, thereby indicating that the explosive power is greatest at YCH = 1.9%. This experiment is a study conducted on a medium-sized simulated horizontal oil tank. From the analysis and comparison of t_{max}, p_{max}, v_{ave}, and E_{max} under different YCH conditions, when ignited at the top position with YCH = 1.9%, the explosion develops most rapidly, thereby generating the highest explosion overpressure and producing the greatest explosive power. This scenario is the most dangerous, and real-time monitoring is required in practical engineering to prevent YCH from reaching values near this threshold.

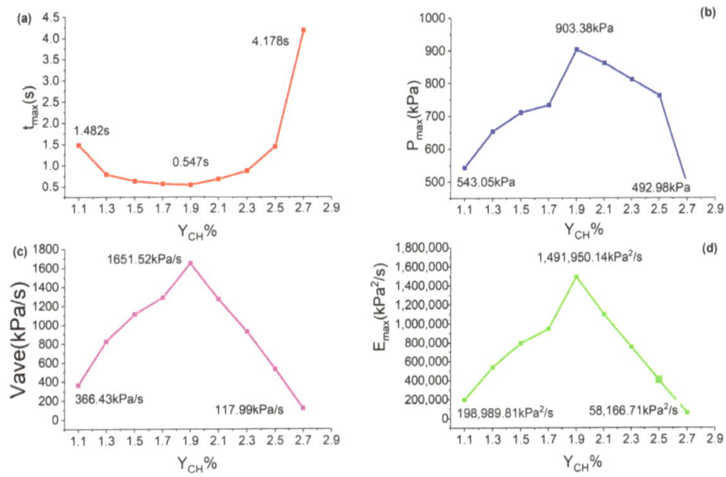

Figure 5. The relationship between explosion overpressure characteristic parameters and YCH when ignited at the top position: (**a**) the trend of t_{max} with respect to YCH; (**b**) the trend of p_{max} with respect to YCH; (**c**) the trend of v_{ave} with respect to YCH; (**d**) the trend of E_{max} with respect to YCH.

3.2. Analysis of the Explosion Overpressure Variation Process

The overpressure parameters were closely related to the volume and shape of the explosion container. The different volumes and shapes of the experimental containers resulted in different explosion characteristics, as presented in Table 3. For the oil–gas mixture with YCH = 1.9%, we obtained p_{max} = 903.38 kPa and (dp/dt)max = 9032.00 kPa/s. However, values obtained from references [5,26–28] are inconsistent, thereby indicating a close correlation between explosion characteristics and the volume and shape of the experimental container. Wang [26] conducted experiments in a 907.50 L tank with YCH = 1.71%, which resulted in p_{max} = 844.00 kPa and (dp/dt)max = 23,041.00 kPa/s. Zhang [5] conducted experiments on a 1150 L cylindrical platform with YCH = 1.7%, thereby yielding p_{max} = 498.12 kPa and (dp/dt)max = 3520.00 kPa/s. Even in experiments with containers of similar volumes but different shapes, considerable differences in p_{max} and (dp/dt)max were observed. Other data can be analyzed and compared based on Table 3. To further study the variation law of explosion overpressure in horizontal oil tanks, the change in the overpressure was divided into stages in this experimental study to determine the optimal explosion suppression timing. This approach has a theoretical reference value in terms of improving the safety and protection of oil materials and engineering practices.

Table 3. Overpressure parameters for gasoline–air mixtures in closed vessels of different sizes.

	Volume/L	YCH/%	P_{max}/kPa	$(\frac{dp}{dt})_{max}$/kPa·s^{-1}
Present work	1000 (tank)	1.90	903.38	8985.28
Wang et al. [26]	907.50 (tank)	1.71	844.00	23,041.00
Zhang et al. [27]	244 (straight tunnel)	1.50	870.00	8190.00
Zhang et al. [5]	5190 (cylindrical)	1.70	932.11	2540.00
	1150 (cylindrical)	1.70	498.12	3520.00
	280 (rectangular)	1.70	398.75	3840.00
	20 (spherical vessel)	1.70	626.52	19,790.00
Qi et al. [28]	20 (spherical vessel)	1.62	835.00	34,381.00

Some researchers [6,26,29] divided the development of the explosion overpressure in confined spaces into stages based on different experimental characteristics, each with its own unique criteria. Research has shown that the magnitude of the instantaneous pressure rise rate v_{ins} does not only describe the characteristics of explosion overpressure change but also facilitates the division of overpressure change into stages. The first–order derivative of the instantaneous pressure rise rate, denoted as u_{ins}, is expressed by Equation (3) and was used to identify the inflection points in the time–pressure curve of the explosion overpressure.

$$u_{ins} = \frac{d}{d(t)}\left(\frac{dp}{dt}\right) \tag{3}$$

Numerous scholars have used the Savitzky–Golay method to smoothen oscillation curves in the Origin software [30], thereby employing a second–order polynomial algorithm for calculation. Figure 6 shows the curves obtained using the Origin software with a second–order polynomial and a 3000–point data window for differentiation when YCH was 1.1%, 1.9%, and 2.7%. Within the "Analysis" tab, the "Mathematics" section was used to calculate the first and second derivatives, resulting in the curve plot. Upon comparison, the pressure–time curve changes minimally after the smoothing process. However, there exist considerable variations in the maximum pressure rise rate (dp/dt)max. Additionally, the variation patterns of (dp/dt)max trends for the three selected representative curves are consistent.

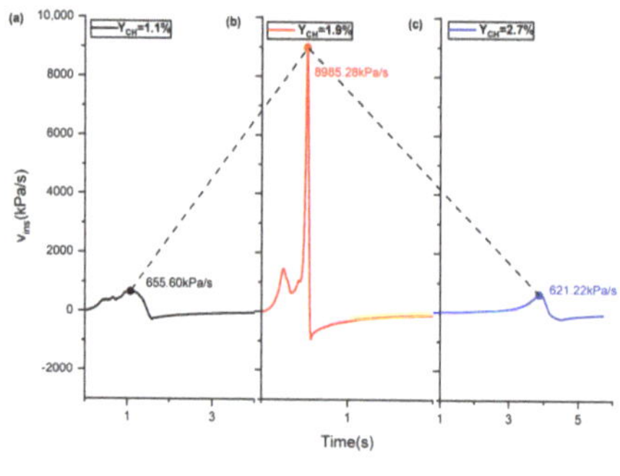

Figure 6. Typical maximum pressure rise rate: (**a**) the maximum pressure rise rate at YCH = 1.1%; (**b**) the maximum pressure rise rate at YCH = 1.9%; (**c**) the maximum pressure rise rate at YCH = 2.7%. The dashed line in the figure represents the trend at three points.

Based on the overall curve pattern, the explosion overpressure variation process was divided into four phases: smooth rise, accelerated rise, fast transitions, and decreasing attenuation [31]. The following is a specific analysis using YCH = 1.9% as an example:

As shown in Figure 7, the division of the explosion stages at YCH = 1.9% is as follows:

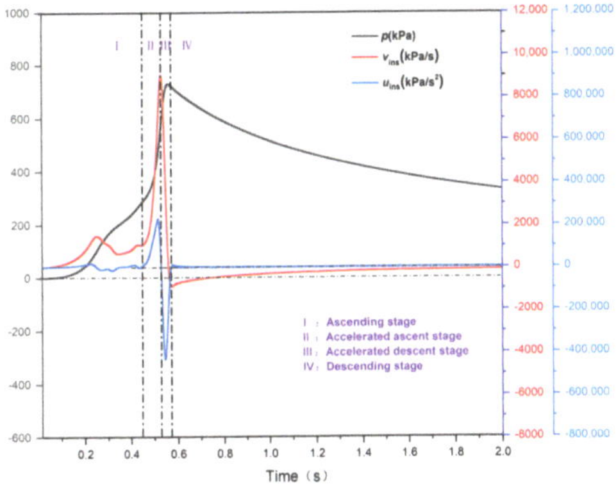

Figure 7. Division of the explosive stage at YCH = 1.9%.

Smooth pressure rise stage (I: 0–0.458 s): In this stage, the v_{ins} and u_{ins} curves exhibit oscillatory development. The v_{ins} values are greater than zero, whereas the u_{ins} values fluctuate around the zero axis. The explosion overpressure steadily increases. However, the overall trend is relatively smooth. This is because after the ignition source ignites the vapor, there is a relatively small amount of fuel and oxygen participating in the reaction. As the flame begins to spread, the heat generated by combustion intensifies molecular collisions within the gasoline–air mixture, thereby resulting in more excited–state radicals. When these radicals accumulate at the flame front, the burning area rapidly increases, thereby resulting in an accelerated increase in the explosion overpressure. This stage ends at 0.458 s with an explosion overpressure of 297.94 kPa, v_{ins} = 1098.37 kPa/s, and u_{ins} = 12,902.95 kPa^2/s.

Accelerated pressure rise stage (II: 0.458–0.528 s): In this stage, the v_{ins} values increase monotonically, while the u_{ins} values change from positive to negative. Here, the combustion intensifies, thereby causing the flame to gradually spread from the ignition source to filling the entire tank. Consequently, the rate of explosion overpressure increased until it reached a value of 569.78 kPa. At this point, v_{ins} = 9030.61 kPa/s and u_{ins} = −18,011.98 kPa^2/s. For gasoline–air mixtures with different initial volume fractions, the explosion overpressure reaches its maximum value in the shortest time when the volume fraction (YCH) is 1.9%.

Fast transitions stage (III: 0.528–0.571 s): In this stage, the v_{ins} values monotonically decrease from positive to negative, while the u_{ins} values are less than 0. Considerable consumption of oil, gas, and oxygen occurred inside the tank. The rates of the gasoline–air mixture combustion reactions and energy release decreased. After reaching the maximum explosive overpressure, the pressure decreased to 720.31 kPa. At this point, v_{ins} = −908.87 kPa/s and u_{ins} = 0.

Decreasing attenuation stage (IV: 0.571–2 s): In this stage, v_{ins} values increase monotonically and gradually approach 0, while u_{ins} values remain positive. As the gasoline–air mixture inside the tank was depleted and oxygen was consumed, the flame was gradually extinguished. Hence, the explosion overpressure decreases progressively, eventually reducing to 0.

4. Explosion Overpressure Process and Characteristic Parameters for Different Initial Gasoline–Air Mixture Volume Fractions with Mid-Position Ignition

Similar to Section 3, this chapter primarily analyzes the developmental patterns of various numerical values, including the maximum overpressure peak (p_{max}), time to reach the maximum overpressure peak (t_{max}), average pressure rise rate (v_{ave}), and explosion power index (E_{max}) under different initial gasoline–air mixture volume fractions with mid–position ignition.

Figure 8 illustrates the variation in explosion overpressure with time for different initial gasoline–air mixture volume fractions (YCH) with mid–position ignition. As YCH increases, the overpressure peak increases and later decreases. As shown in Figure 3, the maximum value is achieved at YCH = 1.9%, thereby reaching 903.38 kPa. As shown in Figure 8, the maximum value is obtained at YCH = 2.1%, thereby reaching 1435.14 kPa, which is approximately a 59% increase from the maximum value in Figure 3. This phenomenon is due to the fact that the aforementioned conditions are close to the stoichiometric concentration, while different ignition positions have a considerable impact on the development of explosion overpressure, thereby resulting in variations in the volume fraction at which combustion is fully reacted.

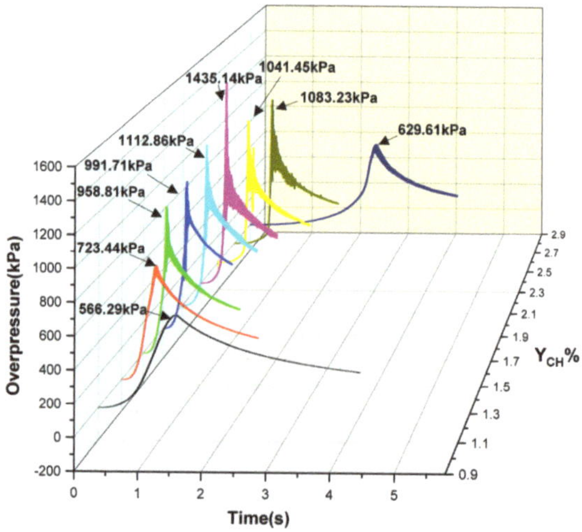

Figure 8. Explosion overpressure time series curves for different YCH when igniting at the middle position.

When YCH was similar, different ignition positions resulted in considerable variations in explosion overpressure. In the YCH range of 1.1% to 1.7%, the increase rate in the explosion overpressure with mid-position ignition compared to top ignition gradually increased at rates of 4.3%, 11%, 34.9%, and 35.1%. When YCH = 2.1%, the increase rate was 66%. This is because, compared to ignition at the top position, when ignition occurred at the middle position, the ignition source was closer to the center in the confined space. This required less time for the flame front to reach both ends of the tank wall. Consequently, the ignition and development of the gasoline–air mixture were faster, while the thermal losses to the tank wall were relatively small. Hence, higher explosive pressures were observed.

The characteristic parameters for mid-position ignition are presented in Table 4. Under a similar initial gasoline–air mixture volume fraction to that presented in Table 2, the explosion overpressure is higher, while the time to reach the maximum explosion overpressure peak is shorter. The general trend in the variation in the characteristic parameters corresponds with the experimental study of Ferrara G, Y. Cao, Guo et al. [15,17,18]. When YCH was 1.1% and 2.7%, it took more than 1.0 s to reach the maximum explosion overpressure. When YCH was 1.3–2.5%, the increase in the explosion overpressure was within 0.80 s. At YCH = 2.1%, p_{max}, v_{ave}, and E_{max} reached their maximum values. Compared to the values presented in Table 2 with top-position ignition, p_{max}, v_{ave}, and E_{max} were larger. Figure 9 shows the variations in t_{max}, p_{max}, v_{ave}, and E_{max} with YCH for mid-position ignition, with the trend exhibited consistent with that shown in Figure 5.

Table 4. The overpressure parameters for different initial gasoline—air mixture volume fractions when igniting at the middle position.

Y_{CH}/%	P_{max}/kPa	t_{max}/s	v_{ave}/kPa·s^{-1}	E_{max}/kPa^2·s^{-1}
1.1	566.29	1.249	453.39	256,750.22
1.3	723.44	0.582	1243.02	899,250.39
1.5	958.81	0.398	2409.07	2,309,840.41
1.7	991.71	0.414	2395.43	2,375,571.89
1.9	1112.86	0.429	2594.07	2,886,836.74
2.1	1435.14	0.465	3086.32	4,429,301.28
2.3	1041.45	0.572	1820.72	1,896,188.84
2.5	1083.23	0.740	1463.82	1,585,653.74
2.7	629.61	2.553	246.62	155,274.42

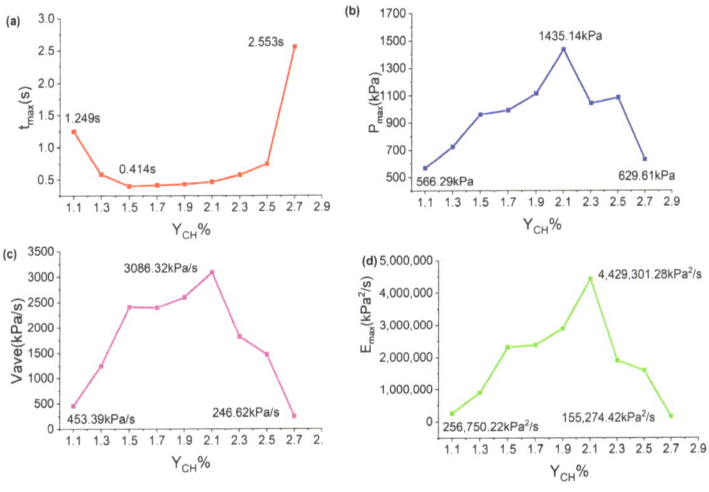

Figure 9. The relationship between explosion overpressure characteristic parameters and YCH when ignited at the middle position: (**a**) the trend of t_{max} with respect to YCH; (**b**) the trend of p_{max} with respect to YCH; (**c**) the trend of v_{ave} with respect to YCH; (**d**) the trend of E_{max} with respect to YCH.

5. Comparison of Ignition-Induced Explosive Overpressure Characteristics between Top and Middle Positions

This chapter compares the typical explosion overpressure characteristic parameters during ignition at the top and middle positions. Emphasis is placed on the comparison of the average rate of pressure rise, v_{ave}, the maximum rate of pressure rise, $\left(\frac{dp}{dt}\right)_{max}$, and the explosion power index, E_{max}. Through comparative analysis, the study further explores the explosion behavior under different operating conditions.

As shown in Figure 10, compared to ignition at the top position, ignition at the middle position resulted in an increased explosion overpressure and a shorter duration. As shown in Figure 10a–c, when the explosion overpressure curves are within the range of 0 to 0.69 s, 0 to 0.35 s, and 0 to 0.39 s, respectively, the explosion overpressure is greater for ignition at the top position than that in the middle position.

As shown in Figure 11, for YCH = 1.1% (1.08 s), the maximum pressure rise rate (dp/dt)max for ignition at the top position reached 655.60 kPa/s; at 0.95 s, the (dp/dt)max for ignition at the middle position was 935.46 kPa/s. Regarding YCH = 1.9%, at 0.53 s, the (dp/dt)max for ignition at the top position reached 8985.28 kPa/s, and at 0.41 s, the (dp/dt)max for ignition at the middle position was 10,089.80 kPa/s. Finally, for YCH = 2.7%, at 3.87 s, the (dp/dt)max for ignition at the top position was 621.22 kPa/s, and at 2.40 s, the (dp/dt)max for ignition at the middle position was 2280.08 kPa/s. Compared to ignition at the top position, that at the middle position resulted in a shorter time as regards reaching the maximum pressure rise rate, with larger numerical values. This was particularly pronounced at YCH = 2.7%, wherein the change rate was the highest, thereby increasing by 267% and reducing by 38%. Based on the analysis of the aforementioned data, a comparison of a typical explosion overpressure characteristic parameters for ignition at the top and middle positions is presented in Table 5.

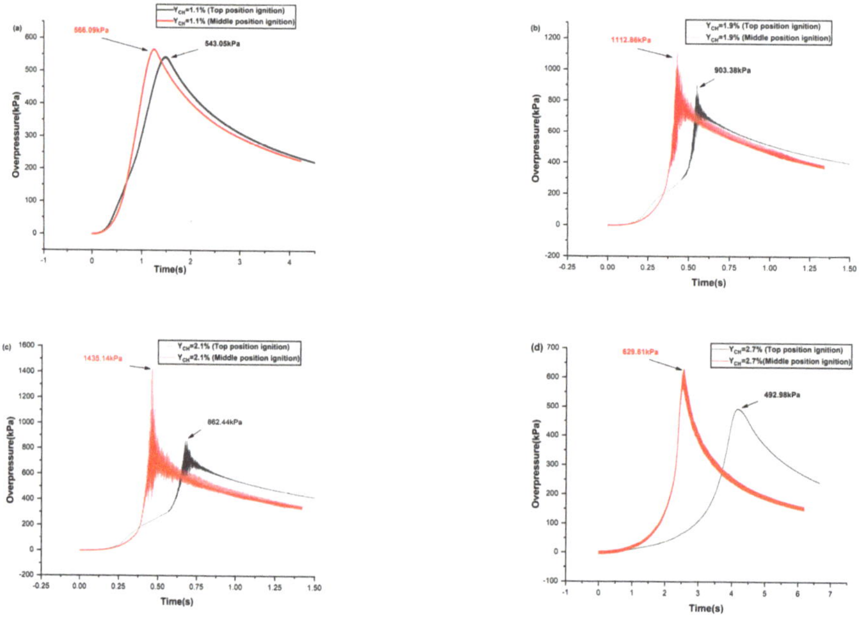

Figure 10. Comparison of typical explosion overpressure-time curves when ignited at the top and middle positions: (**a**) curve comparison charts for YCH = 1.1%; (**b**) curve comparison charts for YCH = 1.9%; (**c**) curve comparison charts for YCH = 2.1%; (**d**) curve comparison charts for YCH = 2.7%.

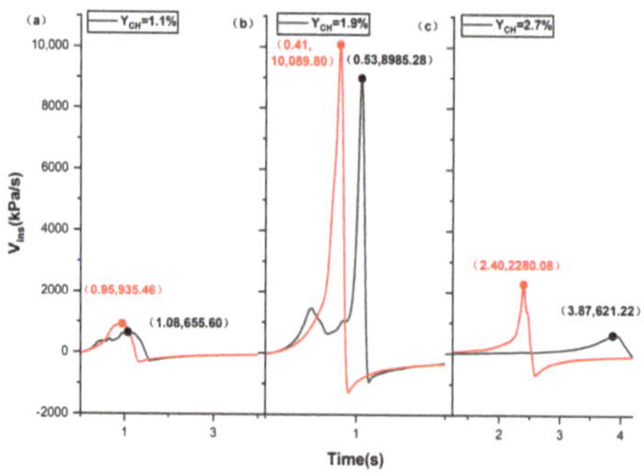

Figure 11. Impact of YCH on the maximum pressure rise rate for ignition at two different positions: (**a**) the maximum pressure rise rate at YCH = 1.1%; (**b**) the maximum pressure rise rate at YCH = 1.9%; (**c**) the maximum pressure rise rate at YCH = 2.7%.

In Figure 10, the overpressure–time sequence curves for the two ignition positions are presented for YCH values of 1.1%, 1.9%, 2.1%, and 2.7%. Here, the explosion overpressure–time sequence curves change with rapid oscillations in the overpressure.

Table 5. Characteristic parameters of explosion overpressure for typical ignition at the top and middle positions.

Ignition Position	YCH/%	Maximum Overpressure Peaks (kPa)	Time to Reach Maximum Overpressure Peaks	Maximum Overpressure Rise Rates (kPa/ s)
Top–position ignition	(a) 1.1	543.05	1.482	655.60
	(b) 1.9	903.38	0.547	8985.28
	(c) 2.1	862.44	0.678	5671.31
	(d) 2.7	492.98	4.178	621.22
Middle–position ignition	(a) 1.1	566.29	1.249	935.46
	(b) 1.9	1112.86	0.429	10,089.80
	(c) 2.1	1435.14	0.465	9324.29
	(d) 2.7	629.61	2.553	2280.08

When Figure 12 is magnified, at YCH = 1.1%, the ignition overpressure curve at the top position exhibits oscillations, whereas the ignition curve at the middle position remains smooth. At YCH = 2.7%, the ignition curve at the middle position shows oscillations, whereas the ignition curve at the top position is smooth. Only when YCH is at 1.9% and 2.1% do the overpressure curves for both ignition positions oscillate back and forth. Furthermore, the oscillations are more pronounced when ignition occurred at the middle position than at the top position [26]. This is because the experimental test rig had relatively large specifications and dimensions, thereby providing sufficient space for the propagation and reflection of overpressure waves. Additionally, it increases the instability of the explosion overpressure and internal fluid turbulence. Further, when YCH approaches a stoichiometric concentration, it can result in the formation of large–scale unstable flames during explosion. The propagation and heat dissipation of these flames accelerated the release of energy, thereby resulting in oscillations.

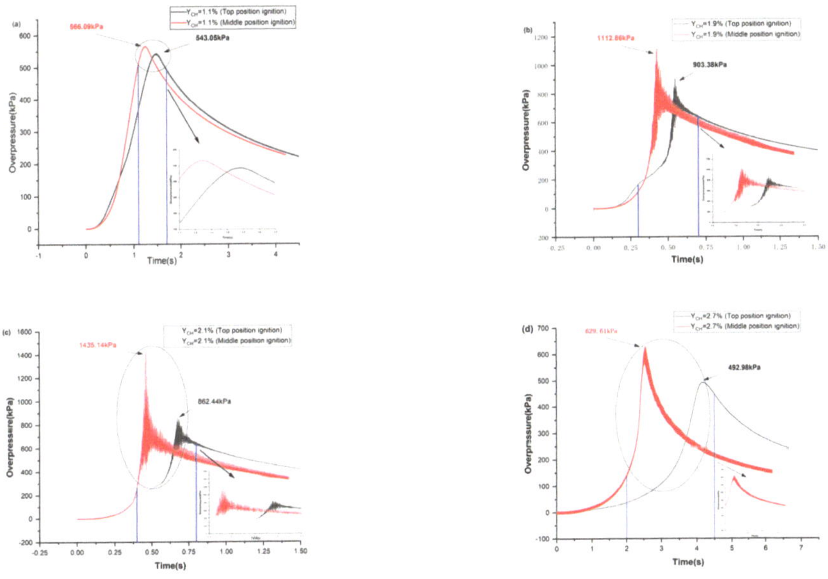

Figure 12. Time sequence curve and localized magnification diagram of typical explosion overpressure oscillations when ignited at the top and middle positions: (**a**) curve comparison charts for YCH = 1.1%; (**b**) curve comparison charts for YCH = 1.9%; (**c**) curve comparison charts for YCH = 2.1%; (**d**) curve comparison charts for YCH = 2.7%.

The variation in overpressure is an important aspect of gasoline–air mixtures explosion research. The average pressure rise rate v_{ave} and maximum explosion power index E_{max} are crucial indicators for evaluating the severity of explosions. Regarding v_{ave} and E_{max}, their values initially increased as YCH increased from 1.1% to 2.7%, as shown in Figure 13. However, they peaked at YCH = 1.9% and YCH = 2.1%, respectively, before decreasing. As shown in Figure 13a, as YCH increases, the overall trend in v_{ave} increases and later decreases. This trend corresponds with the variation pattern of p_{max}. When the YCH values were equal, the v_{ave} for ignition at the middle position was greater than that for ignition at the top position.

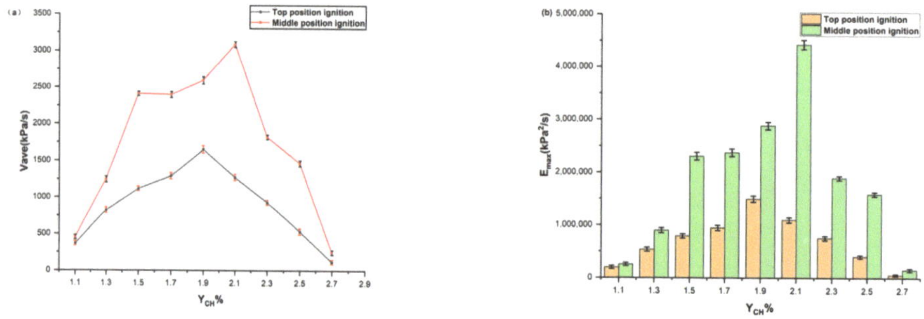

Figure 13. Trend and comparison of the (**a**) average boost rate v_{ave} and (**b**) explosion power index E_{max}.

With the continuous increase in YCH, the value of E_{max} increased and later decreased, as shown in Figure 13b. This trend corresponds with the variation in p_{max}. Furthermore, when YCH was equal, the E_{max} for ignition at the middle position was greater than that for ignition at the top position. At YCH = 2.1, the E_{max} value was the greatest when ignited at the middle position, thereby showing an increase of 304% compared to that when ignited at the top position. At this point, the rate of change in E_{max} was the highest. The results show that when ignited at the top position with YCH = 1.9% and at the middle position with YCH = 2.1%, the chemical reactions of gasoline–air mixtures combustion are most extensive, thereby resulting in the highest energy release and maximum destructive power.

6. Analysis of Typical Flame Morphology

The flame morphology exhibited distinct characteristics based on variations in initial gasoline–air mixture volume fractions and ignition positions. In this chapter, further analysis of flame morphology is conducted through the observation of typical flame propagation images. Based on previous analyses, when the ignition point was located at A7, the peak values of the maximum explosion overpressure p_{max} and explosion power index E_{max} at low, medium, and high initial gasoline–air mixture volume fractions (YCH) were higher than those at the top position. Here, the explosion developed more rapidly and was more prone to generating overpressure oscillations, thereby indicating that gasoline–air mixture explosions ignited at the middle position are more dangerous. In this section, three typical initial gasoline–air mixture volume fractions of YCH at 1.1%, 1.9%, and 2.7% were selected for comparison. After observing the images of the flame variations during the rapid development stage of the explosion and analyzing the changes in the flame morphology and color, we aimed to uncover the process of flame propagation in gasoline–air mixture explosions and the characteristic behavior of flames in such explosions [32,33].

Figure 14 shows an image of the flame at YCH = 1.1%. When the spark from the ignition rod is released, a high–voltage electric arc instantaneously breaks through the gasoline–air mixture between the electrodes. The released electrical energy is converted into thermal energy, thereby igniting the surrounding gasoline–air mixture within a small area and resulting in a high–temperature ignition kernel, as represented by the bright yellow region at 0.002 s in the image. This process is extremely brief and complex, and involves ionization, heat and mass transfer, and chemical reactions in the nucleus. However, a detailed discussion of these processes is not provided here. Once the ignition kernel is formed, heat begins to propagate outward, thereby igniting the surrounding gasoline–air mixture and forming a spherical flame front, as shown at 0.02, 0.05, and 0.10 s in the image. As the heat from the ignition kernel gradually dissipates, the color of the flame changes from bright to dark in this region, thereby forming a less distinct dark red area at 0.10 s. During this process, the darker–colored flame kernel is gradually shifting upwards. This is because the temperature in this region is higher and moves upwards owing to buoyancy effects. Similarly, the temperature in the burned region enclosed by the flame front is higher than that in the unburned region. Hence, in the case of a relatively slow flame propagation, it is possible to clearly observe the overall upward movement of the flame. At 0.23 s, the center of the flame is inclined towards the upper side, and this is owing to the effect of buoyancy. During the first 0.40 s, the flame propagation is relatively unaffected, and the flame front appears smooth without folds. As the flame front gradually approaches the radial container wall, the reflected pressure waves disturb the flame front, causing it to develop into folds. This was observed on the lower side of the flame front at 0.41 s.

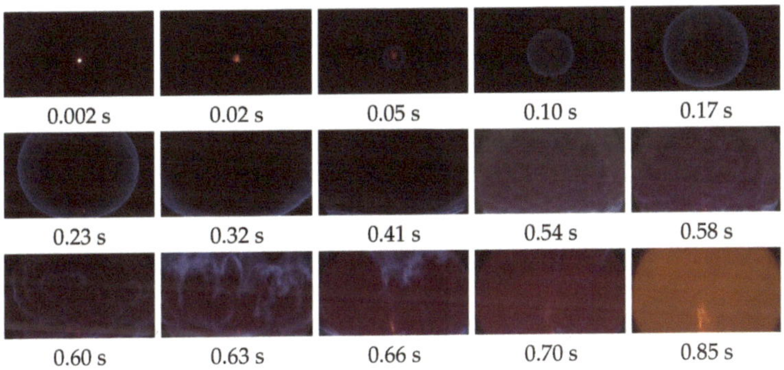

Figure 14. Images depicting the temporal evolution of flame morphology at YCH = 1.1%.

As the flame continues to propagate, the disturbance caused by pressure waves becomes more pronounced, coupled with the folds on the flame front. For better observation, the contrast of the flame images at 0.54 s and 0.58 s was increased. A cellular flame front can be observed in the figure. This phenomenon increases the flame front area, enhances the flow of gases in the front region, intensifies the combustion reaction, accelerates heat release and explosion development, as demonstrated by the flame image at 0.63 s. After a period of development, the unburned gas was gradually consumed, and heat was lost. The flame gradually changed to orange and dark red and eventually extinguished.

As shown in Figure 15, the flame image at YCH = 1.9% exhibits similar early–stage development to that at YCH = 1.1% in the first 0.02 s. However, because the gasoline–air mixture at this point is closer to the stoichiometric concentration, the laminar flame speed is faster. A greater amount of the gasoline air mixture participates in the reaction, thereby releasing heat. At 0.02 s, the bright yellow region inside the flame is larger compared to YCH = 1.1%, and the flame appears more vivid in color. Owing to the combined effects of the greater heat release and faster laminar flame speed, the flame front displacement speed at YCH = 1.9% was considerably higher than that at YCH = 1.1%. As shown in Figure 15, the flame area at 0.05 s is noticeably larger than the flame at 0.05 s in Figure 14. The flame in Figure 14 just reaches the edge of the observation window at 0.17 s, whereas in Figure 15, the flame image has already surpassed the observation window at 0.12 s. A further observation of the flame image in Figure 15 reveals that during the propagation of the flame front, folds gradually form at 0.08 s, 0.12 s, and 0.15 s. At this stage, the flame area was still relatively small, and the folds in the flame front were not formed due to the disturbance from the pressure waves as it gradually contacted the wall but due to the instability of the combustion process. As shown in Figure 14, there are no folds in the flame front during the first 0.23 s. This is because, at this time, the initial gasoline–air mixture volume fraction is lower, thereby resulting in a slower flame propagation. The expansion of the already burned area stretches the flame front, thereby inhibiting the formation of folds. Folds formation in the flame front as shown in Figure 15 further increases the disturbance in the front area, thereby prompting a more intense combustion reaction. Consequently, from 0.15 s to 0.45 s, the turbulence in the flow field of the combustion area intensifies, heat generation becomes more rapid, and the flame gradually changes from pale blue to orange–red. At 0.45 s, it becomes bright white, thereby indicating a highly intense reaction, accompanied by a rapid increase in pressure. As the unburned fuel is consumed and heat is lost through heat transfer to the walls of the oil tank, the flame temperature gradually decreases. Its color changes from bright yellow to dark red at 0.70 s to 1 s until it extinguishes, as shown in the figure.

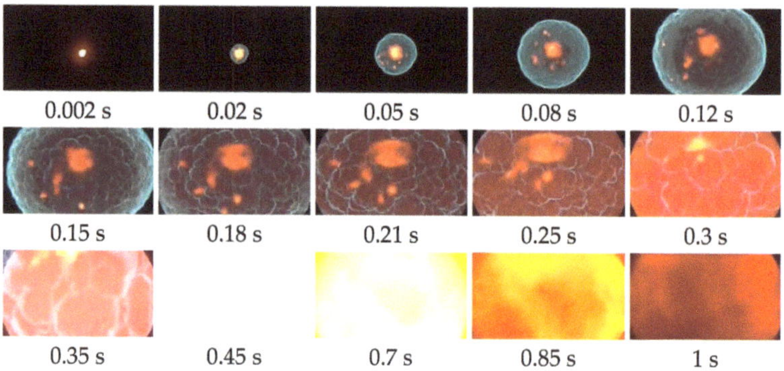

Figure 15. Images depicting the temporal evolution of flame morphology at YCH = 1.9%.

Figure 16 shows the flame image at a gasoline–air mixture volume fraction of YCH = 2.7%. The ignition process in the early stage is similar to that of gasoline–air mixture explosions at other volume fractions. However, owing to the lower air–fuel ratio, the combustion reaction was slower. The flame area at 0.12 s is smaller than the flame at 0.05 s, as shown in Figure 15. However, because of the increased concentration of fuel per unit volume in the reaction, more heat was released, and there were more incomplete reactions. In high-temperature environments, carbon particles precipitate, thereby causing most of the flame region to appear as bright yellow. Additionally, floccule flames formed within the flame front. In the first 0.80 s, owing to the slow flame propagation, the burning area shifts upward under the influence of buoyancy. At 0.80 s, constrained by the upper sidewall, the flame begins to spread downward, as shown in the flame images at 1.20 and 1.50 s. At 1.90 s, a bright white flame formed, thereby indicating an intense explosive reaction. As the gasoline–air mixture is gradually consumed, the flame transitions to a bright yellow and dark red color at 2.60 and 2.80 s, respectively, until the flame is gradually extinguished. Comparing Figures 14–16, when the volume fraction of the gasoline–air mixture is higher than the stoichiometric volume fraction, the flame front becomes more irregular. This is because the heat release is faster, the flame propagation is slower, and the buoyancy effect considerably impacts the flame front.

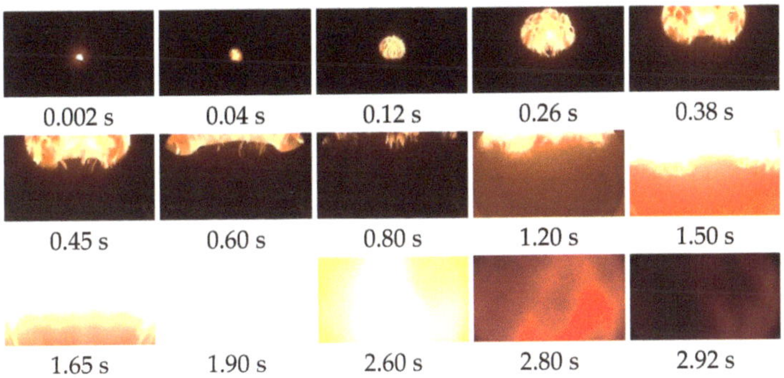

Figure 16. Images depicting the temporal evolution of flame morphology at YCH = 2.7%.

7. Conclusions

To explore the explosion characteristics of a horizontal oil tank within confined spaces, this paper first analyzed the impact of different initial gasoline–air mixture volume fractions and ignition positions on explosion overpressure characteristic parameters. It obtained the most critical gasoline–air mixture volume fractions under various operating conditions. Subsequently, a comparative analysis of the explosion overpressure characteristic parameters was conducted for ignition at two different positions. Finally, flame patterns were studied for typical low, medium, and high initial gasoline–air mixture volume fractions, and a preliminary analysis of flame development characteristics was performed. The research successfully achieved its experimental goals. The main conclusions of the study include the following:

1. With an increase in the initial gasoline–air mixture volume fraction, the peak overpressure increased and later decreased. The development of internal overpressure exhibits four stages: smooth rise, accelerated rise, fast transitions, and decreasing attenuation. When ignited at the top position, the most dangerous gasoline–air mixture volume fraction was 1.9%, with a maximum overpressure p_{max} of 903.38 kPa. When the gasoline–air mixture ignited at the middle position, the most dangerous gasoline–air mixture volume fraction was 2.1%, with a maximum overpressure p_{max} of 1435.14 kPa. In practical engineering, efforts must be made to avoid reaching gasoline–air mixture volume fractions near the aforementioned values.
2. Different ignition positions exhibit similar trends in the variation in the explosion overpressure characteristic parameters. However, when YCH was equal, the values of v_{ave}, (dp/dt)max, and E_{max} for ignition at the middle position were greater than those for ignition at the top position. Hence, ignition at the middle position results in a more powerful explosion, with greater destructive force at the top position.
3. Different initial gasoline–air mixture volume fractions and ignition positions result in distinct flame characteristics. When YCH = 1.9%, i.e., closer to the stoichiometric concentration, the laminar flame speed was faster than when YCH = 1.1%. The flame image exhibited more pronounced folds, thereby shortening the duration. At YCH = 2.7%, wherein the equivalent volume fraction is larger, the flame propagation is relatively slow owing to the rapid heat release, resulting in an irregular flame front.
4. The experimental research in this paper was conducted within a horizontal oil tank. The results of the study not only enriched various experimental setups but also deepened our understanding of the explosive characteristics of gasoline–air mixtures. Due to various constraints in the laboratory, we did not conduct a study with a prototype horizontal oil tank and numerical simulation. In the future, when conditions permit, we will conduct experiments and numerical simulation studies on a prototype experimental test stand.

Author Contributions: Conceptualization, D.Z., R.C. and S.W.; methodology, Y.C.; software, D.Z. and R.C.; validation, Y.C.; formal analysis, P.Z.; investigation, D.H.; resources, X.Q.; data curation, P.Z. and X.Q.; writing–original draft preparation, D.Z.; writing–review and editing, X.J.; visualization, Y.C.; supervision, X.J.; project administration, K.L.; funding acquisition, X.J. and P.Z. All authors have read and agreed to the published version of the manuscript.

Funding: This research was funded by the Key Basic Research Program, grant number 2019–JCJQ–ZD–198–04; Special Project for Technological Innovation and Applied Development in Chongqing City, grant number CSTB2023TIAD–KPX0089; Science and Technology Research Program of Chongqing Municipal Education Commission, grant number KJZD–M202112901; Natural Science Foundation of Chongqing, grant number CSTB2023NSCQ–MSX0127; Army Logistics Academy Program of PLA, grant number LQ–QZ–202227; and Chongqing Graduate Student Research Innovation Program, grant number CYB23297.

Institutional Review Board Statement: Not applicable.

Informed Consent Statement: Not applicable.

Data Availability Statement: The data used to support the findings of this study are available from the corresponding authors upon request.

Conflicts of Interest: The authors declare no conflicts of interest.

Nomenclature

YCH	Initial gasoline–air mixture volume fraction,
P_{max}	Maximum overpressure peak
t_{max}	Time to reach maximum overpressure peak
v_{ave}	Average pressure boost rate
E_{max}	Explosion power index
v_{ins}	Instantaneous pressure rise rate
$\left(\frac{dp}{dt}\right)_{max}$	Maximum overpressure rise rates
u_{ins}	First-order derivative of the instantaneous pressure rise rate

References

1. Wang, S.M. Experimental Research and Numerical Simulation on the Evolution Characteristics of Oil and Gas Explosions in Confined Spaces with Weak Intensity Surface. Ph.D. Thesis, Army Logistics Academy, Chongqing, China, 2018.
2. Guo, W.H.; Li, H. Prevention of Major Hazardous Fire and Explosion Accidents in Petrochemical Enterprises. *Fire Sci. Technol.* **2016**, *35*, 1335–1337.
3. Yu, Z.B.; Lu, Y.L. Analysis of the Causes of Flash Explosion Accidents in Tanker Trucks and Countermeasures. *Pet. Petrochem. Saf. Technol.* **2005**, *21*, 12–13.
4. Wei, S.H.; Du, Y.; Wang, S.M.; Li, M. Experimental Study on Oil and Gas Explosion Characteristics in Different Shaped Confined Spaces. *China Saf. Sci. J.* **2017**, *13*, 41–47.
5. Zhang, P.L.; Wang, J.; Wang, D. Explosions of gasoline vapor/air mixture in closed vessels with different shapes and sizes. *J. Loss Prev. Process Ind.* **2018**, *57*, 327–334. [CrossRef]
6. Cai, Y.X.; Jiang, X.S.; Wang, S.M.; Yu, B.B.; Wang, C.H.; Wang, Z.T.; Li, Y.X. Experimental Study on Oil and Gas Explosion in Simulated Vertical Domed Oil Tanks. *Explos. Impact* **2022**, *42*, 152–168.
7. Huang, Y.; Xi, G.C.; Yang, J.; Xu, N.; Hao, Y.M. Research on the Flame Spread Law of Methane Pipeline Leakage and Explosion. *Safety* **2022**, *43*, 62–68. [CrossRef]
8. Yu, J.L.; Kang, S.L.; Yan, X.Q.; Zhan, X.B. Study on Methane Explosion and Venting Characteristics in Pressurized Vessels. *J. Saf. Environ.* **2023**, 1–9. [CrossRef]
9. Hu, F.F.; Jia, Y.; Wang, W.T.; Jiang, B.Y.; Li, S.Z.; Cheng, Y.F. Study on the Combustion and Explosion Characteristics of Acetylene/Air Premixed Gas in Confined Spaces. *Pyrotechnics* **2022**, *6*, 50–55.
10. Li, D.; Dai, S.J.; Zheng, H.W.; Cui, Z.Y. Research on the Maximum Explosion Pressure and Maximum Rate of Pressure Rise of Premixed Gas in Flameproof Enclosures. *Coal Mine Saf.* **2023**, *54*, 216–221. [CrossRef]
11. Cao, W.G.; Huang, L.Y.; Liang, J.Y.; Miao, N.; Rao, G.N.; Pan, F. Study on the Explosion Characteristics of Coal Powder in Spherical Closed Containers. *J. China Univ. Min. Technol.* **2014**, *43*, 113–119. [CrossRef]
12. Castellanos, D.; Carreto, V.; Skjold, T. Construction of a 36L dust explosion Apparatus and turbulence flow field comparison with a standard 20 L dust explosion vessel. *J. Loss Prev. Process Ind.* **2018**, *55*, 113–123. [CrossRef]

13. Dong, B.Y.; Zha, Y.X.; Zou, Y.; Yin, P.P. Numerical Simulation and Experimental Study on Methane-Hydrogen-Air Explosion Process in Spherical Pressure Vessel. *China Saf. Sci. J.* **2023**, *19*, 157–163.
14. Jiang, X.S.; Xie, W.; Zhao, Y.D.; Li, J.Y.; Li, J.; Yu, B.B. Experimental Study on Oil and Gas Explosion in Long and Narrow Pipes with Different Length-to-Diameter Ratios. *Oil Gas Storage Transp.* **2020**, *39*, 879–884.
15. Ferrara, G.; Willacy, S.K.; Phylaktou, H.N.; Andrews, G.E.; Di Benedetto, A.; Salzano, E.; Russo, G. Venting of gas explosion through relief ducts: Interaction between internal and external explosions. *J. Hazard. Mater.* **2007**, *155*, 358–368. [CrossRef]
16. Bauwens, C.R.; Chaffee, J.; Dorofeev, S. Effect of Ignition Location, Vent Size, and Obstacles on Vented Explosion Overpressures in Propane—Air Mixtures. *Combust. Sci. Technol.* **2010**, *182*, 1915–1932. [CrossRef]
17. Cao, Y.; Guo, J.; Hu, K.; Xie, L.; Li, B. The effect of ignition location on explosion venting of hydrogen—air mixtures. *Shock Waves* **2017**, *27*, 691–697. [CrossRef]
18. Guo, J.; Sun, X.X.; Rui, S.C.; Cao, Y.; Hu, K.L.; Wang, C.J. Effect of ignition position on vented hydrogen–air explosions. *Int. J. Hydrog. Energy* **2015**, *40*, 15780–15788. [CrossRef]
19. Zhang, P.L.; Du, Y. Nitrogen Non-Premixed Inhibition Experiment for Oil and Gas Explosions. *Explos. Impact* **2016**, *36*, 347–348.
20. Wang, S.M.; Yan, Z.H.; Li, X.D.; Li, G.Q.; Guo, H.; Wu, D.J. The venting explosion process of premixed fuel vapor and air in a half—open vessel: An analysis of the overpressure dynamic process and flame evolution behavior. *Fuel* **2020**, *268*, 117385. [CrossRef]
21. Li, H.Z.; Xiao, H.H. Experimental study on the explosion characteristics of NH3/DME/air mixtures. *Fuel* **2023**, *352*, 129069. [CrossRef]
22. Li, R.Z.; Si, R.J. Influence of Gas Concentration on Explosion Pressure and Pressure Rise Rate. *J. Xi'an Univ. Sci. Technol.* **2010**, *30*, 29–33. [CrossRef]
23. Wang, X.F.; Shi, E.H.; Qi, C.; Yan, X.Q.; Zhang, Z.H.; Yu, J.L. Experimental study on aerosol explosion characteristics and flame propagation behavior of aluminum/ethanol nanofluid fuel. *Fuel* **2023**, *352*, 129022. [CrossRef]
24. Zhang, Y.; Li, F.B.; Cai, R. Analysis of Accidental Acetone Explosion Effects in a Closed Container and Safety Measures. *Chem. Eng. Manag.* **2022**, *7*, 88–91. [CrossRef]
25. Li, G.Q.; Zheng, K.; Wang, S.M.; Chen, W.Z. Comparative study on explosion characteristics of hydrogen and gasoline vapor in a semi-confined pipe based on large eddy simulation. *Fuel* **2022**, *328*, 125334. [CrossRef]
26. Wang, S.M.; Wu, D.J.; Guo, H.; Li, X.D.; Pu, X.Y.; Yan, Z.H.; Zhang, P.L. Effects of concentration, temperature, ignition energy and relative humidity on the overpressure transients of fuel-air explosion in a medium-scale fuel tank. *Fuel* **2020**, *259*, 116265. [CrossRef]
27. Zhang, P.L.; Du, Y.; Zhou, Y.; Qi, S.; Wu, S.L.; Xu, J.F. Explosions of gasoline—air mixture in the tunnels containing branch configuration. *J. Loss Prev. Process Ind.* **2013**, *26*, 1279–1284. [CrossRef]
28. Qi, S.; Du, Y.; Zhang, P.L.; Li, G.Q.; Zhou, Y.; Wang, B. Effects of concentration, temperature, humidity, and nitrogen inert dilution on the gasoline vapor explosion. *J. Hazard. Mater.* **2017**, *323*, 593–601. [CrossRef] [PubMed]
29. Wang, B.; Du, Y.; Li, G.Q.; Yuan, G.Q.; Wang, S.M. Study on the explosion characteristics of oil and gas in long and narrow confined pipelines. *Vib. Impact* **2017**, *36*, 80–85+145. [CrossRef]
30. Ge, Y.; Ma, H.H.; Wang, L.Q. On the smoothing explosion pressure curves using Savitzky—Golay method. *J. Loss Prev. Process Ind.* **2022**, *80*, 104929. [CrossRef]
31. Jiang, X.S.; Chen, R.; Zhang, P.L.; Cai, Y.X.; Zhou, D.L.; He, D.H.; Qin, X.Z.; Zhu, S.J. Study on Gasoline—Air Mixture Explosion Overpressure Characteristics and Flame Propagation Behaviors in an Annular Cylindrical Confined Space with a Circular Arch. *Energies* **2023**, *16*, 6944. [CrossRef]
32. Jiang, H.P.; Bi, M.S.; Gao, W. Effect of monoammonium phosphate particle size on flame propagation of aluminum dust cloud. *J. Loss Prev. Process Ind.* **2019**, *60*, 311–316. [CrossRef]
33. Shang, S.; Bi, M.S.; Zhang, T.J.; Jiang, H.P.; Zhang, S.L.; Gao, W. Synthesis of green nanomaterial and discussion on its suppression performance and mechanism to aluminum dust explosion. *Process Saf. Environ. Prot.* **2021**, *151*, 355–364. [CrossRef]

fire

MDPI

Article

Experimental Estimation of Turbulent Flame Velocity in Gasoline Vapor Explosion in Multi-Branch Pipes

Keyu Lin [1], Peili Zhang [1,*], Jimao Duan [1], Shuo Xiang [1], Ting'ao Shen [1] and Chaoshan Yang [2]

[1] Petroleum, Oil and Lubricants Department, Army Logistics Academy, Chongqing 401331, China; duanjimiao@126.com (J.D.)
[2] Department of Military Installations, Army Logistics Academy, Chongqing 401331, China
* Correspondence: zpl612323@163.com; Tel.: +86-180-0836-0107; Fax: +86-8673-1504

Abstract: The overpressure characteristics of gasoline explosions in multi-branch pipes are caused by various factors, with flame velocity as a particularly significant determinant. Overlooking the impact of turbulent flow in the branch pipes can induce a significant discrepancy in the outcome when using laminar flame velocity to determine the maximum rate of overpressure rise. To quantify the impact of turbulent flame velocity on the rate of overpressure rise in the gasoline explosions within branch pipes, the laminar flame velocity was replaced with its turbulent counterpart. Additionally, modifications to the formula for calculating the maximum overpressure rise rate were implemented. Then, experimental data of peak explosion overpressure and overpressure rise rate under different numbers of branches were obtained. Finally, the empirical data were inputted into the modified formula to determine the maximum rate of overpressure rise, thus enabling the calculation of the turbulent flame velocity across varying numbers of branches. The findings reveal a positive correlation between the number of branches and the turbulent flame velocity during tube explosions. When the number of branch pipes increased from 0 to 4, the turbulent flame velocity was found to range from 8.29 to 13.39 m/s. The increase in the number of branches did not consistently enhance the turbulent flame velocity. As the number of branches increased from zero to three, the turbulent flame velocity rose accordingly. Differently, as the number of branches exceeds three, the turbulent flame velocity exhibits fluctuations and peaks at a level approximately 1.8 times higher. The research method of this paper can provide a reference for estimating the turbulent flame velocity in the combustion process of flammable gas explosions in multi-branch tunnels.

Keywords: gasoline–air mixture; branch tube; number of branches; turbulence factor; explosion overpressure

Citation: Lin, K.; Zhang, P.; Duan, J.; Xiang, S.; Shen, T.; Yang, C. Experimental Estimation of Turbulent Flame Velocity in Gasoline Vapor Explosion in Multi-Branch Pipes. *Fire* **2024**, *7*, 37. https://doi.org/10.3390/fire7020037

Academic Editor: Chuyuan Huang

Received: 2 January 2024
Revised: 23 January 2024
Accepted: 24 January 2024
Published: 25 January 2024

1. Introduction

Explosions of combustible gas are a major hazard in both industrial production and daily life. Of particular concern is the explosion of chemical gases, which has the potential to cause severe casualties and property damage. Therefore, explosions of this type have attracted considerable attention. To evaluate the damage caused by gas explosions and improve the design of explosion-proof vessels, it is often necessary to assess various parameters, including maximum explosion pressure, maximum pressure rise rate, and gas explosion index K_G [1–3]. These parameters are widely used in academic and industrial settings. Nevertheless, the aforementioned parameters are significantly affected by the initial conditions of combustion for the combustible mixture, including but not limited to the initial temperature [4], initial pressure [5,6], container size [7] and geometry [8], ignition energy [9] and position [10], and turbulence level [11]. Bauwens et al. [12] found that the increase in turbulence caused by obstacles significantly enhanced the maximum flame area, combustion rate, and external explosion overpressure. Arisova et al. [13] conducted research on the pipe explosion pressure and flame velocity of branch tunnel discharge; their results indicated that the turbulence caused by the branch tunnel can cause multiple different pressure peaks in explosion overpressure and that the flame velocity increases with the distance between the branch tunnel and the ignition position. Similar research conclusions were also found in a relevant study by Ponizy and Henneton [14]. Almerinda [15] analyzed the effect of ignition energy and ignition source on the deflagration index and found that the deflagration index is independent of ignition energy and mainly depends on turbulence. As indicated by Yan Ming [16] and Wang Kai [17], the initial turbulence level is largely affected by the size and shape of the container.

In the context of subterranean oil caverns, oil pipes, and domestic natural gas pipes characterized by a multi-branch configuration, the flame dynamics and overpressure characteristics exhibited by the explosion of flammable gas differ from those observed in a conventional confined space or container during a gas explosion.

Numerous studies [4,18] establish a close correlation between the behavior of explosion flames and the overpressure characteristics of combustible gas in a branch-structured tunnel. Moreover, the intensity of explosions in this context is higher compared to that of gasoline in a confined space. A significant contributor to this phenomenon is a multi-branch structure, which enables the propagation and development of flames and pressure waves in multiple-branch tunnels. Consequently, multiple distinct flame fronts develop, significantly increasing both the total area of flame and the rate of heat release. In turn, the presence of a branch tunnel leads to multiple interacting pressure waves, which further complicates the distribution and evolution of the pressure field within a confined space. In the course of an explosion, a branch-structured tunnel has a significant impact on the turbulence across the flow field. The explosion theory suggests that within a confined container, a positive feedback mechanism exists, involving turbulence, flame velocity, and overpressure [4]. Therefore, a significant characteristic shown by the explosion of flammable gas within a tunnel comprising a complex network of branches is the close correlation between turbulence, flame, and pressure wave phenomena. Recent incidents caused by explosions in a confined space after the structural failure of gasoline branches with great loss of life [18] serve as evidence of the aforementioned findings.

Gas explosions can be exacerbated by the factors that increase turbulence in the gas flow within pipes, which causes a rising level of explosion-induced overpressure. There have been many studies on the impact of turbulence on explosion intensity within tubes. Li et al. [19] conducted an analysis of the different characteristics of flame variation and overpressure in T-shaped pipe branches. This analysis combined experimental investigation and numerical simulation. The findings of this study confirm the argument that intensifying flame folding and turbulence in the T-shaped pipe branches leads to an increase in explosion pressure. The study by Wang Hua et al. [20] examined the impact of initial turbulence on methane–air explosions. According to the experimental results, initial turbulence contributes to an increase in both the maximum explosion pressure and the maximum rate of pressure rise. Additionally, the time to reach the maximum explosion pressure was shortened and the presence of initial turbulence enhanced the explosive intensity and destructive power of the methane–air mixture. According to the research findings presented by Xie Yiyue [21], the explosion range for combustible gas (vapor) is narrower in a turbulent state than in a macroscopic static state. Turbulence induces the folding of the flame front, which leads to an indistinct boundary and a notable increase in flame propagation velocity. The study by Jian Congguang and Lin Baiquan [22] indicated the significant impact of the sudden change in pipe area on turbulence during gas explosions. Abrupt increases or decreases in the cross-sectional area of the pipe give rise to additional turbulence.

The objective of this study is to establish a correlation between the turbulent flame velocity and the maximum rate of overpressure rise through a combination of theoretical analysis and experimentation. On this basis, the turbulent flame velocity of gasoline explosions in enclosed containers will be determined. The findings of this study are expected to support the assessment of damage caused by gasoline explosions, the design of explosion-proof containers, and the implementation of safety measures for pipes.

2. Theoretical Estimation Method for Turbulent Flame Velocity

According to the research conclusions of Pu [23], Dahoe [24], and F. Cammarota et al. [25], the formula for calculating the rise rate of the maximum explosive overpressure of combustible gas in a vertical cylindrical vessel is

$$\left(\frac{dp}{dt}\right)_{MAX} = \frac{A_V}{V} S_L (P_{MAX} - P_0)\left(\frac{P_{MAX}}{P_0}\right)^{\frac{1}{\gamma}} \quad (1)$$

where A_V is the maximum flame front area, m^2; V is volume of explosive vessel, m^3; S_L is the laminar flame velocity, m/s; P_{MAX} is the peak explosion pressure, Pa; P_0 is the initial pressure, Pa; γ is the gas adiabatic index; and $\left(\frac{dp}{dt}\right)_{MAX}$ is the peak boost rate, Pa/s.

The above formula shows that the maximum overpressure rise rate is related to the shape and size of the container (i.e., A_V and V), is proportional to the laminar flame velocity, and is related to the peak overpressure and initial pressure. Wang et al. [17] and Li et al. [19] found that a gas explosion experiment in an over-branching-structure vessel intensified the explosive gas turbulence in the tube, thus leading to an increase in the explosion overpressure rise rate. However, Formula (1) does not consider the influence of turbulence level on the rate of overpressure rise. In fact, due to the short duration of laminar flow in the process of a premixed gas explosion, most of the time is in a turbulent state, and the use of laminar flame velocity to describe the impact on the rate of overpressure rise is bound to bring a large deviation. Therefore, it is necessary to rewrite Formula (1) by replacing laminar flame velocity with turbulent flame velocity. Based on this, Formula (1) is rewritten as follows:

$$\left(\frac{dp}{dt}\right)_{MAX} = \frac{A_V}{V} S_T (P_{MAX} - P_0)\left(\frac{P_{MAX}}{P_0}\right)^{\frac{1}{\gamma}} \quad (2)$$

where S_T is denotes the turbulent flame velocity, m/s. According to Formula (2), it can be calculated using the following formula:

$$S_T = \frac{V}{A_V(P_{MAX} - P_0)}\left(\frac{P_0}{P_{MAX}}\right)^{\frac{1}{\gamma}}\left(\frac{dp}{dt}\right)_{MAX} \quad (3)$$

where V is the explosion vessel volume, m^3; P_{MAX} is peak explosion pressure, Pa; P_0 is initial pressure, Pa; and $(dp/dt)_{MAX}$ is peak pressure rise rate, Pa/s. These parameters can be determined using experimental data, while the maximum flame front area A_V and gas adiabatic index γ can only be obtained via theoretical calculation. Therefore, through Formula (3), the turbulent flame velocity in the gasoline–air explosion process of pipes with different multi-branch structures can be calculated, and the relationship between the flame velocity of turbulence and parameters such as the number and length of branch pipes can be further discussed.

3. Experimental Methods and Data

3.1. Experimental Devices and Methods

In this study, the experimental set-up involved a combined test bench, a dynamic signal acquisition system, a gasoline atomizing circulation system, a volume fraction test system, and an ignition system, as illustrated in Figure 1.

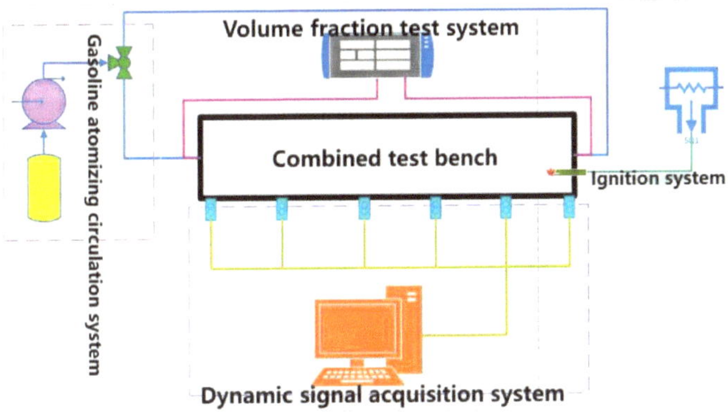

Figure 1. Composition of the experimental system.

The combined test bench primarily consisted of four steel square pipes in different forms and lengths, including a 0.5 m straight pipe, a 1 m straight pipe, a 0.5 m length straight pipe, a 1 m length straight pipe, a tee composed of a 1 m main pipe and a 0.5 m branch pipe, and a tee composed of a 0.5 m main pipe and a 0.5 m branch pipe. These pipes had a wall thickness of 10 mm and a cross-section size of 0.1 m × 0.1 m, with a round hole opened around the square pipe for pressure sensor installation. According to the experimental requirements, four types of square tubes were assembled through flanges, and five types of test benches with 0, 1, 2, 3, and 4 branches, respectively, were constructed. To maintain a constant total volume of the bench, the total length was set to 6 m. Figure 2 shows the structures and the dimensions of the five test benches.

The dynamic signal acquisition system consisted of a high-frequency dynamic pressure sensor and a signal acquisition and analysis system. With six sensors installed on the test platform (as shown in Figure 2), the accuracy level reached 0.25%FS, and the range was 0–1 MPa. During the experiment, pressure data were tested, and the system of dynamic signal acquisition and analysis provided data on pressure changes.

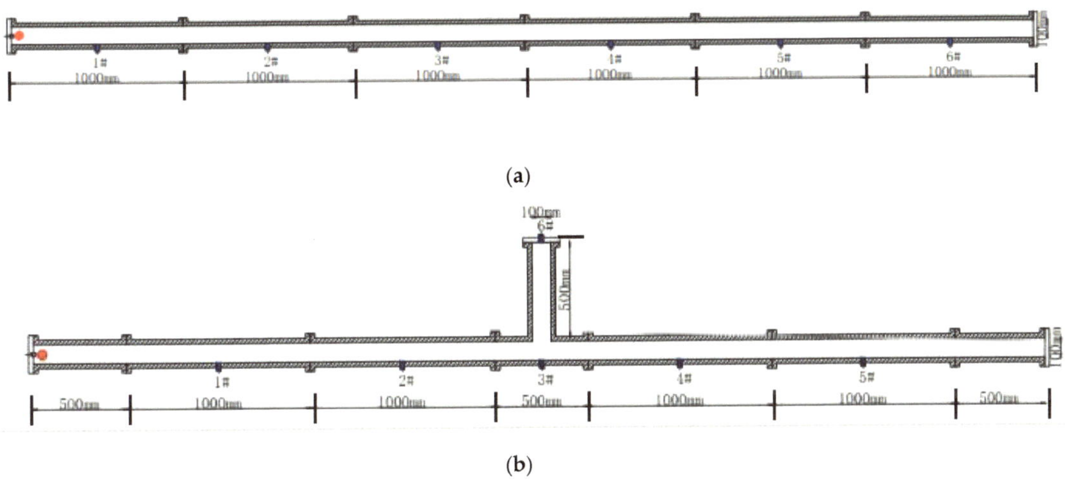

Figure 2. *Cont.*

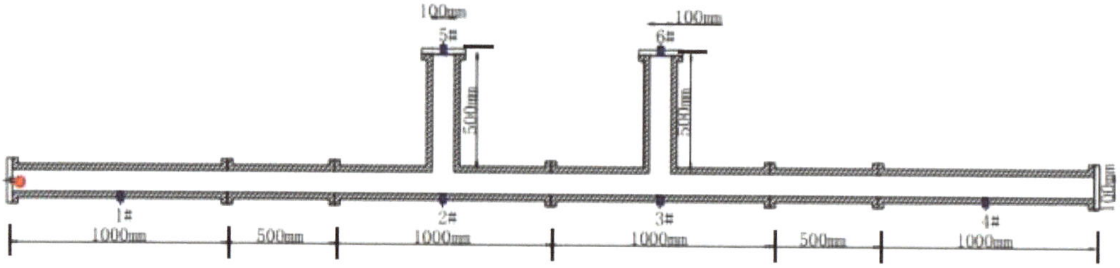

(c)

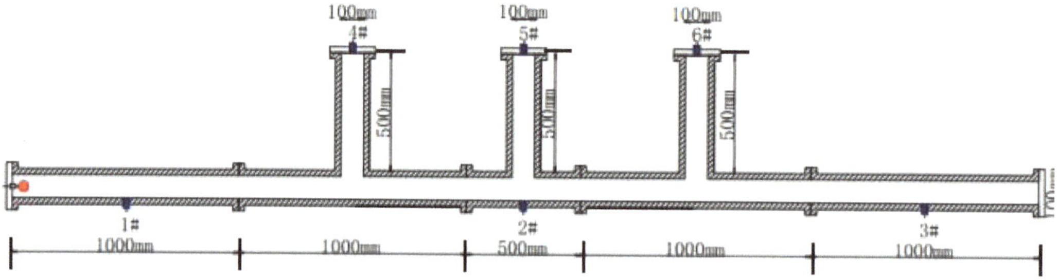

(d)

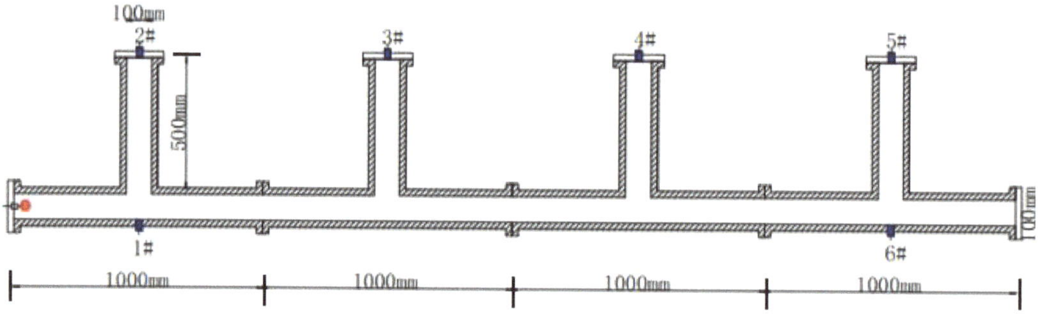

(e)

Figure 2. Scheme of combined test bench and position of transducers (i # pressure transducers). (**a**) Straight pipe, (**b**) straight pipe containing 1 branch, (**c**) straight pipe containing 2 branches, (**d**) straight pipe containing 3 branches, (**e**) straight pipe containing 4 branches.

The gasoline atomizing circulation system consisted of an oil storage bottle, a vacuum pump, an ultrasonic atomizer, a pipe, and a three-way ball valve. Gasoline vapor was generated by the ultrasonic atomizer and pumped into the test bench. The volume concentration of oil vapor was measured using an infrared gas concentration tester. After the target concentration was reached, the three-way valve was adjusted for gas circulation to evenly distribute the gasoline in the test bench. Figure 3 shows the system's structure.

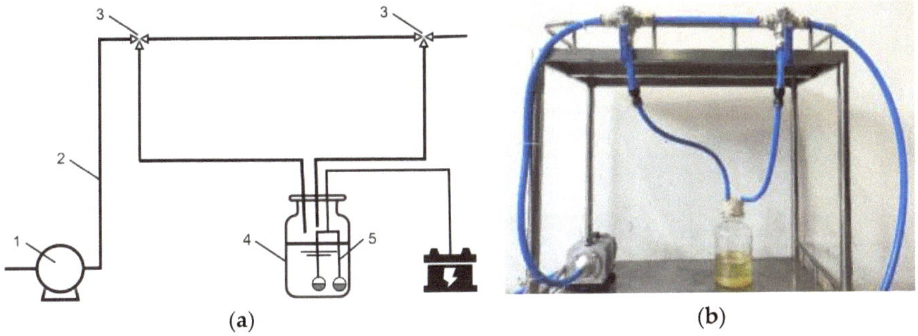

Figure 3. Gasoline atomizing circulation system. (**a**) Specific structure: 1. oil storage bottle; 2. pipe; 3. three-way ball valve; 4. oil storage bottle; 5. ultrasonic atomizer. (**b**) Photo of the gasoline atomizing circulation system.

A high-energy non-interference ignition mode was adopted for the ignition system, with the initial energy of the gasoline explosion provided through a high-energy arc spark.

The initial conditions of the experiment were as follows: the initial temperature was 293 K, the initial pressure was 101 kPa, and the ignition energy was 5 J. No. 92 gasoline was the experimental object; its performance parameters are detailed in Table 1. The explosion characteristics of gasoline are more pronounced within the range of gasoline volume fraction from 1% to 2.7%, with the most evident effects observed between 1.2% and 1.6%. Therefore, the experimental concentration of gasoline was set at 1.2%.

Table 1. Main properties of the 92 RON unleaded gasoline.

Initial Condition	Value of Initial Parameter
Lower flammable limit (v %)	1.0
Octane rating	92
Specific gravity	0.725
Upper flammable limit (v %)	2.7
Average molecular formula	$C_{8.1}H_{14.05}$
Average heating value (kJ/kg)	44,000

The experiment involved three major parts: gasoline premix, ignition experiment, and data acquisition. In the initial phase of oil–gas mixing, the gasoline–air mixture was injected into the experimental pipe through a gasoline atomization system, with the concentration of gasoline measured simultaneously. Once the required concentration was reached, the gasoline atomization system transitioned into the circulation process and lasted 2 min to ensure a thorough and even mixing of gas and air in the pipe. When the premixed gas was stabilized, the gas circulation system was switched off for 30 s. Subsequently, the control ignition system and the pressure acquisition system were activated simultaneously to collect the relevant experimental data. To ensure objectivity and repeatability, the experiment was conducted three times under the same working conditions, and the average of these experiments was analyzed to examine the change in overpressure load.

3.2. *Acquisition of Experimental Data*

Figures 4 and 5 show the impact of varying branch numbers on the sequence of explosion overpressure and boost rate. Obviously, as the number of branches increases, there is a significant rise in both explosion overpressure and pressure boost rate, suggesting a positive influence of pipe branches on the explosion overpressure. Figure 6 illustrates the variations in peak overpressure, the time at which the peak overpressure is reached, and the pressure boost rate with different branch numbers. Additionally, it is evident that the maximum boost rate is higher when branching occurs. In general, as the number of pipe branches increases, both the peak explosion overpressure and pressure boost rate first increase and then decrease. However, it is worth noting that a larger number of branches is correlated with a more significant positive effect. The use of a branch pipe amplifies such phenomena as diffracting, superposition, and excitation associated with the explosion wave generated by the combustion of gasoline and air.

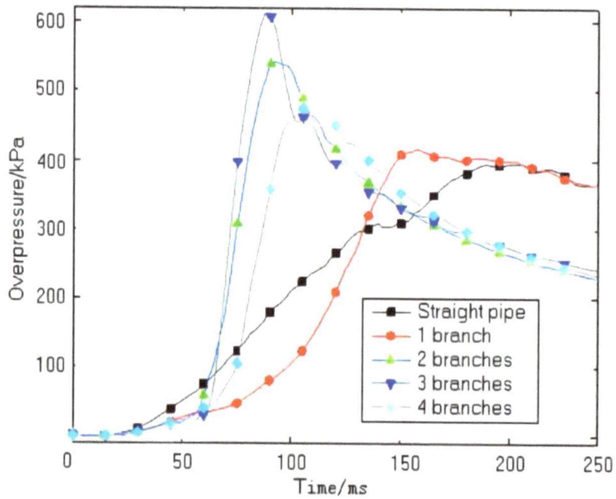

Figure 4. Curves of explosion overpressure history of gasoline–air mixture with different numbers of branches.

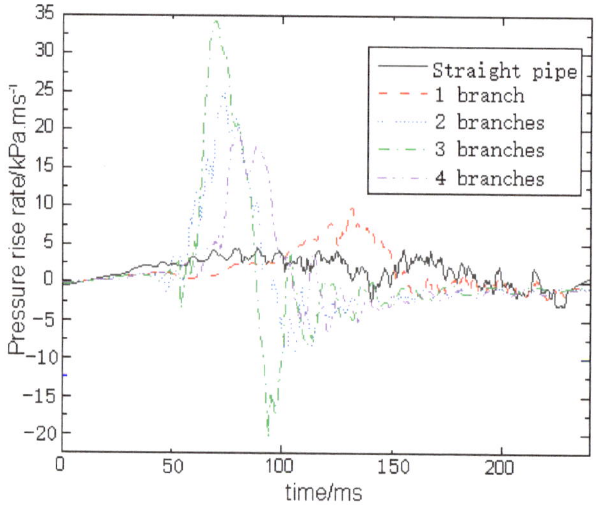

Figure 5. Comparison of overpressure rising rate time-series curves of different branches.

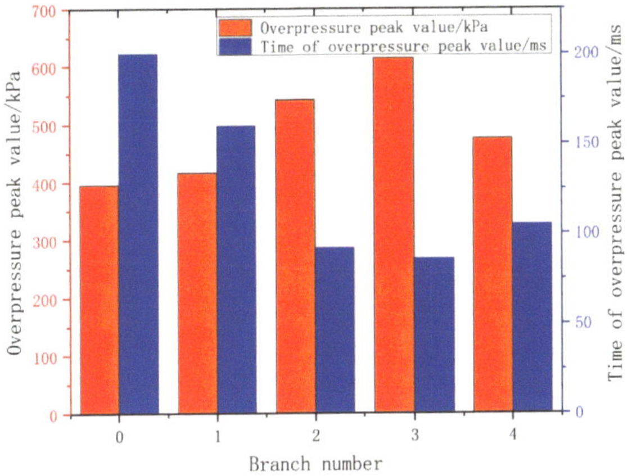

(a)

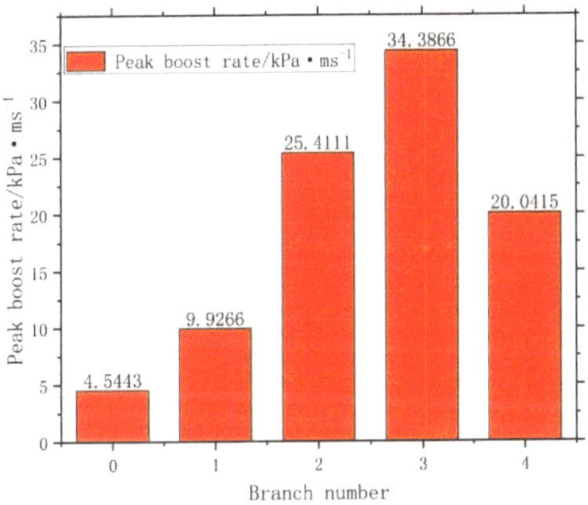

(b)

Figure 6. Peak overpressure and overpressure rising rate of different branches. (**a**) Peak value and time of overpressure. (**b**) Peak rate of overpressure rise.

According to Figures 4–6, the peak values of overpressure and boost rate can be extracted from the experimental values, as shown in Table 2.

Table 2. Peak values of overpressure and boost rate under different working conditions.

Branch Number	Overpressure Peak Value (kPa)	Peak Boost Rate (kPa·ms^{-1})
0	396.00	4.54
1	416.53	9.93
2	542.19	25.41
3	613.74	34.39
4	474.89	20.04

4. Theoretical Calculation of Other Data

4.1. Maximum Flame Front Area

Christophe et al. [26] pointed out that the progression of gas explosion flame propagation in a narrow confined space can be divided into three typical processes: (1) the formation of a hemispherical flame at the initial ignition stage; (2) the axial stretching of the flame, which forms a finger flame; and (3) the deformation of the flame surface into a tulip flame. An analysis of these stages allows for the determination of the maximum flame front area. In the first stage, the premixed combustible mixture is ignited by an electric spark from the ignition electrode, and the flame zone extends in all directions. At this time, the flame range is small, and the wall influence is negligible. An idealized flame front in a pipe has the shape of a half sphere, and the area is calculated according to

$$A_V = 2\pi a^2 = 2 \times \pi \times 0.1^2 = 0.063 \text{ m}^2 \tag{4}$$

In the second stage, with the axial stretching of flame, the flame front became a finger, referred to as the finger phase. At this point, the flame front maintains a basic half-spherical shape along the axial forefront, and in the radial transmission, cold collisions occur as the flame contacts the wall surface. If the container shape remains constant, the maximum flame front area at this stage is identical to that of the first stage. However, in the case of an experimental vessel configured as a multi-branch pipe, an increase in channel area leads to a larger flame area when the flame propagates to the branch. Therefore, the relationship between the maximum flame front area and the number of branches can be approximated.

In the third stage, with the continued spread of flame, due to the return air flow instability and fluid mechanics, the influence of the flame front, initially outward spherical, gradually transforms into a flattened shape with a sagging center. This transformation occurs with a short duration, based on [5], when the explosion total time reached 1.29 times the wall time fireside, a twisted flame occurred and formed a tulip flame. As can be seen from Figure 4, both the explosion overpressure and the pressure boost rate reach their peaks in a very short duration of the explosion. Therefore, in the pipe, the tulip flame exerts little influence on the key parameters of explosion and can be ignored, and the maximum flame front area of this stage is the same as that of the first stage. The above analysis is for a straight pipe; however, because the experimental vessel is a multi-branch pipe, the increase in the channel area will lead to an increase in the flame area when the flame propagates to the branch. When the flame propagates to the branch pipe, the area of the flame front will approximately double. Therefore, for a single branch pipe (Figure 2b), the maximum flame front area is

$$A_{V,1} = 0.063 \times 2 = 0.126 \text{ m}^2$$

For pipes with two, three, and four branches (Figure 2c–e), the distance between adjacent branch pipes is greater than or equal to the length of each branch pipe. Therefore, when the flame front propagates to the next branch pipe, the flame in the previous branch pipe has already spread to the bottom of the branch pipe and been extinguished. It can be approximated that when there are two, three, or four branches, the maximum flame area throughout the explosion process is still twice the maximum flame area of a straight pipe (Figure 2a), which is equal to the maximum flame area when there is one branch, at 0.126 m^2.

In summary, for a straight pipe without any branching, the maximum flame front area is 0.063 m^2. When there are one, two, three, or four branch pipes, the maximum flame area is always 0.126 m^2.

4.2. Laminar Flame Velocity and Gas Adiabatic Index

Metghalhi and Kech [27] carried out a series of experiments in the combustion chambers of reciprocating internal combustion engines and gas turbines and determined the formula for calculating laminar flame propagation velocity of fuel–air mixture through experiments:

$$S_L = S_{L,ref}\left(\frac{T_u}{T_{u,ref}}\right)^{\gamma}\left(\frac{P}{P_{u,ref}}\right)^{\beta} \tag{5}$$

where the subscript *ref* is the value of the reference state (101 kPa, 293 K), γ is the adiabatic index, and β is the pressure index. When the temperature $T_u < 350$ K for the unburned mixture in Formula (5), the laminar flame velocity can be simplified as

$$S_L = C_M + C_2(\Phi - \Phi_M)^2 \tag{6}$$

where Φ is the chemical equivalent ratio of combustible gas. C_M, C_2, and Φ_M are constants determined by the type of combustible gas and are given in Table 3.

Table 3. Φ_M, C_M, C_2 of different combustible gases.

Combustible Gas	Φ_M	C_M (cm·s^{-1})	C_2 (cm·s^{-1})
Methanol	1.11	43.70	−157.22
Propane	1.08	40.11	−186.48
Isooctane	1.13	33.72	−110.82
RMFD-303	1.13	35.58	−140.45

In this study, No. 92 gasoline was selected in the experiment, and the volume concentration of gasoline–air was 1.2%. For the convenience of calculation, No. 92 gasoline was replaced with isooctane. According to Li's study [6], the adiabatic index and pressure index of gas are functions of the chemical equivalent ratio of combustible gas, expressed as

$$\gamma = 2.18 - 0.8(\Phi - 1) \tag{7}$$

$$\beta = -0.16 + 0.22(\Phi - 1) \tag{8}$$

Zhang Peili et al. [28] analyzed the relationship between the equivalent ratio of gasoline combustion and volume fraction in the literature. According to the analysis conclusion of the literature, the chemical reaction formula of isooctane combustion in air is as follows:

$$C_8H_{18} + 12.5(O_2 + 3.76N_2) = 8CO_2 + 9H_2O + 47N_2 \tag{9}$$

The stoichiometric air–fuel ratio is:

$$(\frac{A}{F})_{stoic} = (\frac{m_{air}}{m_{fuel}}) = \frac{12.5 \times 4.76 M_{r_{air}}}{M_{r_{fuel}}} = \frac{59.5 M_{r_{air}}}{M_{r_{fuel}}} \tag{10}$$

And

$$\Phi = \frac{(\frac{A}{F})_{stoic}}{\frac{A}{F}} = \frac{59.5 M_{air}/M_{fuel}}{n_{air}M_{air}/n_{air}M_{fuel}} = 59.5\frac{n_{fuel}}{n_{air}} \tag{11}$$

According to the gas state formula, under the same temperature and pressure conditions,

$$\Phi = \frac{(\frac{A}{F})_{stoic}}{\frac{A}{F}} = 59.5\frac{n_{fuel}}{n_{air}} = 59.5\frac{V_{fuel}}{V_{air}} = 59.5\frac{Y_{fuel}}{Y_{air}} = 59.5\frac{Y_{fuel}}{1 - Y_{fuel}} \tag{12}$$

According to the above formula, when the oil/gas volume fraction $Y_{fuel} = 0.012$, the oil/gas chemical equivalent ratio is

$$\Phi = 59.5\frac{Y_{fuel}}{1 - Y_{fuel}} = 0.72$$

The gas adiabatic index and pressure index are

$$\gamma = 2.18 - 0.8(\Phi - 1) = 2.4$$

$$\beta = -0.16 + 0.22(\Phi - 1) = -0.22$$

According to Table 2, the laminar flame velocity of isooctane can be calculated using Formula (6):

$$S_L = 33.72 - 110.82 \times (0.72 - 1.13)^2 = 15.09 \text{ cm/s}$$

5. Turbulent Flame Velocity Estimation Results and Discussion

By substituting the experimental data and theoretical calculation data obtained under the above working conditions into Formula (3), the turbulent flame velocity can be calculated in different working conditions. The calculation results are shown in Table 3, where the branch number of 0 represents the working condition of the straight vessel.

The data in Table 4 are plotted in Figure 7.

Table 4. Turbulence factors under different conditions.

Branch Number	Turbulent Flame Velocity (m/s)	Laminar Flame Velocity (m/s)	$S_{T,i}/S_{L,i}$	$S_{T,i}/S_{T,0}$
0	8.29	0.15	54.94	1.00
1	8.32	0.15	55.04	1.01
2	13.62	0.15	90.23	1.64
3	15.06	0.15	99.80	1.82
4	13.39	0.15	88.74	1.62

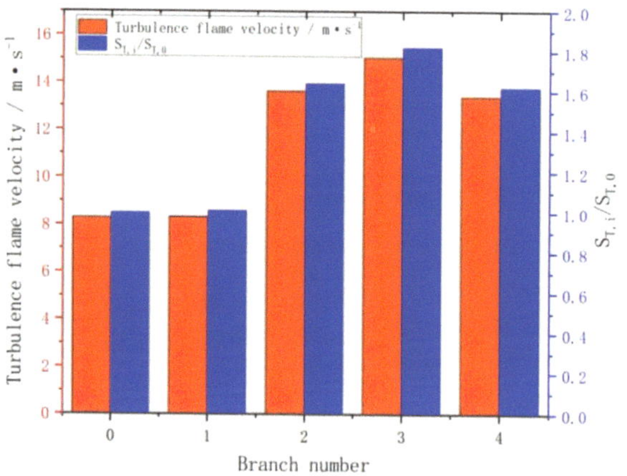

Figure 7. Scatter diagrams of turbulent flame velocity under different numbers of branches.

The analysis shown in Figure 7 reveals a notable increase in turbulent flame velocity compared to the laminar flame. Additionally, the turbulent flame exhibits exponential growth during the explosion as the number of branches increases. When the number of branches is one, the turbulent flame velocity is basically equivalent to that of the straight tube as the overpressure rate and flame area are synchronous. However, as the number of branches continues to rise to three, the influence of the overpressure rate exceeds that of the flame area, with a substantial increase in turbulent flame velocity of up to 1.8 times. However, beyond the number of three branches, the turbulent flame velocity begins to decline. Existing literature examines the correlation between flame velocity in gasoline explosions across multiple branches and the number of branches. According to these studies, an increase in the number of branches does not necessarily result in a proportional increase in flame velocity [29]. Instead, the flame velocity decreases when the number of branch tunnels exceeds a specific threshold.

The escalation in turbulence during the branch pipe explosion process is mainly attributed to the propagation of the flame into the branch. At this stage, the flame front undergoes serious creasing, bending, and deformation, which results in an expanded burning area of gas flame, which, in turn, enhances the entrainment effect of mixed gas and increases the combustion reaction rate, accelerating the active substances' involvement in combustion reaction and enhancing the heat transport rate. As a result, the flame travels at a faster velocity, giving rise to a rise in the intensity of the compression wave generated by the explosion. Meanwhile, the precursor shock wave effectively compresses and preheats the unburned gas in front of the flame, causing a significant disturbance in the airflow. This disturbance contributes to an increased gradient in the flow field and further amplifies the combustion rate and turbulent kinetic energy. From the perspective of hydrodynamics, as the number of branches increases from 0 to 3, the increase in the turbulent flame velocity caused by increased turbulence is greater than the decrease in the turbulent flame velocity caused by heat loss from the flame to the pipe wall. Therefore, the overall effect is an increase in the turbulent flame velocity. However, when the number of branch tunnels becomes four, the increase in the turbulent flame velocity caused by increased turbulence is less than the decrease in flame velocity caused by heat loss from the flame to the pipe wall. As a result, the overall effect is a decrease in turbulent flame velocity. This may be the primary reason for the observed decrease in the turbulent flame velocity when the number of branches reaches four, as shown in Figure 7.

6. Conclusions

In order to quantitatively estimate the turbulent flame velocity of a gasoline vapor explosion in a multi-branch pipe, a formula that can quantitatively calculate the turbulent flame velocity was first obtained through theoretical analysis, and then other overpressure parameters were obtained through experiments. Finally, the turbulent flame velocity of the gasoline vapor explosion process in a multi-branch pipe was quantitatively calculated and analyzed. The main conclusions of this present paper can be summarized as follows:

1. The current formula for the maximum overpressure rise rate does not take into account the influence of turbulence, which inevitably leads to deviation when this formula is applied to multi-branch-structured pipe. Considering the influence of turbulence, it is feasible to estimate the turbulent flame velocity of explosions in multi-branch pipes by modifying the formula.
2. Compared with straight pipelines, the turbulent flame velocities of explosions in pipes with one, two, three, and four branches are 1.01, 1.64, 1.82, and 1.62 times higher than those in a straight pipe.
3. The turbulent flame velocity increases from 8.29 to 15.06 m/s as the number of branch pipes increases from 0 to 3. However, an increase in the number of branch pipes does not always lead to an increase in turbulent flame velocity. When the number of branches increased from three to four, the turbulent flame velocity decreased from 15.06 to 13.39 m/s.

Notably, the turbulent flame velocity determined in this study represents the approximate turbulent flame velocity in a multi-branch pipe during the gasoline–air explosion process. The occurrence of complex structural pipe explosions caused by overpressure during the gasoline–air process is influenced by various factors, including the rate of heat release from fire and the rate of heat loss from solid boundaries. Consequently, it is necessary to conduct a thorough evaluation of the collective impact of these factors on the characteristics of overpressure. Despite the challenges in quantifying the velocity of turbulent flames during gasoline–air explosions within an enclosed container, the research methodology and findings presented in this paper remain useful for investigations into the overpressure characteristics of gas explosions in enclosed containers.

Author Contributions: Writing-original draft, K.L.; Project administration, P.Z.; Methodology, J.D.; Software, C.Y.; Data curation, S.X. and T.S. All authors have read and agreed to the published version of the manuscript.

Funding: This research was funded by the Science and Technology Research Program of Chongqing Municipal Education Commission (No. KJZD-M202112901 and KJZD-M202212901), Natural Science Foundation of China (No. 52272338), Natural Science Foundation of Chongqing (CSTB2023NSCQ-MSX0127), Technological Innovation and Application Development Project of Chongqing (2023TIAD-KPX0113), Chongqing Graduate Scientific Research Innovation Project (No. CYS21527), Natural Science Foundation of China (52302422 and 52272338), a Major Project of the Science and Technology Research Program of the Chongqing Education Commission of China (KJZD-M202212901).

Institutional Review Board Statement: Not applicable.

Informed Consent Statement: Not applicable.

Data Availability Statement: The data used to support the findings of this study are available from the corresponding authors upon request.

Conflicts of Interest: The authors declare no conflicts of interest.

References

1. Jiang, X.S.; Zhou, L.; Cai, Y.X.; Zhou, D.L.; Chen, R.; Qin, X.Z. Explosion suppression experiment of oil storage vessel based on porous explosion suppression materials. *Oil Gas Storage Transp.* **2023**, *42*, 1158–1165.
2. Zhang, B.Y.; Cui, J.R. Experimental study on explosion propagation barrier properties of metal foamed with different ex-plosion surface structures for methane gas. *Explos. Shock. Waves* **2023**, *43*, 168–178.
3. Liang, H.; Wang, T.; Luo, Z.; Wang, X.; Kang, X.; Deng, J. Risk Assessment of Liquefied Petroleum Gas Explosion in a Limited Space. *ACS Omega* **2021**, *38*, 24683–24692. [CrossRef]
4. Gao, N. Effect of Initial Temperature on Free Radicals of Gas Explosion in Restricted Space. *Adv. Mater. Res.* **2013**, *2657*, 798–799. [CrossRef]
5. Cui, Y.H.; Jiang, J.C.; Yu, Y.; Zhang, Q.W. Initial Pressure Influence on Explosion Pressures of Methane-Air Deflagrations in Linked Vessels. *Adv. Mater.Res.* **2014**, *3187*, 936. [CrossRef]
6. Li, Y.; Xu, H.L.; Wang, X.S. Experimental Study on the Influence of Initial Pressure on Explosion of Methane-coal Dust Mixtures. *Procedia Eng.* **2013**, *62*, 980–984. [CrossRef]
7. Heidari, A.; Wen, J.X. Numerical simulation of flame acceleration and deflagration to detonation transition in hydrogen-air mixture. *Int. J. Hydrog. Energy* **2014**, *39*, 21317–21327. [CrossRef]
8. Pan, Y.; Luke, A.L. Experimental and numerical studies on the response of stiffened plates subjected to gas explosions. *J. Constr. Steel Res.* **1999**, *52*, 171–193. [CrossRef]
9. Choi, K.; Sakasai, H.; Nishimura, K. Minimum ignition energies of pure magnesium powders due to electrostatic discharges and nitrogen's effect. *J. Loss Prev. Process Ind.* **2016**, *41*, 141–146. [CrossRef]
10. Ferrara, G.; Willacy, S.; Phylaktou, H.; Andrews, G.; Di Benedetto, A.; Salzano, E.; Russo, G. Venting of gas explosion through relief ducts: Interaction between internal and external explosions. *J. Hazard. Mater.* **2008**, *155*, 358–368. [CrossRef]
11. Tan, Y.X.; Xie, Y.Y.; Huo, Y.J.; Han, Y. Influence of different turbulence states on flame propagation process of ethanol vapor near explosive lower limit. *J. North Univ. China (Nat. Sci. Ed.)* **2018**, *39*, 89–98.
12. Bauwens, C.R.; Chaffee, J.; Dorofeev, S. Effect of Ignition Location, Vent Size, and Obstacles on Vented Explosion Overpressures in Propane-Air Mixtures. *Combust. Sci. Technol.* **2010**, *182*, 1915–1932. [CrossRef]
13. Arisova, V.N.; Trykov Yu, P.; Ponomareva, I.A. Structural changes in magnesium-aluminum composite obtained by explosion welding after bending and thermal treatment. *Russ. J. Non-Ferr. Met.* **2014**, *55*, 46–50. [CrossRef]
14. Ponizy, B.; Henneton, N.; Claverie, A.; Veyssiere, B. Detailed investigation of flame transmission from a vessel to a discharge duct. *Combust. Flame* **2014**, *161*, 1348–1364. [CrossRef]
15. Di Benedetto, A.; Garcia-Agreda, A.; Russo, P.; Sanchirico, R. Combined Effect of Ignition Energy and Initial Turbulence on the Explosion Behavior of Lean Gas/Dust-Air Mixtures. *Ind. Eng. Chem. Res.* **2011**, *51*, 7663–7670. [CrossRef]
16. *Numerical Simulation of Flame Propagation Process of Premixed Gas in Bifurcation Pipeline*; Harbin Engineering University: Harbin, China, 2015.
17. Wang, K.; Su, M.; Wei, L.; Chen, S.; Kong, X.; Fang, Y. Effect of initial turbulence on explosion behavior of stoichiometric methane-ethylene-air mixtures in confined space. *Process Saf. Environ.Prot.* **2022**, *161*, 583–593. [CrossRef]
18. Yang, S.G.; Cai, J.W.; Yang, Y.; Sun, W.S.; Men, J.M. Disaster effect of combustible Gas explosion in urban underground shallow buried pipe ditch (II): Influencing factor analysis and consequence assessment. *Explos. Shock. Waves* **2023**, *43*, 155–166.
19. Li, G.Q.; Du, Y.; Bai, J. Numerical simulation of flame propagation characteristics of gas vapor explosion in T-shaped branch pipeline. *China Saf. Sci. Technol.* **2016**, *12*, 120–127.

20. Wang, H.; Chen, T. Experimental and Numerical Study of the Impact of Initial Turbulence on the Explosion Behavior of Methane-Air Mixtures. *Chem. Eng. Technol.* **2021**, *44*, 1195–1205. [CrossRef]
21. Xie, Y.Y. *Effect of Turbulence on Explosion Limit of Combustible Gas (Vapor) and Flame Propagation Process*; North University of China: Taiyuan, China, 2017.
22. Jian, C.G.; Lin, B.Q.; Song, Z.C.; Zhou, S.N. Induction of turbulence and its effect on flame and explosion wave in gas explosion process. *J. Exp. Mech.* **2004**, *19*, 39–44.
23. Pu, Y.K.; Jarosinski, J.; Johnson, V.G.; Kauffman, C.W. Turbulence effects on dust explosions in the 20-liter spherical vessel. *Symp. Combust.* **1991**, *23*, 843–849. [CrossRef]
24. Dahoe, A.E.; Zevenbergen, J.F.; Lemkowitz, S.M.; Scarlett, B. Dust explosions in spherical vessels: The role of flame thickness in the validity of the 'cube-root law'. *J. Loss Prev. Process Ind.* **1996**, *9*, 33–44. [CrossRef]
25. Cammarota, F.; Benedetto, A.D.; Russo, P.; Salzano, E. Experimental analysis of gas explosions at non-atmospheric initial conditions in cylindrical vessel. *Process Saf. Environ. Prot.* **2010**, *88*, 341–349. [CrossRef]
26. Christophe, C.; Geoffrey, S. On the "tulip flame" phenomenon. *Combust. Flame* **1996**, *105*, 225–238.
27. Metghalchi, M.; Keck, J.C. Burning velocities of mixtures of air with methanol, isooctane, and indolene at high pressure and temperature. *Combust. Flame* **1982**, *48*, 191–210. [CrossRef]
28. Zhang, P.L.; Qi, S.; Wang, S.M.; Cheng, S.G. Calculation and analysis of adiabatic combustion temperature of No. 93 gasoline vapor. *J. Logist. Eng. Coll.* **2017**, *33*, 33–39.
29. Zhang, P.L.; Xiao, J.; Wang, J.; Li, Y.X. Branch pipe length and the number of the influence of the gas vapor explosion overpressure features. *J. Saf. Environ.* **2022**, *48*, 2412–2419.

fire

Article

Investigating the Influence of Flue Gas Induced by Coal Spontaneous Combustion on Methane Explosion Risk

Sijia Hu [1,2,3], Yanjun Li [1,2,3,*], Chuanjie Zhu [2,3], Baiquan Lin [2,3], Qingzhao Li [2,3], Baolin Li [1,2,3] and Zichao Huang [4,*]

1 School of Environment and Safety Engineering, North University of China, Taiyuan 030051, China; tb19120011b2@cumt.edu.cn (S.H.); 20220060@nuc.edu.cn (B.L.)
2 Key Laboratory of Gas and Fire Control for Coal Mines, Ministry of Education, China University of Mining and Technology, Xuzhou 221116, China; anq021@cumt.edu.cn (C.Z.); lbq21405@126.com (B.L.); qingzhaolee@163.com (Q.L.)
3 School of Safety Engineering, China University of Mining and Technology, Xuzhou 221116, China
4 Industrial Safety Research Branch, China Coal Technology and Engineering Group Chongqing Research Institute, Chongqing 400037, China
* Correspondence: 20230116@nuc.edu.cn (Y.L.); passion_hzc@163.com (Z.H.)

Abstract: During the process of coal spontaneous combustion (CSC), a plethora of combustible gases alongside inert gases, such as CO_2, are copiously generated. However, prior investigations have regrettably overlooked the pivotal influence of inert gas production on the propensity for methane explosions during CSC. To investigate the impact of the flue gas environment generated by CSC, containing both combustible and inert gases, on the risk of methane explosion, a high-temperature programmed heating test system for CSC was employed to analyze the generation pattern of flue gas. It was found that CO, CO_2, and CH_4 were continuously generated in large quantities during the process of CSC, which are the main components of CSC flue gas. The effect of the concentration and component ratio (C_{CO2}/C_{CO}) of the flue gas on the methane explosion limit was tested. It was found that the CSC flue gas led to a decrease in the methane explosion limit, and that the explosion limit range was facilitated at $0 < C_{CO2}/C_{CO} < 0.543$ and suppressed at $C_{CO2}/C_{CO} > 0.543$. As the temperature of CSC increases, the risk of methane explosion is initially suppressed. When the coal temperature exceeds 330~410 °C, the explosion risk rapidly expands.

Keywords: coal spontaneous combustion; gas generation; methane explosion; explosion limit; risk assessment

Citation: Hu, S.; Li, Y.; Zhu, C.; Lin, B.; Li, Q.; Li, B.; Huang, Z. Investigating the Influence of Flue Gas Induced by Coal Spontaneous Combustion on Methane Explosion Risk. *Fire* **2024**, *7*, 105. https://doi.org/10.3390/fire7040105

Academic Editors: Chuyuan Huang, Haipeng Jiang and Lijuan Liu

Received: 12 February 2024
Revised: 15 March 2024
Accepted: 21 March 2024
Published: 22 March 2024

1. Introduction

Energy stands as the bedrock of human advancement, with coal wielding a significant sway in global energy consumption [1–3]. Throughout the coal mining process, various disaster risks are encountered, among which methane explosions pose the gravest threat, resulting in profound human casualties and extensive property devastation [4–7]. The residual coal left within mining sites, owing to the inherent limitations in extraction techniques, lays the groundwork for the emergence of coal spontaneous combustion (CSC) [8,9]. When the CSC develops to a high temperature sufficient to detonate methane, it is possible to form a serious methane explosion accident [10]. In the genesis of this high-temperature source, copious volumes of mixed flue gas, comprising combustible and inert constituents, are incessantly generated, exerting a tangible influence on methane explosion risks [11–13]. Delving into the generation law of flue gas during coal spontaneous combustion and its consequential impact on methane explosion limits not only enriches our comprehension of the perils posed by methane explosion disasters stemming from coal spontaneous combustion in goaf but also fortifies our arsenal of preventive and mitigation measures against such calamitous events [14,15]. Hence, a plethora of scholarly endeavors have been dedicated to scrutinizing the methane explosion risk under the aegis of CSC conditions [16].

The genesis of combustible gases poses a palpable escalation in the risk profile of methane explosions [17–19]. In a study by Zhang et al., the impact of CO on methane explosion limits was examined across varying temperature regimes, employing a 20 L explosive sphere apparatus. Their findings elucidated a conspicuous augmentation in gas explosion risk concomitant with heightened initial temperatures and CO concentrations [20]. Hao et al. conducted investigations into the influence of H_2 on methane explosion risks, uncovering a notable expansion in the explosion limit range of methane due to the presence of H_2 [19]. Further delving into this intricate nexus, Ma et al. probed the effects of CO-H_2 mixtures on methane explosion limits. Their comprehensive analysis unveiled a discernible decrease in the lower explosion limit (LEL) and a noteworthy expansion in methane explosion limits, thereby substantially amplifying methane explosion risks, with CO-H_2 mixtures demonstrating a more pronounced effect compared to pure CO [14]. Wang et al. undertook experiments centered on the spontaneous combustion gas emissions of long-flame coal to scrutinize the impact of CO-C_2H_4-C_2H_6 mixtures on methane explosion limits. Their discernments highlighted a heightened explosion hazard attributed to the addition of these combustible gases, with C_2H4 and C_2H_6 exerting more significant effects than CO [21]. Building upon the foundations laid by Wang et al., Luo et al. expanded the purview of inquiry by incorporating the influence of H_2, revealing a parabolic increase in methane explosion hazards within CSC mixtures [22].

While significant strides have been made in scrutinizing the impact of gases engendered during coal spontaneous combustion (CSC) on methane explosion risks, these endeavors are not devoid of certain lacunae. Primarily, extant investigations predominantly center on gas analyses at temperatures below 500 °C [23,24], whereas the temperature required for CSC to ignite methane often needs to exceed 650 °C [25]. Moreover, prevailing studies tend to concentrate on the effects of individual combustible gases or their mixtures generated during CSC on methane risk, neglecting the distinctive characteristic of yielding copious volumes of mixed gas comprising both combustible and inert constituents during coal spontaneous combustion. This oversight inevitably engenders inaccuracies in analysis outcomes [13]. Therefore, in this study, the variation law of the concentration and composition of gases generated during the process of CSC temperature rising to 730 °C was analyzed. The impact of flue gas, characterized by varying concentrations and composition ratios, on methane explosion limits was systematically evaluated, thereby elucidating the dynamics of methane explosion risk under the influence of flue gas in relation to coal temperature.

2. Flue Gas Generation Characteristics during CSC

2.1. The Flue Gas Generation Characteristic Test Experiment of CSC

2.1.1. Experimental System

The temperature-programmed experiment serves as a cornerstone method for elucidating the gas generation law of CSC [26]. Given that CSC-induced methane explosions necessitate attaining temperatures exceeding 600 °C, a high-temperature programmed experimental system was built, as illustrated in Figure 1 The system was composed of a gas supply system, programmed temperature control system, temperature acquisition system, and gas analysis system. The gas supply system consisted of gas cylinders, flow controllers, gas mixing tanks, check valves, and gas purifiers. In order to simulate the air conditions in the goaf of coal mine, O_2 and N_2 were used as the gas source. The input gas flow was meticulously regulated by high-precision gas flow controllers, maintaining an N_2/O_2 ratio of 4. The programmed temperature control system was composed of an SG-GL1400 vacuum tube furnace (heating range 0~1400 °C, maximum heating rate 10 °C/min) and a coal sample container of 316 stainless steel. The coal samples were broken and screened to obtain particles with a size range of 40–80 mesh. Subsequently, these particles were placed into coal sample containers, sealed, and then placed horizontally in a vacuum tube furnace to ensure uniform heating of the coal sample container. The temperature acquisition system was composed of thermocouples and a temperature acquisition instrument. The measurement accuracy of the thermocouple was 0.1 °C. The temperature changes inside the container were collected in real-time. The gases generated during the coal heating process were collected in gas sampling bags and analyzed using a GC-4000A gas chromatograph.

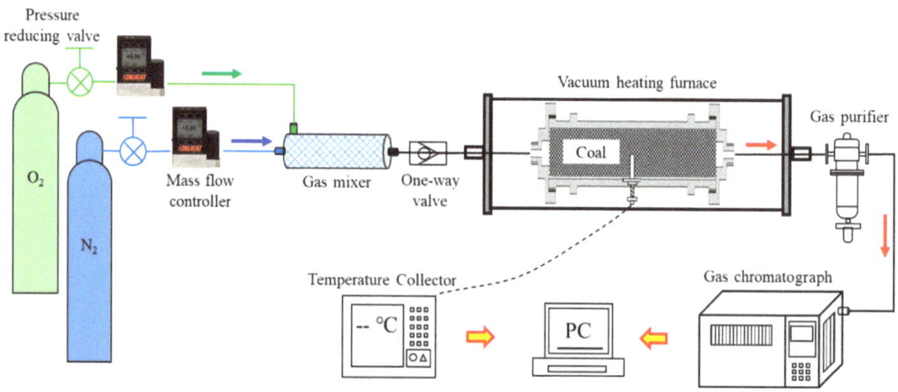

Figure 1. Schematic diagram of high-temperature programmed experiment system.

2.1.2. Experimental Process

(1) Coal samples screening: The fresh coal samples taken from the site were removed by sandpaper, and then the coal samples were broken into different sizes by a crusher. Then, the coal samples were screened by a sieve with a screening range of 0.18–0.38 mm to obtain a coal sample with a particle size range of 40~80 mesh. Finally, 300 g coal samples were selected and placed in a vacuum-drying oven at 50 °C for 24 h to remove external moisture.

(2) Heating pretreatment: After the dry coal samples were loaded into the coal sample container, N_2 was first introduced into the furnace at a flow rate of 120 mL/min for about 10 min to drain the air in the tank. Subsequently, the furnace temperature was set at 30 °C and maintained for 30 min to make the initial temperature conditions of the experiment consistent.

(3) Temperature programming: The flow controller was adjusted to supply gas to the coal sample tank at 300 mL/min, and the proportion of O_2 in the supplied gas was 21%. At the same time, the flow and composition of the outlet gas were monitored. After the outlet gas was stable, the programmed temperature control system was turned on, and the heating rate was set to 2 °C/min, and the heating range was 30~710 °C.

(4) Data collecting: After the temperature started to rise, the temperature acquisition instrument was opened, and the acquisition frequency was set to 1 time/s. At the same time, for every 20 °C increase in coal temperature, flue gas generated from coal spontaneous combustion was collected and stored separately in gas bags with a volume of 10 mL each. Subsequently, the flue gas in each gas bag was tested using the chromatograph.

2.1.3. Experimental Samples

Three types of coal samples with different degrees of metamorphism were selected for testing to analyze the generation law of combustible and inert gases during the CSC process. These coal samples were obtained from the Shengli Coal Mine in the Inner Mongolia Autonomous Region (lignite), Baikuang Coal Mine in Pingdingshan, Henan Province (bituminous coal), and the Linhua Coal Mine in Guizhou Province (anthracite). The industrial analysis and vitrinite reflectance test results of the coal samples are shown in Table 1.

Table 1. Reflectance and industrial analysis of coal sample maceral groups.

Coal Samples	Proximate Analysis (*w*/*w* %)				Vitrinite Reflectance $R_{0,\ max}$ (%)
	M_{ad}	A_{ad}	V_{daf}	FC_{ad}	
Lignite	11.37	14.63	53.41	20.59	0.32
Bituminous	1.16	9.37	24.60	64.87	1.22
Anthracite	2.18	15.45	6.64	75.73	3.33

2.2. Flue Gas Generation Law of CSC

Figure 2 portrays the dynamic interplay between gas concentrations and coal temperature throughout the heating process. Across the spectrum of temperatures spanning from 30 to 710 °C, the gases emanating from the heating of three distinct coal types predominantly encompass CO, CO_2, CH_4, C_2H_4, C_2H_6, and C_3H_8. The initial concentrations of CO and CO_2 for lignite, bituminous coal, and anthracite were 0.0029% and 0.27%, 0.0022% and 0.22%, and 0.0017% and 0.16%, respectively. Notably, the emergence of CO and CO_2 pervaded the entirety of the coal's thermal evolution, steadfastly constituting the principal constituents of the generated gases. The advent of CH_4 burgeoned notably once the coal temperature surpassed the threshold of 290 °C, exhibiting a sustained propensity for generation even amidst the heightened temperatures. This phenomenon underscores the potential for localized explosion hazards in regions with diminished methane content in the goaf, where evaluation in coal temperature can instigate methane release. In contrast, the generation of C_2H_4, C_2H_6, and C_3H_8 predominates within the temperature band of 350 to 550 °C, yielding relatively diminutive quantities, each remaining below 0.5. Consequently, the influence of these gases on methane explosions remains circumscribed. Thus, in instances of (CSC) within the goaf, the gases prone to amass in the combustion zone primarily comprise CO, CO_2, and CH_4. Among them, CH_4 is the inherent gas in the goaf, while CO and CO_2, as gases generated in large quantities during the coal heating process, may affect methane explosions. Accordingly, this investigation delves into the ramifications of CSC flue gas, predominantly constituted by CO and CO_2, on the risk of methane explosions.

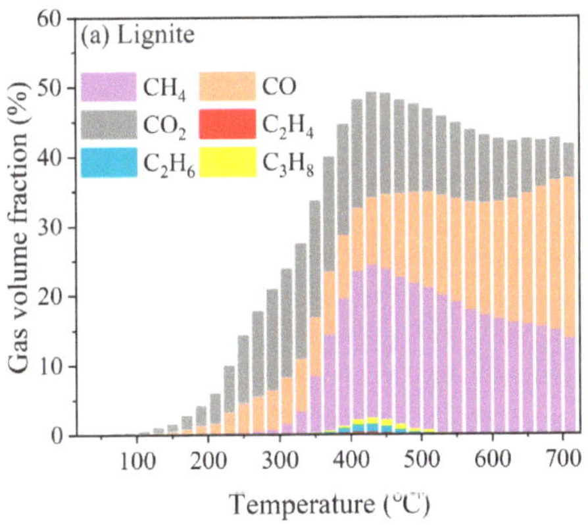

Figure 2. *Cont.*

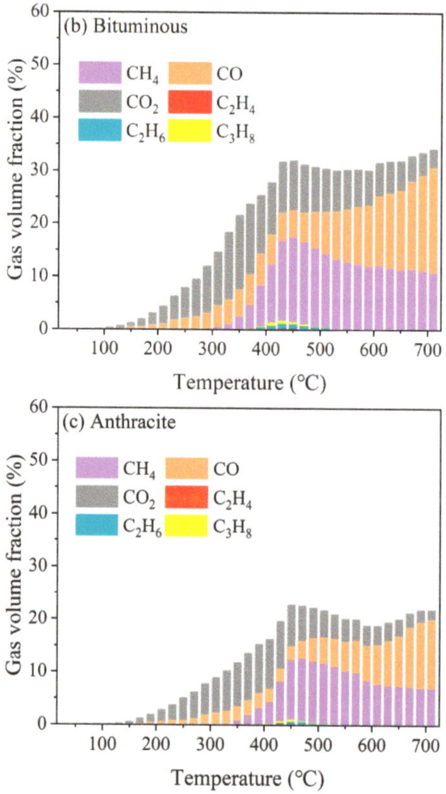

Figure 2. Generation law of CSC flue gas.

3. Influence of CSC Flue Gas on Methane Explosion Limit

The determination of the explosion limits of combustible gases was achievable through both experimental and theoretical calculation methodologies [27,28]. Despite the array of theoretical approaches available for calculating these limits, experimental validation remains the gold standard [22,29,30]. Consequently, this study endeavored to employ experimental testing as the cornerstone for investigating the explosion limits of methane in the presence of flue gas.

3.1. Explosion Limit Test Experiment of Multi-Component Mixed Gas

3.1.1. Experimental System

In order to investigate the effect of CSC flue gas on the explosion limits of methane, a multi-component gas explosion limit testing experimental system was constructed as shown in Figure 3, mainly composed of an explosion room, gas supply system, ignition system, and data acquisition system. The explosion room was a cylindrical chamber made of 304 stainless steel, with a volume of 16 L and was capable of withstanding an explosion pressure of 10 MPa. The gas supply system mainly consisted of gas cylinders, control valves, driers, vacuum pumps, and high-precision vacuum gauges. The gas was supplied using Dalton's law of partial pressures. During the supply process, precise control of gas concentration as achieved through a high-precision vacuum gauge, with an accuracy of up to 0.01%. The ignition system consisted of ignition electrodes and an ignition controller, with a gap of 2 mm between the positive and negative electrodes. The ignition controller was a KTD-W adjustable igniter capable of releasing ignition energy ranging from 50 to 500 mJ. The data acquisition system consisted of a high-frequency pressure sensor and a TST6300 dynamic signal acquisition instrument. The highest sampling frequency for explosion overpressure was 1 MHz, with a maximum sampling width of 2 M and a sampling accuracy of up to ±0.001 FS. According to the European Union standard BSEN 1839:2012, the criterion for an explosion is whether the maximum explosion pressure exceeds 5% of the initial pressure [31].

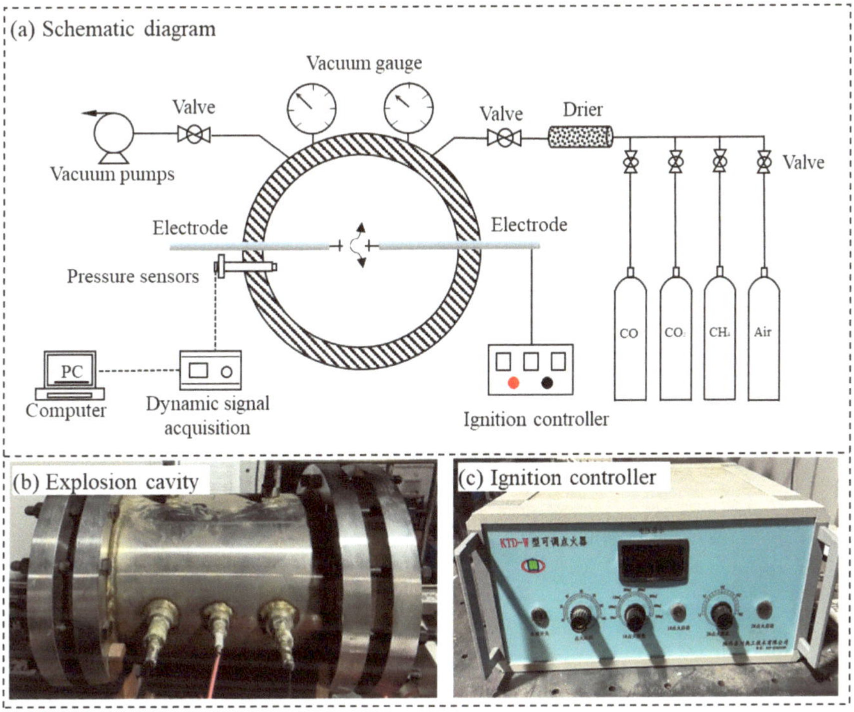

Figure 3. Experimental system for testing the explosion characteristics of multi-component gases.

3.1.2. Experimental Process

(1) Experimental preparation: Before the experiment, the line connections and the air tightness of the device were checked separately. The ignition system and explosion overpressure acquisition system were debugged under unpressurized conditions to ensure the normal operation of the testing apparatus.
(2) Vacuuming: The intake valve was closed, the vacuum pump was turned on, the high-precision vacuum pressure gauge was observed, and once the pressure decreased to a stable level, the vacuuming valve was closed.
(3) Gas supply: The partial pressure of each gas according to Dalton's law was calculated, then sequentially the gas valve was opened and the concentration of each gas was controlled using the vacuum pressure gauge. After completion, the gases were allowed to stand for 10 min to ensure thorough mixing.
(4) Ignition: The explosion overpressure signal acquisition was turned on, then the mixed gas was ignited using the igniter controller.
(5) Exhaust the waste gas: After the explosion ended, the exhaust valve was opened and dry air was introduced into the explosion room via the intake pipe for 30 s to purge any remaining explosive gases inside the chamber.

3.1.3. Experimental Design

In accordance with the generation law of CSC flue gas, CO and CO_2 emerged as prominent constituents capable of sustained and copious generation, each bearing distinct implications for methane explosion limits [32–34]. Therefore, this study took the concentration of flue gas (C_{flue}) and the concentration ratio of CO_2 to CO in flue gas (α) as influencing factors to analyze the influence of CSC flue gas on the methane explosion limit. To analyze the variation trends of C_{flue} and α with coal temperature, corresponding values of C_{flue} and α were calculated based on the CO and CO_2 concentrations obtained at each temperature adjustment as depicted in the experimental results of Figure 2. The results are presented in Figure 4. From Figure 4a, it is discernible that during the initial phases of coal temperature elevation, the concentration of flue gas remains relatively modest, with none of the three coal samples surpassing 1%. Subsequently, upon attaining 150 °C, the rate of flue gas generation escalates swiftly, albeit not exceeding 30% at its zenith. Figure 4b delineates the rapid initial decline followed by a gradual stabilization of α as coal temperature escalates. Prior to reaching 150 °C, the flue gas concentration remains relatively subdued, exerting limited impact on methane explosion limits. Consequently, this study directs its focus towards scrutinizing the effect of CSC flue gas on methane explosion limits subsequent to coal temperature surpassing 150 °C. During this phase, the α value of the flue gas predominantly oscillates within the range of 0 to 5.

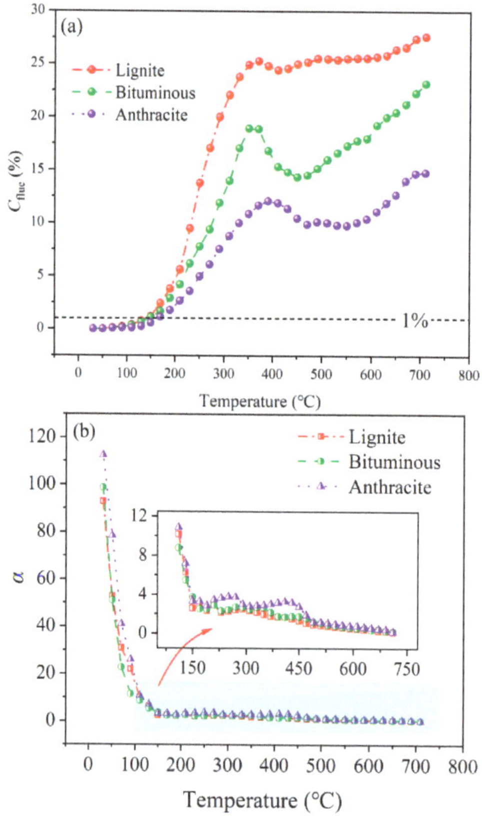

Figure 4. Variations of C_{flue} and α with the increase in coal temperature.

Based on the preceding analysis, the concentration of flue gas generated by CSC predominantly fluctuates between 1% and 30%, while the α value primarily ranges from 0 to 5. In light of this, to scrutinize the impact of flue gas on methane explosion limits, six distinct concentration conditions of flue gas were devised, namely 5%, 10%, 15%, 20%, 25%, and 30%. Additionally, 5 α value conditions were delineated, encompassing 0.2, 0.5, 1, 3, and 5.

3.2. Experiments

Figure 5 delineates the repercussions of CSC flue gas, characterized by varying concentrations (C_{flue}) and α values, on methane explosion limits. Notably, the presence of CSC flue gas precipitates a decline in both the lower explosion limit (LEL) and upper explosion limit (UEL) of methane. This trend is similar to the experimental findings in reference [35], but the values are slightly higher than the test values in reference [35]. This may be due to different explosion limit standards and experimental conditions. The reduction in UEL is ascribed to the competitive influences of CO and CO_2 on O_2 molecules, coupled with the dilution, insulation, and cooling effects of CO_2 on reactants, alongside the inhibitory impact of CO_2 on the reaction process. Furthermore, as the proportion of CO_2 in the flue gas mixture escalates, the role of CO_2 becomes more conspicuous, yielding a more pronounced decrease in the UEL. However, the reasons for the decline of LEL are different from those of UEL. CO tends to diminish the LEL of methane, whereas CO_2 tends to elevate it. The reason for the decrease in LEL of methane is that the effect of CO in the flue gas (α = 0~5) is stronger than that of CO_2. Consequently, while the decrease in LEL may engender fresh explosion hazards in regions where methane concentrations were initially sub-threshold, the continuous escalation of C_{flue}, particularly in scenarios such as α = 0.2, can drive the LEL of methane to 0%, rendering the flue gas itself explosively susceptible, thereby heightening the likelihood of ignition or explosion.

The effects of different flue gases with varying C_{flue} or α on the methane explosion limits are different. When α remains constant, both the LEL and UEL of methane decrease linearly with the increase in C_{flue}. However, when C_{flue} remains constant, the impact of flue gases with different α on the UEL and LEL of methane varies. Specifically, the UEL decreases with increasing α, while the LEL increases with increasing α.

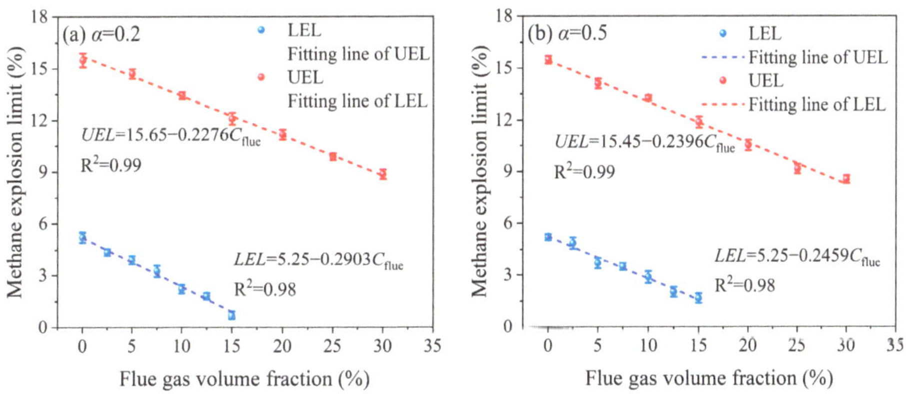

Figure 5. *Cont.*

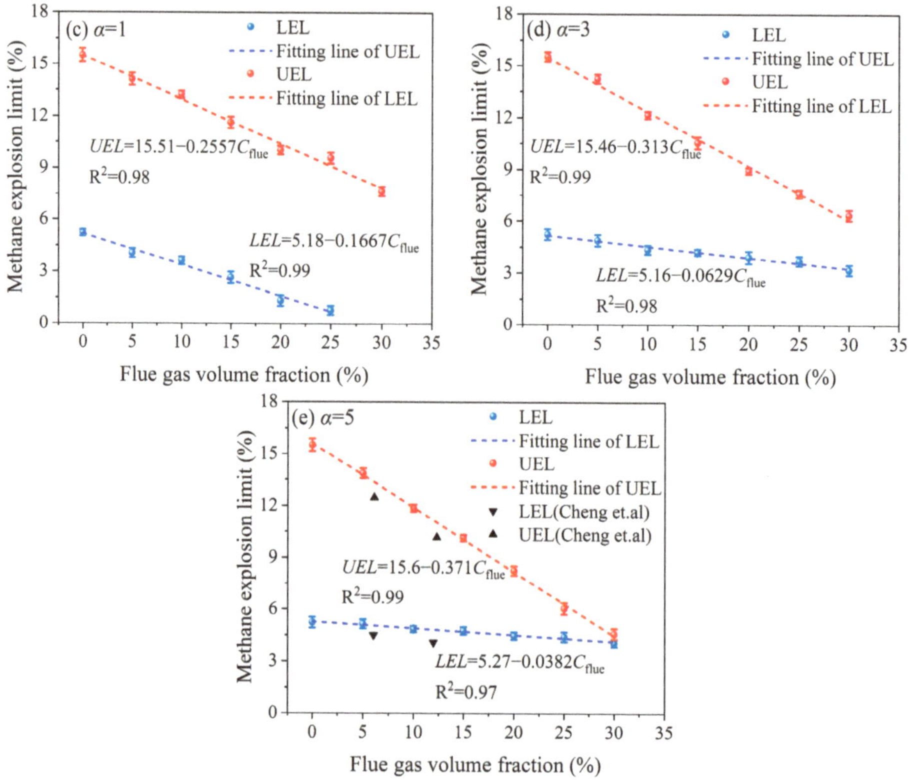

Figure 5. Influence of flue gases of different C_{flue} and α on the methane explosion limit [35].

4. Influence of CSC Flue Gas on the Risk of Methane Explosion

4.1. Influence of Flue Gas on Methane Explosion Limit Range

The explosion limit range stands as a pivotal parameter for assessing explosion risks [36,37]. In the context of this study, the measured methane explosion limit spans from 5.22% to 15.51%, yielding a range of 10.29%. Figure 6 elucidates the impact of flue gas characterized by diverse C_{flue} and α values on the range of methane explosion limits. It is discernible that flue gas with varying α values exerts disparate effects on the explosion limit range of methane. Specifically, when α assumes values of 0.2 or 0.5, the flue gas augments the explosion limit range of methane. Notably, when $C_{flue} \geq 25\%$, there appears to be a decline in the explosion limit range of methane. However, this decline does not signify a mitigation of methane explosion risks. Rather, it signals that flue gas at elevated concentrations possesses explosive characteristics, thereby posing a heightened risk of direct ignition. Conversely, when α is 1, 3, or 5, the flue gas manifests an inhibitory effect on the explosion limit range of methane. Moreover, this inhibitory influence amplifies with escalating C_{flue}.

The fluctuations in the range of methane explosion limits are intricately influenced by both LEL and UEL. Specifically, when the LEL of methane exhibits a slower decline with increasing C_{flue} compared to the UEL, the range of the gas explosion limit contracts. When the decrease rate of the methane LEL with increasing C_{flue} is faster than that of UEL, the explosion limit range of methane expands. Conversely, if the rate of decrease in the methane LEL with increasing C_{flue} surpasses that of the UEL, the explosion limit range of methane expands.

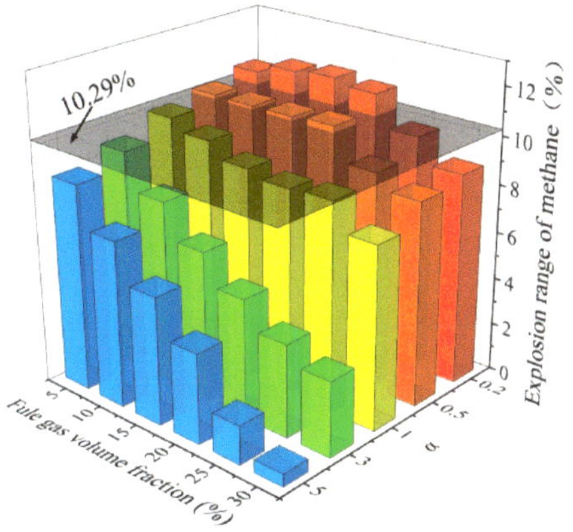

Figure 6. Influence of flue gas on the explosion range of methane.

This rate of decrease can be quantified by the absolute value of the slope ($|k|$) of the linear reduction in the methane explosion limit. In essence, when $|k|_{LEL} > |k|_{UEL}$, flue gas augments the explosion limit range of methane. Conversely, when $|k|_{LEL} < |k|_{UEL}$, flue gas constrains the explosion limit range of methane.

As illustrated in Figure 7, the variation of $|k|$ with α is delineated. It can be observed that $|k|_{LEL}$ decreases as α increases, while $|k|_{UEL}$ increases with α. Notably, at $\alpha = 0.543$, $|k|_{LEL} = |k|_{UEL}$, signifying a point where the impact of flue gas on both the UEL and LEL of methane is equivalent. Consequently, the methane explosion limit range is consistent with the condition without flue gas. In other words, the value of 0.543 denotes the critical threshold at which flue gas transitions from promoting to inhibiting the methane explosion limit range.

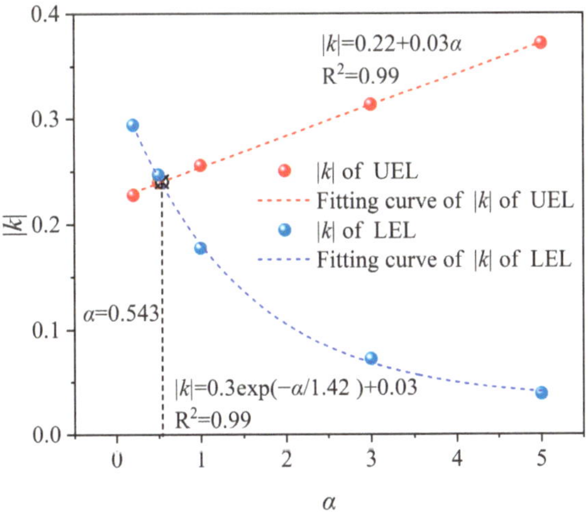

Figure 7. The variation law of $|k|$ with α.

In summation, flue gas characterized by low α values tends to enhance the explosion limit range of methane, whereas flue gas with high α values tends to curtail it. Considering the actual conditions in coal mine goafs, C_{flue} remains relatively subdued during the initial stages of temperature elevation, thereby exerting minimal influence on the explosion limit range of methane. However, as the coal temperature continues to rise, C_{flue} steadily escalates while α gradually diminishes, signaling a trajectory of heightened methane explosion risk. Particularly during the high-temperature phases, when the α of flue gas plummets below 0.5, the flue gas itself acquires explosive properties, precipitating a sudden upsurge in the risk of explosion within the goaf area.

4.2. Variation Law of Methane Explosion Risk during the Temperature Increasing Process of CSC

4.2.1. Fitting of Methane Explosion Limits under the Influence of Flue Gas

The elevated temperatures generated by CSC in the goaf serve as crucial ignition sources for initiating methane combustion [38,39]. Concurrently, flue gas undergoes continuous generation throughout the coal temperature escalation process within the goaf, culminating in the formation of a dynamic flue gas environment characterized by fluctuating C_{flue} and α values. This, in turn, exerts a notable influence on the explosion limits of methane.

To assess the risk of methane explosion amidst coal temperature elevation, the methane explosion limit within the flue gas milieu was modeled. as shown in Figure 8. The specific outcomes of this fitting process are as follows:

$$LEL = 5.22 - 0.3C_{flue}\mathrm{EXP}(-\alpha/1.42) - 0.03C_{flue} \quad (1)$$

$$UEL = 15.52 - 0.22C_{flue} - 0.03\alpha C_{flue} \quad (2)$$

where LEL is the lower explosion limit of methane, %; UEL is the upper explosion limit of methane, %; C_{flue} is the flue gas concentration, %; and α is the ratio of CO_2 and CO concentrations in the flue gas.

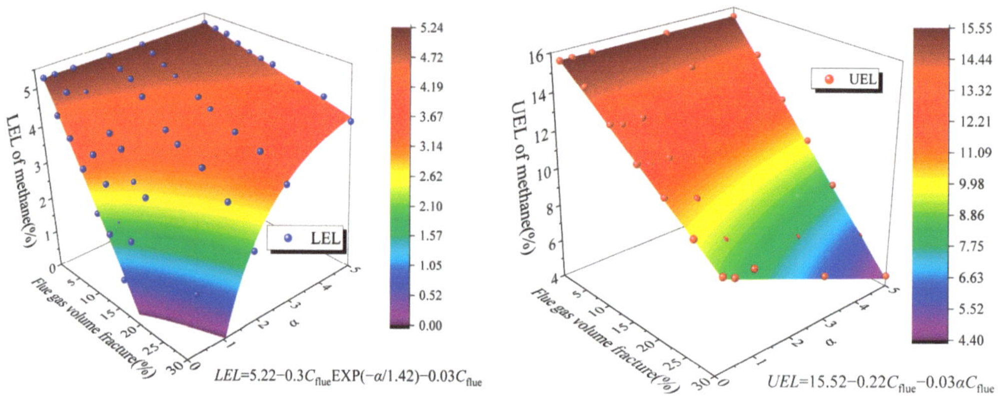

Figure 8. Methane explosion limit fitting.

4.2.2. Variation Law of Methane Explosion Limits during the Temperature Increasing Process of CSC

To assess the explosion risk of methane during the process of coal temperature elevation, it is necessary to calculate the methane explosion limits at various temperatures. By substituting the experimental results of the temperature-programmed experiment into the aforementioned fitted formula for explosion limits calculation, the variation pattern of methane explosion limits with coal temperature was obtained, as depicted in Figure 9. Notably, from the pattern of LEL variation illustrated in Figure 9a, it becomes apparent that under the influence of flue gas, the LEL of methane undergoes a continuous decline with increasing coal temperature. In the oxidation zone of the goaf, the methane concentration typically remains at a low level owing to air dilution, thereby mitigating methane explosion risks [40,41]. However, the presence of flue gas serves to diminish the LEL of methane, thereby introducing new explosion hazards in areas that were initially non-explosive. Something that is particularly noteworthy is the scenario wherein the temperature of bituminous and anthracite coal reaches 550 °C and 650 °C, respectively, resulting in the LEL of methane plummeting to 0%. At this juncture, the risk of methane explosion escalates significantly due to the combustibility of the flue gas itself.

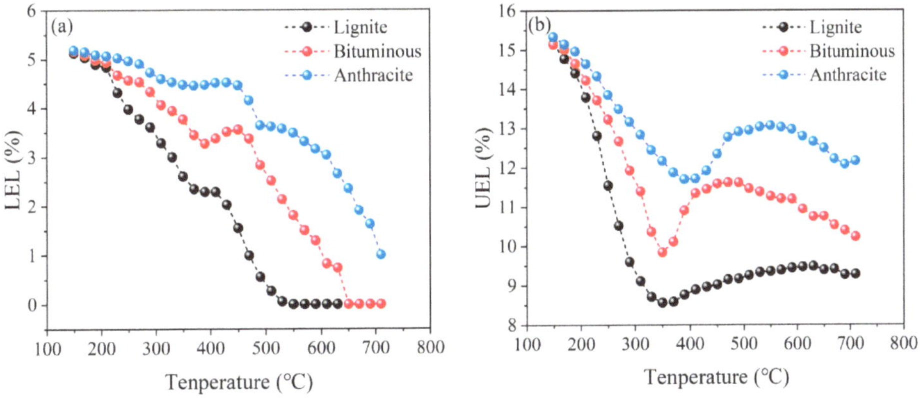

Figure 9. Variation law of methane explosion limit during the temperature rising process of coal.

Similarly, Figure 9b shows that the UEL of methane experiences a decrease under the influence of flue gas. However, unlike the pattern observed in LEL variation, fluctuations in UEL emerge after the temperatures of lignite, bituminous coal, and anthracite reach 350 °C, 350 °C, and 390 °C, respectively. In the event of a CSC disaster in the goaf, the reduction in LEL can potentially mitigate the risk of methane explosion. However, this inhibitory effect is constrained by the low methane concentration prevalent in the oxidation zone. Furthermore, it is noteworthy that the impact of flue gas generated by different coal types on the methane explosion limits varies. Lignite exerts the greatest influence, followed by bituminous coal, with anthracite exerting the least impact. This observation implies that the CSC flue gas generated by low metamorphic grade coal is more likely to escalate the risk of methane explosion.

4.2.3. Variation Law of Methane Explosion Limits Range during the Temperature Increasing Process of CSC

In order to further analyze the impact of flue gas on methane explosion risk during the coal temperature elevation process, it is necessary to analyze the variation of methane explosion limit range with coal temperature. This value can be calculated based on the difference between the LEL and the UEL in Figure 9. The results are illustrated in Figure 10. Notably, it becomes evident that as coal temperature rises, the range of methane explosion limits initially diminishes before rebounding. Relative to conditions devoid of flue gas, the presence of flue gas engenders a reduction in the range of methane explosion limits, reaching a nadir when the temperatures of lignite, bituminous coal, and anthracite reach 330 °C, 350 °C, and 410 °C, respectively. At this juncture, although the decrease in methane LEL remains relatively modest, the inhibitory effect on the explosion limit range peaks, thereby yielding the lowest risk of methane explosion. Subsequently, with the swift expansion of the methane explosion limit range and the continual decline in LEL, the risk of methane explosion progressively escalates.

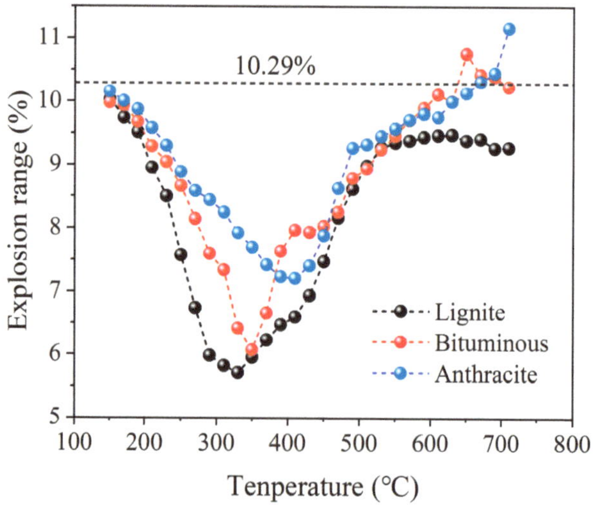

Figure 10. The variation law of methane explosion range during the temperature rising process of coal.

5. Conclusions

(1) Throughout the CSC temperature elevation process, CO, CO_2, and CH_4 are consistently generated in substantial quantities, constituting the primary components of CSC flue gas. They represent the predominant gases that influence the risk of methane explosions. Conversely, C_2H_4, C_2H_6, and C_3H_8 are produced only in minor quantities within the temperature range of 350~550 °C, and their impact on the risk of methane explosions remains limited.

(2) Both the LEL and UEL of methane demonstrate a linear decline with increasing C_{flue}. The UEL of explosion diminishes with escalating α, whereas the LEL increases with α.

(3) Flue gas characterized by lower α values facilitates an expansion of the range of methane explosion limits, whereas flue gas with higher α values constrains the range of methane explosion limits. $\alpha = 0.543$ marks the critical threshold at which flue gas transitions from promoting to inhibiting the explosion limit range of methane.

(4) At lower coal temperatures, the influence of flue gas on methane LEL is minimal, while the explosion range is inhibited, culminating in a reduced risk of methane explosion. However, once the coal temperature surpasses 330~410 °C, the risk of methane explosion continues to escalate under the dual effects of a rapid decline in LEL and a continuous increase in the explosion limit range.

Author Contributions: Conceptualization, C.Z. and Y.L.; methodology, B.L. (Baiquan Lin); formal analysis, S.H.; investigation, Z.H.; resources, Q.L.; data curation, Y.L.; writing—original draft preparation, S.H.; writing—review and editing, Y.L.; visualization, C.Z.; supervision, Q.L.; project administration, B.L. (Baolin Li); funding acquisition, Y.L. All authors have read and agreed to the published version of the manuscript.

Funding: This research was funded by the National Science and Technology Major Project grant number 2020YFA0711803, the National Natural Science Foundation of China grant number 52174214, 52304260, and Shanxi Province Basic Research Program grant number 202203021222031. And The APC was funded by 202203021222031.

Data Availability Statement: The data in this article is presented in the text and no additional database is required.

Conflicts of Interest: The authors declare no conflict of interest.

References

1. Alshehry, A.; Belloumi, M. The Symmetric and Asymmetric Impacts of Energy Consumption and Economic Growth on Environmental Sustainability. *Sustainability* **2024**, *16*, 205. [CrossRef]
2. Fouquet, R. The Digitalisation, Dematerialisation and Decarbonisation of the Global Economy in Historical Perspective: The Relationship Between Energy and Information Since 1850. *Environ. Res. Lett.* **2024**, *19*, 014043. [CrossRef]
3. Wang, X.; Lu, Y.; Chen, C.; Yi, X.; Cui, H. Total-factor Energy Efficiency of Ten Major Global Energy-consuming Countries. *J. Environ. Sci.* **2024**, *137*, 41–52. [CrossRef]
4. Zhang, X.; Jian, S.; Lin, B.; Zhu, C. Study on the Influence of Different-voltage Plasma Breakdowns on Functional Group Structures in Coal. *Energy* **2023**, *284*, 128768. [CrossRef]
5. Liu, T.; Li, M.; Zou, Q.; Li, J.; Lin, M.; Lin, B. Crack Instability in Deep Coal Seam Induced By the Coupling of Mining Unloading and Gas Driving and Transformation of Failure Mode. *Int. J. Rock Mech. Min. Sci.* **2023**, *170*, 105526. [CrossRef]
6. Li, B.; Wang, E.; Li, Z.; Cao, X.; Liu, X.; Zhang, M. Automatic Recognition of Effective and Interference Signals Based on Machine Learning: A Case Study of Acoustic Emission and Electromagnetic Radiation. *Int. J. Rock Mech. Min. Sci.* **2023**, *170*, 105505. [CrossRef]
7. Zheng, Y.; Li, S.; Xue, S.; Jiang, B.; Ren, B.; Zhao, Y. Study on the Evolution Characteristics of Coal Spontaneous Combustion and Gas Coupling Disaster Region in Goaf. *Fuel* **2023**, *349*, 128505. [CrossRef]
8. Duan, H.; Zhao, L.; Yang, H.; Zhang, Y.; Xia, H. Development of 3D Top Coal Caving Angle Model for Fully Mechanized Extra-thick Coal Seam Mining. *Int. J. Min. Sci. Technol.* **2022**, *32*, 1145–1152. [CrossRef]
9. Liu, Q.; Lin, B.; Zhou, Y.; Li, Y. Permeability and Inertial Resistance Coefficient Correction Model of Broken Rocks in Coal Mine Goaf. *Powder Technol.* **2021**, *384*, 247–257. [CrossRef]
10. Zheng, Y.; Li, Q.; Zhu, P.; Li, X.; Zhang, G.; Ma, X.; Zhao, Y. Study on Multi-field Evolution and Influencing Factors of Coal Spontaneous Combustion in Goaf. *Combust. Sci. Technol.* **2023**, *195*, 247–264. [CrossRef]
11. Ma, D.; Qin, B.; Li, L.; Gao, A.; Gao, Y. Study on the Methane Explosion Regions Induced By Spontaneous Combustion of Coal in Longwall Gobs Using a Scaled-down Experiment Set-up. *Fuel* **2019**, *254*, 115547. [CrossRef]
12. Gao, A.; Qin, B.; Zhang, L.; Ma, D.; Li, L. Experimental Study on Gas Migration Laws at Return Air Side of Goaf Under High-temperature Conditions. *Combust. Sci. Technol.* **2023**, *195*, 1930–1944. [CrossRef]
13. Bai, G.; Li, X.; Zhou, X.; Wang, J.; Linghu, J. Evaluation of Lignite Combustion Characteristics and Gas Explosion Risks Under Different Air Volumes. *Energy Sources Part A Recovery Util. Environ. Eff.* **2020**, *1–15*, 1770375. [CrossRef]
14. Ma, D.; Qin, B.; Zhong, X.; Sheng, P.; Yin, C. Effect of Flammable Gases Produced From Spontaneous Smoldering Combustion of Coal on Methane Explosion in Coal Mines. *Energy* **2023**, *279*, 128125. [CrossRef]
15. Zhuo, H.; Qin, B.; Qin, Q. The Impact of Surface Air Leakage on Coal Spontaneous Combustion Hazardous Zone in Gob of Shallow Coal Seams: A Case Study of Bulianta Mine, China. *Fuel* **2021**, *295*, 120636. [CrossRef]
16. Li, J.; Li, X.; Liu, C.; Zhang, N. Study on the Air Leakage Characteristics of a Goaf in a Shallow Coal Seam and Spontaneous Combustion Prevention and Control Strategies for Residual Coal. *PLoS ONE* **2022**, *17*, e0269822.
17. Yang, Y.; Fei, J.; Luo, Z.; Wen, H.; Wang, H. Experimental Study on Characteristic Temperature of Coal Spontaneous Combustion. *J. Therm. Anal. Calorim.* **2023**, *148*, 10011–10019. [CrossRef]
18. Wang, Y.; Sun, Y.; Dai, L.; Wang, K.; Cheng, G. Mechanisms of CO and CO_2 Production During the Low-temperature Oxidation of Coal: Molecular Simulations and Experimental Research. *Fire* **2023**, *6*, 475. [CrossRef]
19. Hao, Q.; Luo, Z.; Wang, T.; Xie, C.; Zhang, S.; Bi, M.; Deng, J. The Flammability Limits and Explosion Behaviours of Hydrogen-enriched Methane-air Mixtures. *Exp. Therm. Fluid Sci.* **2021**, *126*, 110395. [CrossRef]
20. Zhang, X.; Zhou, X.; Bai, G.; Li, C. Influence of CO on Explosion Limits and Characteristics of the CH_4/air Mixture. *ACS Omega* **2022**, *7*, 24766–24776. [CrossRef]
21. Wang, H.; Liu, Y.; Shan, Q. Influence of Gas From Long-flame Coal Spontaneous Combustion on Gas Explosion Limit. *Int. J. Chem. Eng.* **2023**, *2023*, 5096109. [CrossRef]
22. Luo, Z.; Liang, H.; Wang, T.; Cheng, F.; Su, B.; Liu, L.; Liu, B. Evaluating the Effect of Multiple Flammable Gases on the Flammability Limit of CH_4: Experimental Study and Theoretical Calculation. *Process Saf. Environ. Prot.* **2021**, *146*, 369–376. [CrossRef]
23. Li, H.; Chen, X.; Shu, C.M.; Wang, Q.; Ma, T.; Bin, L. Effects of Oxygen Concentration on the Macroscopic Characteristic Indexes of High-temperature Oxidation of Coal. *J. Energy Inst.* **2019**, *92*, 554–566. [CrossRef]
24. Zhao, J.; Wang, T.; Deng, J.; Shu, C.M.; Zeng, Q.; Guo, T.; Zhang, Y. Microcharacteristic Analysis of CH_4 Emissions Under Different Conditions During Coal Spontaneous Combustion with High-temperature Oxidation and in Situ Ftir. *Energy* **2020**, *209*, 118494. [CrossRef]
25. Robinson, C.; Smith, D.B. The Auto-ignition Temperature of Methane. *J. Hazard. Mater.* **1984**, *8*, 199–203. [CrossRef]
26. Wang, K. An Approach for Evaluation of Grading Forecasting Index of Coal Spontaneous Combustion By Temperature-programmed Analysis. *Environ. Sci. Pollut. Res.* **2023**, *30*, 3970–3979. [CrossRef] [PubMed]
27. Ji, W.; Wang, Y.; Yang, J.; He, J.; Wen, X. Methods to Predict Variations of Lower Explosion Limit Associated with Hybrid Mixtures of Flammable Gas and Dust. *Fuel* **2022**, *310*, 122138. [CrossRef]
28. Wang, G.; Shi, G.; Yang, Y.; Liu, S. Experimental Study on the Exogenous Fire Evolution and Flue Gas Migration During the Fire Zone Sealing Period of the Coal Mining Face. *Fuel* **2022**, *320*, 123879. [CrossRef]

29. Nie, B.; Peng, C.; Gong, J.; Yin, F.; Wang, K. Explosion Characteristics and Energy Utilisation of Coal Mine Ultra-lean Methane. *Combust. Theory Model.* **2021**, *25*, 73–95. [CrossRef]
30. Jiao, F.; Zhang, H.; Li, W.; Zhao, Y.; Guo, J.; Zhang, X.; Zhang, Y. Experimental and Numerical Study of the Influence of Initial Temperature on Explosion Limits and Explosion Process of Syngas-air Mixtures. *Int. J. Hydrogen Energy* **2022**, *47*, 22261–22272. [CrossRef]
31. European Committee for Standardisation. *Determination of Explosion Limits of Gases and Vapours: BS EN 1839:2012 [S]*; BIS Standards Publication: Brussels, Belgium, 2012; p. 7.
32. Zhu, X.; Zhou, J.; Guo, Z.; Zhang, X. CO_2 Suppression of Methane/air Explosion Flame and Pressure Coupling Analysis. *Fresenius Environ. Bull.* **2022**, *31*, 4995–5003.
33. Luo, Z.; Su, Y.; Li, R.; Chen, X.; Wang, T. Effect of Inert Gas CO_2 on Deflagration Pressure of CH_4/CO. *ACS Omega* **2020**, *5*, 23002–23008. [CrossRef] [PubMed]
34. Deng, J.; Cheng, F.; Song, Y.; Luo, Z.; Zhang, Y. Experimental and Simulation Studies on the Influence of Carbon Monoxide on Explosion Characteristics of Methane. *J. Loss Prev. Process Ind.* **2015**, *36*, 45–53. [CrossRef]
35. Cheng, F.; Deng, J. Experimental study on the influence of carbon monoxide on carbon dioxide-inerted methane explosion. *Xian Univ. Sci. Technol.* **2016**, *36*, 315–319.
36. Wang, T.; Zhou, Y.; Luo, Z.; Wen, H.; Zhao, J.; Su, B.; Deng, J. Flammability Limit Behavior of Methane with the Addition of Gaseous Fuel at Various Relative Humidities. *Process Saf. Environ. Prot.* **2020**, *140*, 178–189. [CrossRef]
37. Sun, Y.; Qian, X.; Yuan, M.; Zhang, Q.; Li, Z. Investigation on the Explosion Limits and Flame Propagation Characteristics of Premixed Methanol-gasoline Blends. *Case Stud. Therm. Eng.* **2021**, *26*, 101000. [CrossRef]
38. Zhang, L.; Wang, H.; Chen, C.; Wang, P.; Xu, L. Experimental Study to Assess the Explosion Hazard of CH_4/coal Dust Mixtures Induced by High-temperature Source Surface. *Process Saf. Environ. Prot.* **2021**, *154*, 60–71. [CrossRef]
39. Li, L.; Liu, T.; Li, Z.; Chen, X.; Wang, L.; Feng, S. Different Prevention Effects of Ventilation Dilution on Methane Accumulation at High Temperature Zone in Coal Mine Goafs. *Energies* **2023**, *16*, 3168. [CrossRef]
40. Deng, L.; Lei, C.; Xiao, Y.; Cao, K.; Ma, L.; Wang, W.; Bin, L. Determination and Prediction on "three Zones" of Coal Spontaneous Combustion in a Gob of Fully Mechanized Caving Face. *Fuel* **2018**, *211*, 458–470. [CrossRef]
41. Ma, D.; Qin, B.; Gao, Y.; Jiang, J.; Feng, B. An Experimental Study on the Methane Migration Induced By Spontaneous Combustion of Coal in Longwall Gobs. *Process Saf. Environ. Prot.* **2021**, *147*, 292–299. [CrossRef]

fire

MDPI

Article

Characteristics of Carbon Monoxide and Ethylene Generation in Mine's Closed Fire Zone and Their Influence on Methane Explosion Limits

Dong Ma [1], Leilin Zhang [2,*], Tingfeng Zhu [3] and Zhenfang Shi [3]

1 State Key Laboratory for Fine Exploration and Intelligent Development of Coal Resources, China University of Mining and Technology, Xuzhou 221008, China
2 Engineering Technology Research Centre for Safe and Efficient Coal Mining, Anhui University of Science and Technology, Huainan 232001, China
3 School of Safety Engineering, China University of Mining and Technology, Xuzhou 221116, China
* Correspondence: 2014017@aust.edu.cn

Abstract: Methane explosions often occur during the closure process of mine fire zones, during which the concentration of combustible gases such as monoxide and ethylene produced by coal combustion dynamically changes, which changes the risk of methane explosion. Therefore, studying the gas concentration distribution and methane explosion limits during the process of mine closure is of great significance for disaster prevention and control. In this paper, a three-dimensional physical model of gob was built, and the distribution of monoxide and ethylene in the process of fire zone closure was investigated. Further, the explosion limits of methane enriched with CO and C_2H_4 in the closed fire zone of gob were analyzed. The results indicate that CO and C_2H_4 would form a small-scale accumulation phenomenon near the fire zone after the closure of the fire zone, and when the fire zone is closed for more than 15 min, the mixed combustible gases in the environment lose their explosiveness.

Keywords: coal mine fires; methane explosion; closed fire zone; explosion limit

Citation: Ma, D.; Zhang, L.; Zhu, T.; Shi, Z. Characteristics of Carbon Monoxide and Ethylene Generation in Mine's Closed Fire Zone and Their Influence on Methane Explosion Limits. *Fire* **2024**, 7, 168. https://doi.org/10.3390/fire7050168

Academic Editors: Chuyuan Huang, Haipeng Jiang and Lijuan Liu

Received: 13 March 2024
Revised: 11 May 2024
Accepted: 12 May 2024
Published: 14 May 2024

1. Introduction

Mine fires have always been a major safety hazard, causing significant casualties and economic losses [1–3]. In order to prevent the expansion of a mine fire and control the disaster, the timely closure of the affected zone is an important emergency measure [4–6]. After the sealing, the oxygen supply to the fire zone is cut off from outside, so the heat generated inside the fire zone is less than the heat dissipated, which reduces the internal temperature of the fire zone and effectively controls the development of the fire [7–9]. Furthermore, once the fire zone is sealed, the oxygen concentration also decreases so the coal enters a negative combustion state. This generates a significant amount of flammable gases, including CO and C_2H_4. These gases accumulate in the closed fire zone, making it susceptible to methane explosions [10]. For instance, the Jilin Babao mine experienced a massive methane explosion during the closure process in 2013, resulting in the tragic death of 53 workers [11]. Similarly, a methane explosion in the Soma mine in Turkey caused the death of 301 workers in 2014 due to improper operation of a closed fire zone [12].

To better understand the internal gases' composition and combustion state after the fire zone is sealed, the indicator gases in the process of fire zone sealing are often used for risk prediction and assessment. Carbon monoxide and ethylene, as predictors of coal combustion, are the main combustible gases produced in the process of coal combustion [13]. Understanding the coal's combustion state is crucial for comprehending the mechanism of explosive accidents during hole sealing and preventing future incidents.

In practice, it is difficult to study the gas distribution in the fire zone using on-site measurements. Therefore, researchers tend to use numerical simulation and experimental platform methods for their study. Zhu et al. [14] used numerical simulation to study the effect of N_2 injection in a sealed fire zone on the development of a fire and found the transport law of the gas mixture in the fire zone. Wang et al. [15] used numerical simulation to study the distribution pattern of CH_4 and smoke after the fire zone was closed, as well as the influencing factors of the methane explosion's range. It was found that fire smoke will affect the methane explosion range, and the explosion zone first appeared in the return air side. Brune and Saki [16] used a computational fluid dynamics model to study the dynamic sealing of a progressive sealed longwall gob with nitrogen. They found that this approach can separate methane and air, thereby reducing the risk of explosion. However, although numerical simulation can solve some of the research problems, the use of numerical simulation for complex mine environments, where parameters such as coal-rock porosity and oxygen consumption are often assumed, can lead to unavoidable errors in the experimental results. Therefore, building a physical experimental platform is the most appropriate method for studying the distribution of flue gas before and after the sealing. Zhou et al. [5] used an experimental platform to study the gas changes in the fire zone after opening and found an instantaneous elevation of CO after the rekindling of the opening. Wen et al. [17] used a homemade high-temperature coal spontaneous combustion experimental setup to study the great influence of oxygen on coal spontaneous combustion at different temperatures in the sealed fire zone. Shi et al. [6] studied the impact of sealed fire zones on methane explosions and found that sealing could lead to changes in temperature and CH_4 concentration, as well as cause methane migration, which can concentrate in the fire zone. Zheng et al. [18] used a physical model to study the coupling effect of gas and temperature in the mining airspace and concluded that the temperature of the airspace leakage area should be monitored to prevent causing methane explosion accidents. Wang et al. [19] studied the effects of different sealing methods on the generation and flow of smoke in fire zones by building gob models and found that sealing methods had a great impact on the development of fires, which may lead to methane explosion accidents.

So far, much research has been carried out regarding the closed fire zone, but there is a lack of research related to CO and C_2H_4 in the process of fire zone closure. Therefore, in this paper, an experimental platform for the modeling of the mining zone was established, and the change law of the distribution of the indicator gases in the process of fire zone closure was investigated. In the study, the dynamics of CO and C_2H_4 transport and distribution were delineated during the thermal escalation of coal under conditions of spontaneous combustion. Further, the explosion limits of methane enriched with CO and C_2H_4 mixtures were tested. From the insights gleaned, the distribution pattern of CO and C_2H_4 gases at different temperatures in the closed fire zone were conducive to predicting the combustion state of enclosed fire zones.

2. Materials and Methods

2.1. Gob Platform

To investigate the gas concentration changes during spontaneous coal combustion and closure in the mining zone, a three-dimensional physical model of gob was constructed, as shown in Figure 1. The schematic diagram of the physical model is shown in Figure 2. The experimental platform comprises the gob, the working face, the intake and return air roadways, the coal combustion chamber, the ventilation system, the temperature monitoring system and the gas collection and analysis system. This ventilation system is a variable-frequency mixed-flow duct fan, as the fan supplies air to the working face. Additionally, the ventilator was connected to the working face by an air duct, and a hot-wire anemometer measures the internal ventilation of the working face before each test. The temperature monitoring system was made up of a heating plate, a temperature controller, a temperature sensor and a data acquisition card. Considering the temperature control and monitoring of the fire source in the spontaneous combustion zone of coal, the heat source of the experiment is the coal powder heated by the heating plate. Thermocouples were used to connect to the TES1310 pyrometer (Furuida Instrument Technology Co., Ltd.,

Dongguan, China) for real-time monitoring of the internal temperature of the coal. As a result, the experiment focuses on the internal temperature and concentration fields of the gob. Therefore, through the gas sample collection and analysis system, the simulated internal flue gas component and the concentration distribution law of the air mining zone can be quantitatively derived. Table 1 displays the specific dimensions of the gob.

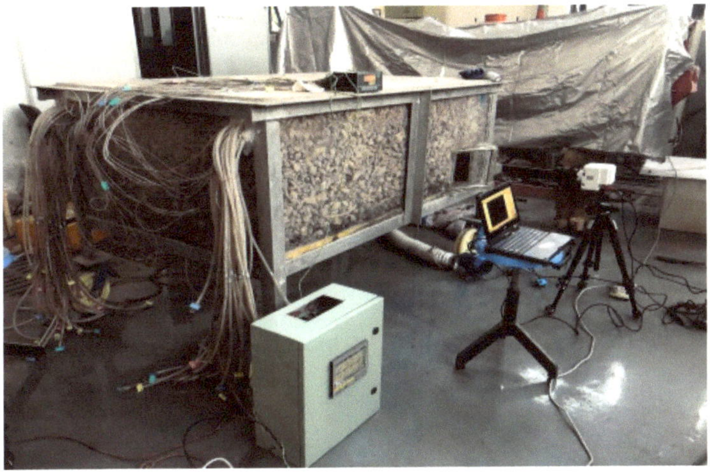

Figure 1. Experimental platform.

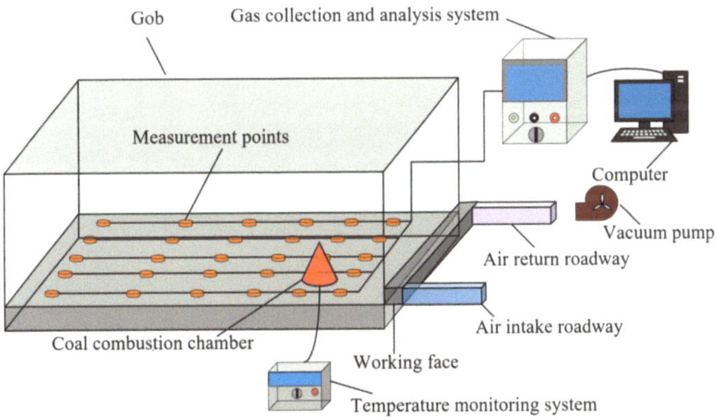

Figure 2. Schematic diagram of the three-dimensional physical model of gob.

Table 1. The dimensions of the gob.

Part	Length/cm	Width/cm	Height/cm
Mining face	120	5	3
Air intake roadway	20	4	3
Air return roadway	20	4	3
Mine gob	180	120	80

2.2. Methods

The experimental platform is set up with a layer of measurement points, with the distance from the measurement points to the ground being 1 cm. The layer of measurement points is arranged in 5 rows along the strike direction of the gob and 5 columns along the inclination direction, a total of 25 measurement points. As shown in Figure 3, the focus of this study is the area near the combustion chamber of the coal, which includes a total of 9 measurement points. The analysis and research mainly rely on the experimental data from the monitoring points on two monitoring lines, L1 and L2. Prior to each experiment, the ventilation velocity was measured with a hot wire anemometer to ensure that the wind speed conditions of the working face reached 0.6 m/s. At the inception of the experiment, 100 g of coal powder was placed on the heating plate with the highest temperature of 400 °C, and the thermocouple was embedded in the coal. The thermocouple was connected to the TES1310 temperature tester to monitor the temperature of the coal. Subsequently, an automatic negative-pressure gas sampler was used to collect gas generated during the fire zone closure process. Finally, a gas chromatograph was used to detect the composition of the flue gas to obtain its distribution and concentration at different temperatures and different positions. There were 4 aspects of uncertainty in the experiment: deviation of gas concentration during gas distribution (A1), error arising during pumping tests (A2), error in ventilation velocity at the working face (A3) and error from ambient temperature (A4). The deviation of the gas concentration during gas distribution (A1) and the error arising during the pumping test (A2) originate from the accuracy of the instrument, with an error of about 0.3%. The error in the ventilation velocity at the working face (A3) and the error from ambient temperature (A4) come from the testing instruments and manual readings, with an error of about 3.2%. Thus, the total experimental uncertainty $\sqrt{A1^2 + A2^2 + A3^2 + A4^2}$ is less than 5%.

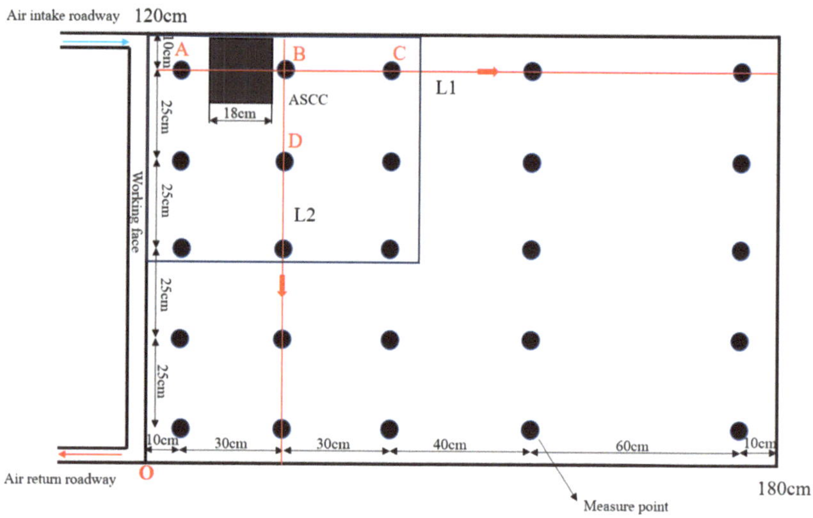

Figure 3. Layout of model measuring points.

3. Results and Discussion

3.1. The Distribution of CO in the Closed Fire Zone of Gob

As discernible from Figure 4, when the fire zone was not sealed, the combustion intensified with the rise in coal temperature, leading to substantial production of CO and a sharp increase in its concentration. As shown in Figure 4a, when the temperature increased to 100 °C, a portion of the CO was discharged along the working face into the air return roadway. Additionally, due to the presence of gob leakage, another portion of CO flowed towards the deeper and upper parts of the gob. When the coal temperature reached 250 °C, the area of high CO concentration moved further away from the working face, as shown in Figure 4b. Moreover, owing to the gas vortex phenomenon in coal spontaneous combustion fire zone, it resulted in a much higher CO concentration at the accumulation than in the surrounding area [20]. Since then, the fire zone was sealed. It can be seen from Figure 4 that the temperature in the fire zone showed a trend of increasing and then decreasing.

When the coal temperature reached 400 °C, the CO concentration increased dramatically, and its distribution area moved further into the deeper part of the gob. During the cooling process, due to the absence of intake airflow, the coal spontaneous combustion flue gas was only influenced by the thermal buoyancy movement of the gas, and the influence range was smaller than the heating process. As shown in Figure 4d, when the coal temperature was reduced to 300 °C, the area of high concentration of CO was mainly near the high-temperature point, deep in the gob, and near the air return roadway. As shown in Figure 4e, when the temperature was lowered to 250 °C, the region of high CO concentration was mainly near the high temperature point, and the distribution of CO was further balanced. When the temperature was further reduced, the distribution of CO was more balanced and basically distributed in all locations of the gob, as shown in Figure 4f.

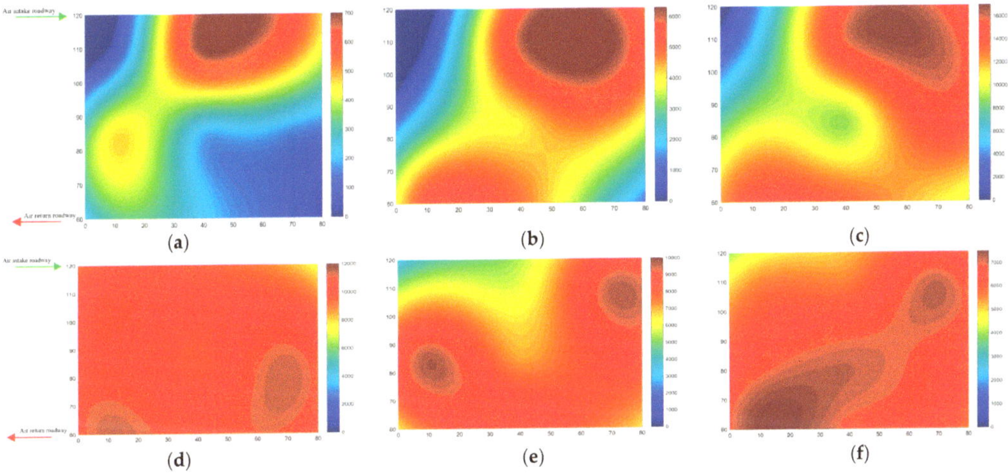

Figure 4. CO concentration under different coal temperature conditions. (**a**) Coal temperature 100 °C; (**b**) coal temperature 250 °C; (**c**) coal temperature 400 °C; (**d**) coal temperature 300 °C; (**e**) coal temperature 250 °C; (**f**) coal temperature 100 °C.

The variation of CO concentration on the two monitoring lines, L1 and L2, are shown in Figures 5 and 6. During the temperature rise and fall periods, the CO concentration at different temperatures shows an initially increasing and then decreasing strike direction of the gob. However, the inclination direction of the gob showed fluctuating changes. As shown in Figure 5a, when the coal temperature was 250 °C, the CO concentrations of the measuring points on the monitoring line L1 were 32 ppm, 4246 ppm, 6355 ppm, 3500 ppm and 100 ppm, respectively, and the highest concentration was near the middle of the strike direction of the gob. As shown in Figure 6a, when the coal temperature was 400 °C, the CO concentrations of the measuring points on the monitoring line L2 were 800 ppm, 3000 ppm, 14,386 ppm, 9435 ppm and 13,127 ppm, respectively, and the highest concentration was near the middle of the inclination direction of the gob. The CO concentrations in the strike direction and the inclination direction showed an increase with coal temperature increase. For example, as shown in Figure 5a, the CO concentrations at measurement point B were 542 ppm, 4246 ppm and 13,127 ppm at coal temperatures of 100 °C, 250 °C and 400 °C, respectively.

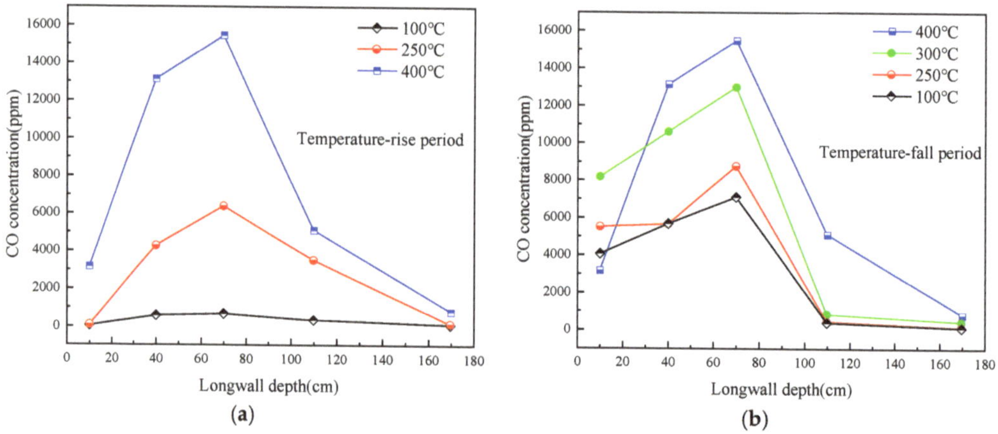

Figure 5. Changes in CO concentration at different temperatures in the L1 monitoring line. (**a**) Temperature-rise period; (**b**) temperature-fall period.

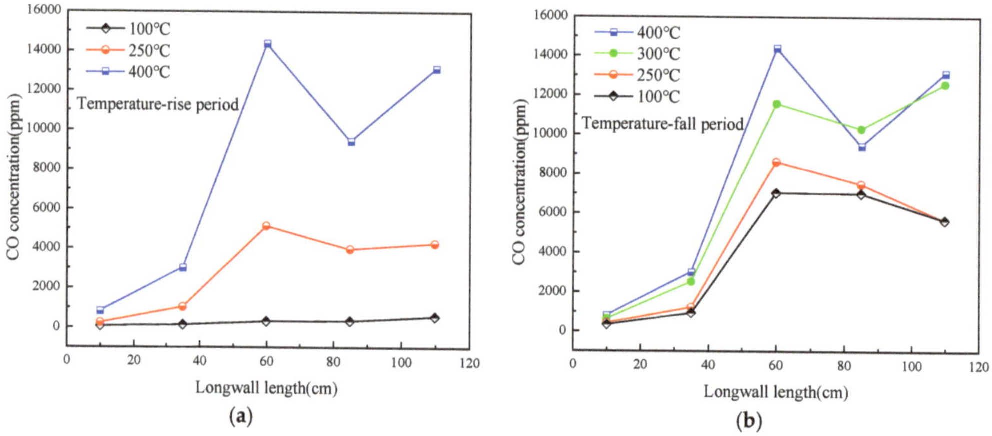

Figure 6. Changes in CO concentration at different temperatures in the L2 monitoring line. (**a**) Temperature-rise period; (**b**) temperature-fall period.

3.2. The Distribution of C_2H_4 in the Closed Fire Zone of Gob

Figure 7 shows the distribution of C_2H_4 concentration. As can be seen in Figure 7a, when the coal temperature reached up to 100 °C, the C_2H_4 concentration generated by the combustion of the coal was relatively small. Although it was influenced by the intake airflow, most of the deep part of the gob was not detected in a higher concentration of C_2H_4, with an average concentration of less than 10 ppm. With the increase in the temperature of the coal, there is a significant rise in the concentration of C_2H_4 generated in the inner part of the gob. As the temperature of the coal reached up to 250 °C, as shown in Figure 7b, there was a small accumulation of C_2H_4 concentration near the high-temperature point, and the maximum concentration was less than 150 ppm. Subsequently, as the fire zone was closed, as shown in Figure 7c, the concentration of C_2H_4 continued to increase, which gradually spread to the deep part of the gob and the air return side under the influence of the intake airflow. Due to the closure of the fire zone, resulting in a gradual decrease in the temperature of the gob, C_2H_4 concentration generation was reduced and uniformly distributed in the deep part of the gob, the coal temperature dropped to 300 °C and the C_2H_4 concentration of the various measurement points was between 50 and 60 ppm, as shown in Figure 7d. However, when the coal temperature dropped to 250 °C, as in Figure 7e, the C_2H_4 concentration on the intake side appeared to drop roughly 20 ppm. Additionally, the air return side and the deep part of the gob still showed a wide range of C_2H_4 accumulation. Finally, when the coal temperature reached up to 100 °C, the C_2H_4 concentration showed a significant decrease, and the maximum concentration did not exceed 50 ppm.

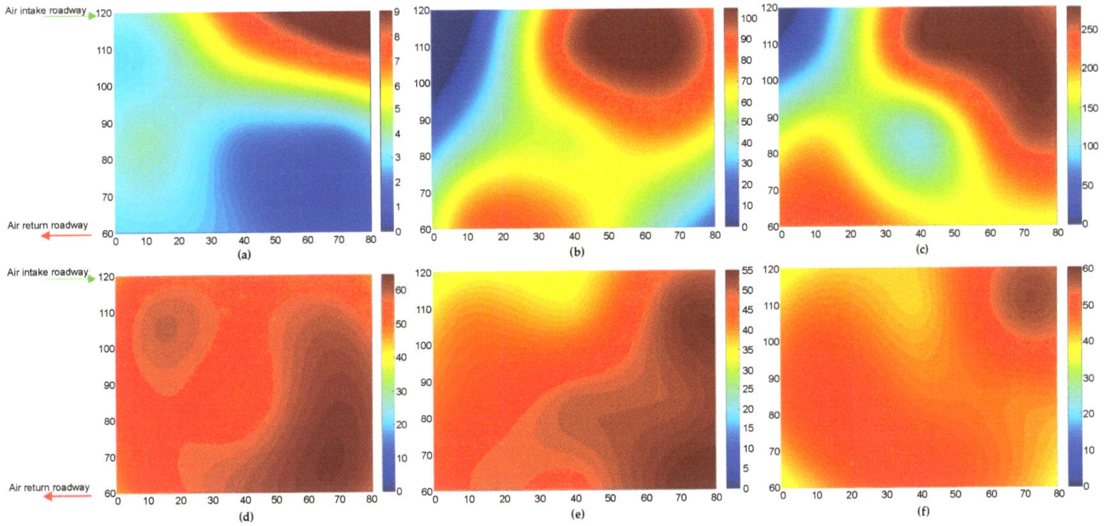

Figure 7. C_2H_4 concentration under different coal temperature conditions. (**a**) Coal temperature 100 °C; (**b**) coal temperature 250 °C; (**c**) coal temperature 400 °C; (**d**) coal temperature 300 °C; (**e**) coal temperature 250 °C; (**f**) coal temperature 100 °C.

The changes in C_2H_4 concentration along the two monitoring lines, L1 and L2, are shown in Figures 8 and 9. As can be seen from Figure 8a, along the direction of the monitoring line L1, C_2H_4 shows a trend of increasing and then decreased under different temperature conditions. For example, when the coal temperature was 250 °C, the C_2H_4 concentration increased from 9 ppm to 110 ppm and then decreased to 20 ppm. When the coal temperature was 400 °C, the C_2H_4 increased from 30 ppm to 260 ppm and then decreased to 5 ppm. After the closure of fire zone, as shown in Figure 8b, the C_2H_4 concentration was basically the same as that in the temperature-rise period. As the temperature decreased to 300 °C, the C_2H_4 concentration increased from 52 ppm by 5 ppm and then decreased to 8 ppm. Comparing the C_2H_4 concentration at different temperatures, it can be seen that there was a good linear relationship between the coal temperature and the C_2H_4 concentration. Moreover, as shown in Figure 9b, when the temperature of the coal decreased to 300 °C, the concentration of C_2H_4 increased from 4 ppm to 59 ppm and then decreased to 55 ppm. Overall, the C_2H_4 concentration decreased with the coal temperature decreases in the temperature-fall period, and the peak C_2H_4 concentration ranged from 110 ppm to 50 ppm from 400 °C to 100 °C.

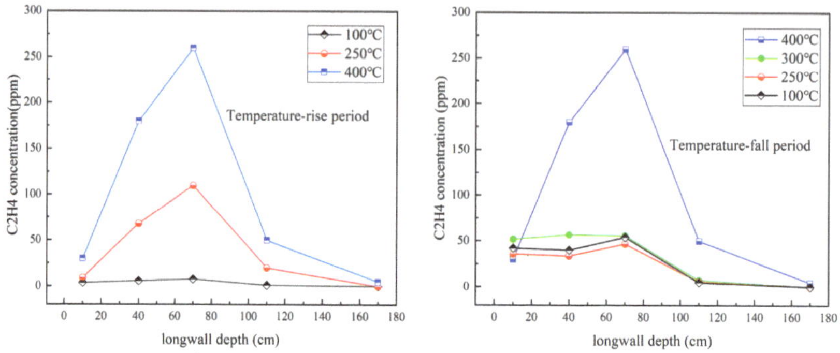

(a) (b)

Figure 8. Changes in C_2H_4 concentration in the L1 monitoring line. (**a**) Temperature-rise period; (**b**) temperature-fall period.

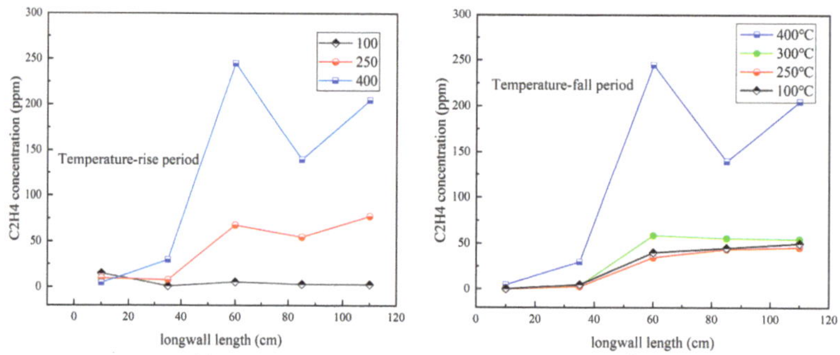

(a) (b)

Figure 9. Changes in C_2H_4 concentration in the L2 monitoring line. (**a**) Temperature-rise period; (**b**) temperature-fall period.

3.3. The Methane Explosion Concentration in the Closed Fire Zone of Gob

Combustible mixed gases mixed with oxygen may explode if they encounter a sufficiently high-energy ignition source. When the concentration of air or oxygen is high enough, the concentration of combustible mixed gases is low and causes a lean combustion state, and there is not enough fuel to maintain the explosion reaction. On the contrary, if the concentration of combustibles is high and the concentration of air or oxygen is low, the combustibles are in a lean oxygen state, and there is not enough oxygen to assist the explosion reaction. Combustible mixed gases must be within the range of the lower explosive limit and upper explosive limit in order to explode.

The testing of methane explosion limits is based on a 20 L spherical explosion system, as introduced in reference [21]. The system mainly includes a 20 L approximately spherical explosive reaction vessel, ignition device, gas distribution system, vacuum pumping system, data acquisition system, controller, high-speed camera. The inner diameter of a 20 L approximately spherical explosion reaction vessel was about 34 cm, and the maximum pressure of the reaction vessel can reach 4 MPa, which can ensure the safe conduct of mixed gas explosion experiments. The ignition device consists of an ignition electrode located at the center of a spherical reaction vessel and an external electric spark generator. After the gas balance in the reactor, the ignition device was automatically controlled by a controller. In the experiment, the ignition energy was 60 J, and the ignition delay time was 120 ms. The gas distribution system mainly consisted of gas cylinders, air compressors, flow meters and vacuum pumps. The gas cylinders used in the experiment were 99.99% pure CH_4, CO, C_2H_4, O_2 and N_2 cylinders. The mixed gas was configured using the partial pressure method, and the gas distribution accuracy was 0.1%. During the experiment, the ambient temperature was 18 °C~25 °C, the air humidity was 55%~65% and the pressure was 0.1 MPa.

According to the EU BSEN 1839:2012 standard [22], this paper takes whether the maximum explosion pressure exceeded (5 ± 0.1)% of the initial pressure as the standard for explosion occurrence. Each operating condition experiment should be repeated at least three times to ensure good accuracy and repeatability of the test results.

The oxygen concentration changes at four monitoring points near the fire area are shown in Figure 10. It can be seen that after the fire zone is closed, the oxygen concentration at the measuring points showed a rapid decreasing trend and eventually approached complete depletion. Among them, the oxygen concentration at point B decreased the fastest. When the fire zone was closed for 20 min, the oxygen concentration at point B was consumed to nearly 0%, while points A, C and D decreased to nearly 0% after 22 min, 25 min and 26 min, respectively. Based on the O_2 concentration, CO concentration and C_2H_4 concentration of these four measurement points, the methane explosion limit of the monitoring points during the fire zone closure process were tested, as shown in Figure 11.

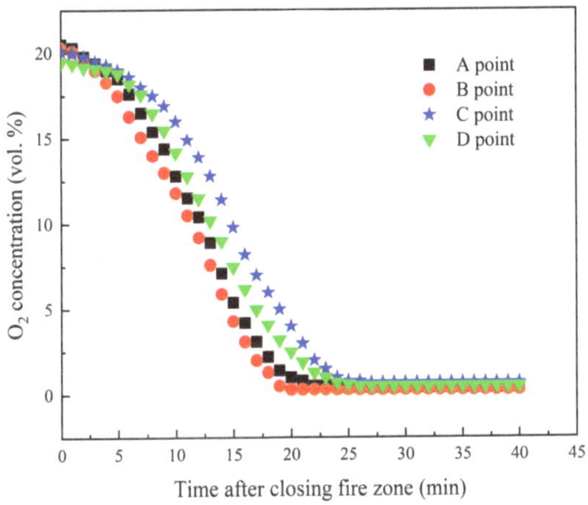

Figure 10. The variation of O_2 concentration in the zone near the fire.

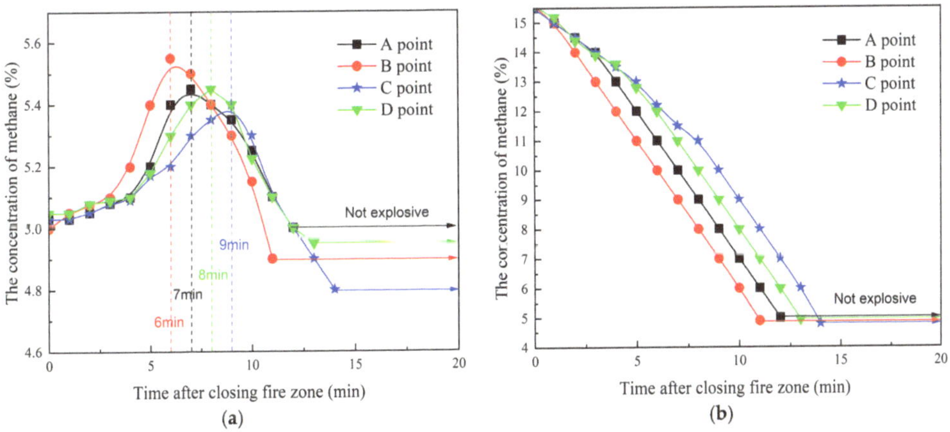

Figure 11. The variation of the minimum and maximum methane explosion for explosions in the fire zone. (**a**) The minimum methane concentration for explosions; (**b**) the maximum methane concentration for explosions.

The change in the minimum methane concentration for explosions after the closure of the fire zone is shown in Figure 11a. It can be seen that with the extension of closure time, the minimum methane concentration at each measuring point showed a trend of first increasing and then decreasing. The peak value appeared at four measurement points, A, B, C and D, after being closed for 7 min, 6 min, 9 min and 8 min, respectively. This was because in the initial stage of closure, coal combustion consumed a large amount of oxygen concentration, leading to a decrease in oxygen concentration near the fire zone, thereby causing an increase in the minimum methane concentration for explosions [23]. Among them, point B, which is closer to the fire zone, had the fastest decrease in O_2 concentration, resulting in the largest increase in the minimum methane concentration. Subsequently, as the combustion in the fire zone continued, higher concentrations of CO and C_2H_4 gas were generated, leading to a gradual decrease in the minimum methane concentration. However, when the closure time of the fire zone reached 10–15 min, the mixed combustible gases in the environment lose their explosiveness due to the O_2 concentration in the reaction system being lower than the critical oxygen concentration required for gas explosion [24]. The change in the maximum methane concentration for explosions after the closure of the fire zone is shown in Figure 11b. It can be seen that after the fire zone was closed, the maximum methane concentration for explosions at each measuring point showed a rapid downward trend. This was mainly attributed to the decrease in oxygen concentration in the fire zone environment and the generation of combustible gases, which together led to a decrease in the maximum methane concentration for explosions. Additionally, when the fire zone was closed for more than 15 min, the mixed combustible gases in the environment lost their explosiveness, which was consistent with the test results for the minimum methane concentration.

4. Conclusions

In the study, a three-dimensional physical model of gob was built, and the dynamics of CO and C_2H_4 distribution were delineated during the thermal escalation of coal under conditions of spontaneous combustion. Further, the explosion limits of methane enriched with CO and C_2H_4 mixtures were tested. The main conclusions of this paper are as follows:

- A three-dimensional physical test model of the gob was constructed according to the characteristics of the coal gob on a scale of 1:100, and the experimental platform mainly consisted of the air intake and return roadways, the working face, the gob, the temperature monitoring system, the ventilation system, the coal spontaneous combustion fire zone and the collection system.
- With the increase in coal temperature, CO and C_2H_4 mainly spread to the depth of the mining zone and the vicinity of the air return side. After sealing the fire zone, CO and C_2H_4 first formed a small-scale accumulation phenomenon near the fire zone. With the lowering of the temperature, the C_2H_4 and CO concentration dispersed towards the interior of the gob, but a small scale of C_2H_4 and CO still accumulated on the side of the return air corner.
- After the fire zone was sealed, the minimum methane concentration for explosions near the fire zone showed a trend of first increasing and then decreasing, while the maximum methane concentration for explosions decreased significantly. When the fire zone was closed for more than 15 min, the mixed combustible gases in the environment lost their explosiveness.

Author Contributions: D.M.: methodology, software, investigation, validation, visualization, data curation, writing—original draft. L.Z.: supervision, conceptualization, resources, writing—review and editing, project administration. T.Z.: supervision, resources. Z.S.: software, review and editing. All authors have read and agreed to the published version of the manuscript.

Funding: This work was supported by the Engineering Technology Research Centre for Safe and Efficient Coal Mining (Anhui University of Science and Technology) (NO. SECM2208) and the National Natural Science Foundation of China (NO. 52204250).

Institutional Review Board Statement: Not applicable.

Informed Consent Statement: Informed consent was obtained from all subjects involved in the study.

Data Availability Statement: The data are not publicly available due to commercial confidentiality, as they contain information that could compromise the privacy of research participants.

Conflicts of Interest: The authors declare no conflicts of interest.

References

1. Song, Z.; Kuenzer, C. Coal fires in China over the last decade: A comprehensive review. *Int. J. Coal Geol.* **2014**, *133*, 72–99. [CrossRef]
2. Ling, Y.; Luo, H. Current situation and development trend for coal mine fire prevention and extinguishing techniques in China. *Mei T'an Hsueh Pao (J. China Coal Soc.)* **2008**, *33*, 126–130.
3. Singh, R.V.K. Spontaneous heating and fire in coal mines. *Procedia Eng.* **2013**, *62*, 78–90. [CrossRef]
4. Geng, J.; Sun, Q.; Zhang, Y.; Gong, W.; Du, S. Non-destructive testing and temperature distribution of coal mine roadway lining structure under exogenous fire. *J. Loss Prev. Process Ind.* **2018**, *55*, 144–151. [CrossRef]
5. Zhou, F.B.; Li, J.H.; He, S.; Liu, Y.S. Experimental modeling study on the reignition phenomenon when opening a sealed fire zone. *Procedia Earth Planet. Sci.* **2009**, *1*, 161–168. [CrossRef]
6. Shi, G.Q.; Wang, G.Q.; Ding, P.X.; Wang, Y.M. Model and simulation analysis of fire development and gas flowing influenced by fire zone sealing in coal mine. *Process Saf. Environ. Prot.* **2021**, *149*, 631–642. [CrossRef]
7. Zuo, Q.; Li, J. Simulation of fire smoke disaster in a goaf during the closure process. *Therm. Sci.* **2021**, *25 Pt A*, 3399–3407. [CrossRef]
8. Zhu, C.G.; Wang, J.; Xie, W.X.; Zheng, T.T.; Lv, C. Improving strontium nitrate-based extinguishing aerosol by magnesium powder. *Fire Technol.* **2015**, *51*, 97–107. [CrossRef]
9. Taraba, B.; Pavelek, Z. Investigation of the spontaneous combustion susceptibility of coal using the pulse flow calorimetric method: 25 years of experience. *Fuel* **2014**, *125*, 101–105. [CrossRef]
10. Ma, T.; Larrañaga, M. Theoretical flammability diagram for analyzing mine gases. *Fire Technol.* **2015**, *51*, 271–286. [CrossRef]
11. Zhu, Y.; Wang, D.; Shao, Z.; Xu, C.; Zhu, X.; Qi, X.; Liu, F. A statistical analysis of coalmine fires and explosions in China. *Process Saf. Environ. Prot.* **2019**, *121*, 357–366. [CrossRef]
12. Duzgun, H.S.; Yaylaci, E.D. An evaluation of Soma underground coal mine disaster with respect to risk acceptance and risk perception. In Proceedings of the 3rd International Symposium on Mine Safety Science and Engineering, Montreal, QC, Canada, 13–19 August 2016; pp. 368–374.
13. Liang, Y.; Zhang, J.; Wang, L.; Luo, H.; Ren, T. Forecasting spontaneous combustion of coal in underground coal mines by index gases: A review. *J. Loss Prev. Process Ind.* **2019**, *57*, 208–222. [CrossRef]
14. Zhu, Y.; Zhou, X.; Wang, H.; Wang, B. *Numerical Simulation on Migrating Laws of Mixed Gas after Nitrogen Injection in Sealing Fire Zone of Coal Mine*; Coal Industry Publ House: Fuxin, China, 2008.
15. Wang, G.-Q.; Shi, G.-Q.; Wang, Y.-M.; Shen, H.-Y. Numerical study on the evolution of methane explosion regions in the process of coal mine fire zone sealing. *Fuel* **2021**, *289*, 119744. [CrossRef]
16. Brune, J.F.; Saki, S.A. Prevention of gob ignitions and explosions in longwall mining using dynamic seals. *Int. J. Min. Sci. Technol.* **2017**, *27*, 999–1003. [CrossRef]
17. Wen, H.; Guo, J.; Jin, Y.; Wang, K.; Zhang, Y.; Zheng, X. Experimental study on t influence of different oxygen concentrations on coal spontaneous combustion characteristic parameters. *Int. J. Oil Gas Coal Technol.* **2017**, *16*, 187–202. [CrossRef]
18. Zheng, Y.; Li, Q.; Zhang, G.; Zhao, Y.; Zhu, P.; Ma, X.; Li, X. Study on the coupling evolution of air and temperature field in coal mine goafs based on the similarity simulation experiments. *Fuel* **2021**, *283*, 118905. [CrossRef]
19. Wang, G.; Shi, G.; Yang, Y.; Liu, S. Experimental study on the exogenous fire evolution and flue gas migration during the fire zone sealing period of the coal mining face. *Fuel* **2022**, *320*, 123879. [CrossRef]
20. Li, L.; Qin, B.; Ma, D.; Zhuo, H.; Liang, H.; Gao, A. Unique spatial methane distribution caused by spontaneous coal combustion in coal mine goafs: An experimental study. *Process Saf. Environ. Prot.* **2018**, *116*, 199–207. [CrossRef]
21. Ma, D.; Qin, B.; Zhong, X.; Sheng, P.; Yin, C. Effect of flammable gases produced from spontaneous smoldering combustion of coal on methane explosion in coal mines. *Energy* **2023**, *169*, 128125. [CrossRef]
22. *BS EN 1839:2012*; Determination of Explosion Limits of Gases and Vapours. European Committee for Standardisation, BIS Standards Publication: Brussels, Belgium, 2012; p. 7.

23. Luo, Z.; Su, B.; Wang, T.; Cheng, F.; Wang, Y.; Liu, B.; Xie, C. Effects of propane on the flammability limits and chemical kinetics of methane–air explosions. *Combust. Sci. Technol.* **2020**, *192*, 1785–1801. [CrossRef]
24. Su, B.; Luo, Z.; Wang, T.; Zhang, J.; Cheng, F. Experimental and principal component analysis studies on minimum oxygen concentration of methane explosion. *Int. J. Hydrogen Energy* **2020**, *45*, 12225–12235. [CrossRef]

fire

MDPI

Article

Study on the Characteristics of Combustible Mixed Gas Production during Lignite Oxidation Process

Dong Ma [1,*], Tingfeng Zhu [2], Puchun Yuan [2] and Leilin Zhang [3]

1 State Key Laboratory for Fine Exploration and Intelligent Development of Coal Resources, China University of Mining and Technology, Xuzhou 221008, China
2 School of Safety Engineering, China University of Mining and Technology, Xuzhou 221116, China; ts23120206p31@cumt.edu.cn (T.Z.); 16223141@cumt.edu.cn (P.Y.)
3 Engineering Technology Research Centre for Safe and Efficient Coal Mining, Anhui University of Science and Technology, Huainan 232001, China; 2014017@aust.edu.cn
* Correspondence: madongcumt@163.com

Abstract: CO, H_2, and other combustible gases will be produced during coal oxidation in coal mines, which will increase the risk of explosion when mixed with methane. Therefore, it is very important to understand the production characteristics of combustible gas during coal oxidation. In this paper, a programmed temperature gas test system is built to study the impact of lignite on the production of gases at different particle sizes and temperatures, and the release characteristics of gases are also analyzed. The result shows that the production of combustible gas is influenced by the coal particle size significantly when the temperature is above 200 °C, and it decreases as the particle size increases. CO is the main gas during the early stage of coal spontaneous combustion, and the release of CH_4 and H_2 increases after 300 °C. The fitted equations of gas generation and temperature are consistent with the experimental results. The research results are helpful in understanding the hazards of coal spontaneous combustion and have a certain guiding significance for coal mine monitoring and prevention of coal spontaneous combustion.

Keywords: coal spontaneous combustion; particle size; combustible gas; lignite

Citation: Ma, D.; Zhu, T.; Yuan, P.; Zhang, L. Study on the Characteristics of Combustible Mixed Gas Production during Lignite Oxidation Process. *Fire* **2024**, *7*, 367. https://doi.org/10.3390/fire7100367

Academic Editors: Chuyuan Huang, Haipeng Jiang and Lijuan Liu

Received: 24 September 2024
Revised: 13 October 2024
Accepted: 14 October 2024
Published: 16 October 2024

1. Introduction

During the oxidation process of coal, oxygen is absorbed through physical and chemical adsorption, simultaneously releasing gaseous products. The primary gaseous products during coal self-ignition include carbon oxides and hydrocarbons [1,2]. The mine fires and gas explosions caused by coal spontaneous combustion are major safety hazards in coal mines, resulting in serious casualties and property losses. In recent years, there have been multiple accidents caused by coal spontaneous combustion in major coal-producing countries around the world. In 2021, a coal spontaneous combustion occurred at Shenmu Coal Mine in Shaanxi, China, causing a fire and gas explosion, resulting in the death of 5 workers. In 2022, a coal mine in the Indian state of Jharkhand caught fire due to spontaneous combustion, causing the mine to shut down for several weeks. Based on the frequent occurrence of coal spontaneous combustion accidents, a large number of scholars have conducted research on the process of coal spontaneous combustion.

In the low-temperature oxidation stage of coal, the quantity of gaseous products increases with the rise in coal oxidation temperature, and the concentration of gas products exponentially grows [3–5]. Furthermore, since some coals inherently adsorb gases like CO_2, CH_4, and C_2H_6, these gases gradually desorb as the coal's temperature rises, resulting in relatively high emissions at the early stages of coal oxidation [6–8]. The concentration of gaseous products produced during coal oxidation correlates with changes in coal temperature, allowing for predictions of coal self-ignition states based on the release patterns of gaseous products [9,10]. Yuan et al. [11] found that the CO/CO_2 ratio tends to a constant value of 0.2 in different coal samples, regardless of the properties of the coal. The rate of CO production is highly correlated with coal temperature. When the initial temperature of the coal body is 90 °C, the concentration of CO increases rapidly and can reach a maximum value of 1300 ppm. Green et al. [12,13] studied the generation characteristics of CO and CO_2 during the coal self-ignition process and further analyzed the pathways of their gaseous products' formation. They found that the thermal decomposition of oxygen-containing functional groups to produce CO does not depend on the coal grade, and the production of CO_2 depends on the coal grade; the decomposition of air oxygen by chemical adsorption presents a completely opposite situation to the former. Li Zenghua et al. [14] measured the production patterns of H_2 during coal oxidation, finding that H_2 is detected in the temperature range of 200–250 °C, and its concentration subsequently increases exponentially with temperature. To study the gaseous product generation patterns at high temperatures, Xi'an University of Science and Technology independently developed a testing system for the self-ignition characteristics of coal under high-temperature conditions and examined the generation patterns of gaseous products during high-temperature oxidation stages of coal self-ignition [15–18]. Dong et al. [19] used programmed temperature experiments to study the gas production patterns of different coal samples during heating and oxidation processes. They discovered that the order of gas production was CO, C_2H_6, C_2H_4, and C_3H_8, and that the amount of gas produced was approximately exponentially related to the temperature. Ren and colleagues [20] studied the relationship between temperature and exothermic intensity with the amount of gas produced under different oxygen concentrations. They found that CO, CH_4, C_2H_4, and C_2H_6 increased with rising temperature. The exothermic intensity had the greatest correlation with CO_2, and under low oxygen conditions, the exothermic intensity had a stronger correlation with CO and CH_4. Sun et al. [21] studied the CO gas production characteristics of low metamorphic coals, CYM, RNM, and QM, through programmed temperature experiments. They found that CYM had the fastest CO concentration and gas production rate, reaching a CO concentration of 1000×10^{-11} mol cm^{-3} s^{-1} at 170 °C. Yan et al. [22] studied the relationship between the indicator gases and their ratios during the coal oxidation process with coal temperature using grey correlation analysis. They concluded that between 90 °C and 150 °C, the grey relational degrees of C_3H_8/CH_4, C_2H_6/CH_4, and C_3H_8/C_2H_6 were 0.9701, 0.9562, and 0.9240, respectively.

The particle size of coal also affects the release patterns of gaseous products during the self-ignition process. It was found that the critical particle size of coal is in the range of 38–74 μm (D50 = 50.56 μm), and below this particle size, the oxidation rate does not change significantly. CO_2/CO was proposed as the main indicator for quantitative coal spontaneous combustion under the influence of coal particle size [23–25]. Singh et al. [26] tested the characteristics of gaseous products produced by coal below 200 °C, finding that the particle size of coal significantly impacts the production of CO while having a lesser effect on H_2 production. Song et al. [27] studied the characteristics of gas production from weathered coal of different particle sizes under conditions of high temperature and hypoxia. They found that during low-temperature oxidation (<250 °C), the concentration of indicator gases increased with decreasing particle size. During high-temperature oxidation (>250 °C), weathered coal of different particle sizes exhibited their own characteristics, with a 3 mm particle size being the critical size for high-temperature oxidation.

These studies illustrate the characteristics of the primary gaseous products in coal self-ignition under various environmental conditions and can provide guidance for predicting and preventing coal self-ignition at coal mine sites. However, regarding the underground fire areas in coal mines, it is crucial to understand the generation pattern of combustible gas products, which may be mixed with methane to form multi-component gases and increase the risk of gas explosion. At present, both domestically and internationally, the relationship between coal temperature and gas-phase products is mainly focused on low-temperature oxidation (<200 °C), and most studies focus on the relationship between single CO or H_2 and coal temperature. There is a lack of research on the relationship between multi-component products and coal temperature changes at high temperatures (>200 °C). The systematic research on the generation law of combustible gases generated by coal spontaneous combustion is not sufficient, so it is impossible to grasp the explosive characteristics of mixed gases in the coal spontaneous combustion environment. This article systematically analyzes the generation patterns of multiphase products at different coal temperatures (<500 °C), and explores the effects of particle size differences to further reveal the dynamic changes of products during coal spontaneous combustion. Unlike conventional experiments, this study simulated an actual environment closer to the spontaneous combustion of coal in the fire zone and examined the synergistic effect of temperature and particle size, two key factors. The research results can provide important references for early warning of coal spontaneous combustion in fire zones, and also propose new ideas for fire prevention and extinguishing strategies in actual coal mines.

In this work, lignite from Datong Ermuwan Coal Mine is taken as a study object, and the coal spontaneous combustion programmed temperature gas test system is used to study the production of gas with different coal particle sizes and temperatures. The amount of combustible gas is fitted by segment, and the generation characteristic is analyzed. The results can predict the degree of coal spontaneous combustion, and further provide guidance on avoiding coal fires.

2. Materials and Methods

2.1. Coal Samples

In the experiment, lignite from Datong Ermuwan Coal Mine in Shanxi Province, China (referred to as EMW), was chosen as the coal sample. Fresh coal samples collected from underground were sealed in multi-layer plastic bags and transported to the laboratory. The oxidized layer of the coal was removed, and the samples were crushed under vacuum conditions. Coal samples of three different particle sizes—0.096~0.18 mm, 0.18~0.38 mm, and 0.38~0.83 mm—were sieved and set aside. Prior to experimentation, the coal samples were placed in a vacuum drying oven at 40 °C for 48 h to remove extraneous moisture. An automatic industrial analyzer, model 5E-MAG6600, was used to conduct the industrial analysis of five coal samples. The results of the analysis are presented in Table 1.

Table 1. Industrial analysis of coal samples.

Coal Sample	Coal Rank	Industrial Analysis Results			
		Moisture M_{ad} (%)	Ash A_d (%)	Volatile V_{daf} (%)	Fixed Carbon FC_{ad} (%)
EMW	lignite	13.80	10.98	35.08	40.14

2.2. Experimental Equipment

The coal spontaneous combustion programmed temperature gas test system is illustrated in Figure 1. The system comprises a quartz reactor, a programmable heating system, a gas distribution system, a data acquisition system, and a gas chromatograph. The quartz reactor has an inner diameter of 100 mm and a length of 680 mm. The programmable heating system includes a heating wire and a temperature control module, which allows for the adjustment of the heating rate, as well as a real-time temperature display. The gas distribution system consists of two gas flow meters with a range of 0 to 10 L/min and two gas cylinders of oxygen and nitrogen, enabling intelligent control of the gas composition and flow rate during coal spontaneous combustion. The data acquisition system primarily includes temperature sensors, a multi-channel data collector, and a PLC main control screen, facilitating real-time recording and storage of the coal temperature during the combustion process. Gases produced during the coal combustion are analyzed using a GC-4000A gas chromatograph.

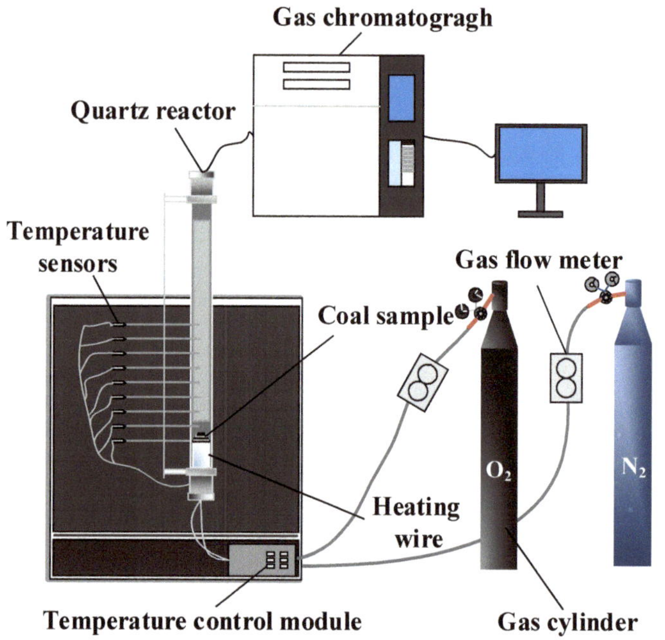

Figure 1. Coal spontaneous combustion programmed temperature gas test system.

2.3. Experimental Procedure

Prior to the commencement of the experiment, nitrogen gas was introduced into a quartz reaction tube at a flow rate of 5 L/min for 10 min to expel any atmospheric air present, thereby mitigating potential interference with the experimental outcomes. Subsequently, both the electric heating system and the automatic gas delivery system were activated. The target temperature for the electric heating program's control module was set at 300 °C, employing a linear temperature increase approach. Concurrently, the intake of oxygen and nitrogen was adjusted to mimic the proportions found in air, with a total inflow rate of 500 mL/min. Once the temperature of the molybdenum heating element reached 300 °C, the temperature control system was deactivated, and heating was discontinued to allow the coal to continue oxidizing and burning in a ventilated environment. When the coal temperature dropped below 200 °C, an increase of 10 °C prompted gas analysis via gas chromatography. For temperatures exceeding 200 °C, gas analysis was conducted for every 20 °C increment. Real-time data on the temperatures of both the molybdenum wire and the coal were recorded by the data acquisition system.

3. Results and Discussion

3.1. Influence of Coal Particle Size on the Generation of Combustible Gas

The concentration of combustible gas produced during the heating process with different particle sizes is illustrated in Figure 2. Figure 2a reveals that larger coal particles produce lower concentrations of CO. This phenomenon is attributed to the larger particles having a smaller specific surface area and lower porosity, which results in fewer active sites for coal surface interaction with oxygen, thus reducing the CO concentration during the oxidation process. Furthermore, it is observed that the peak temperature of CO concentration significantly shifts towards higher values with increasing coal particle size. For example, for coal particles sized between 0.096 and 0.18 mm, the peak CO concentration occurs at 400 °C, whereas for particles sized between 0.38 and 0.83 mm, it occurs at 440 °C.

Figure 2b illustrates the variation in CH_4 under different particle size conditions as a function of temperature. The graph indicates that prior to reaching 300 °C, the methane concentrations generated from coal samples of varying particle sizes are relatively low. However, once the temperature of the coal exceeds 300 °C, the methane concentrations significantly increase due to the oxidative cleavage of aromatic rings and cycloalkanes within the coal molecules. The methane production shows exponential growth with increasing temperature until the end of the experiment. A comparison among coal samples of different particle sizes reveals that the methane concentration produced during the coal oxidation process increases as the particle size decreases. For instance, at a coal temperature of 500 °C, coal samples with a particle size range from 0.096 to 0.18 mm generate a methane concentration of 0.58%, while those with a particle size range from 0.38 to 0.83 mm produce a methane concentration of 0.41%.

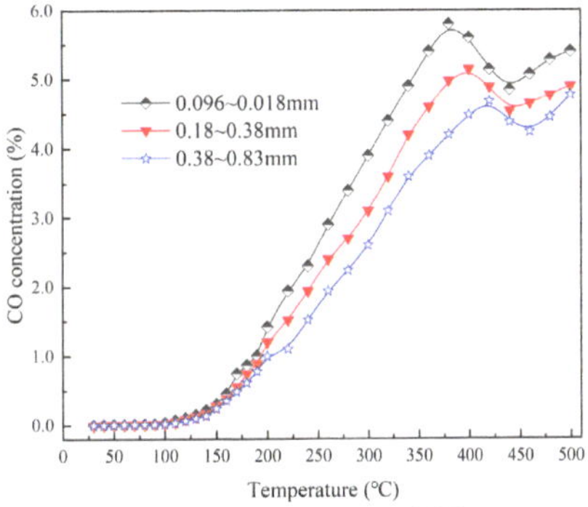

(**a**) The concentration of CO

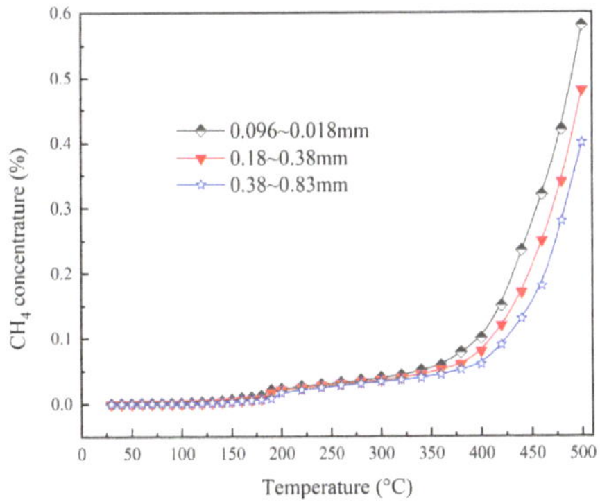

(**b**) The concentration of CH_4

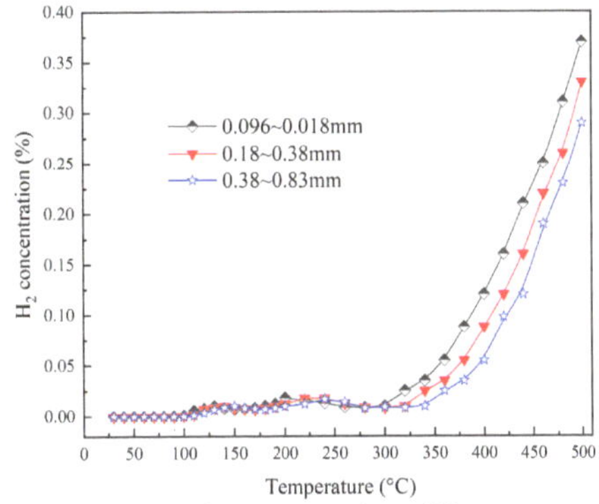

(**c**) The concentration of H_2

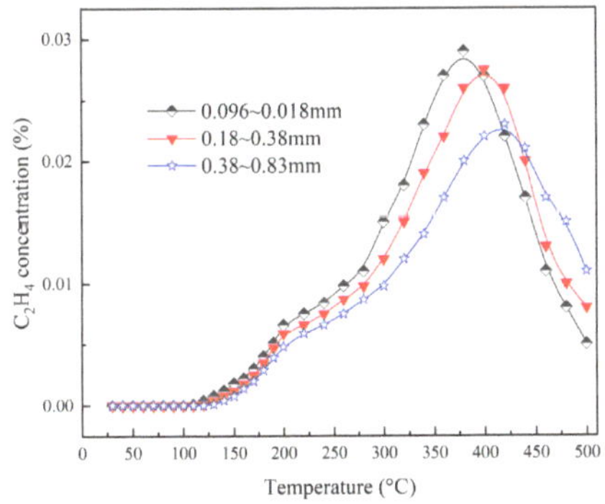

(**d**) The concentration of C_2H_4

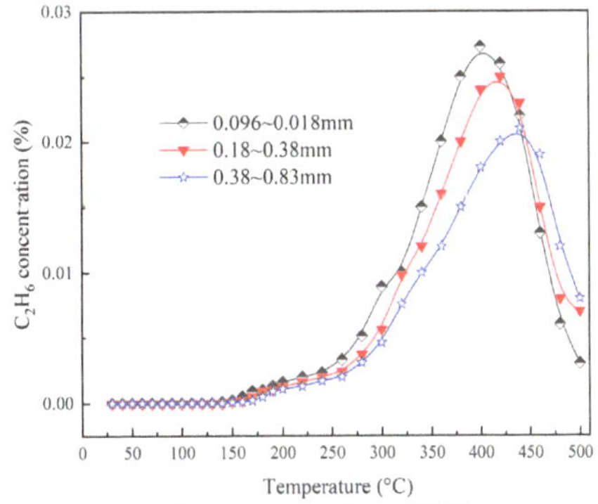

(**e**) The concentration of C_2H_6

Figure 2. The concentration of gas with temperature during the heating process.

Figure 2c presents H_2 under varying particle sizes. The graph illustrates a similar pattern across different coal particle sizes. Prior to reaching a temperature of 300 °C, the concentration of H_2 produced by coal samples of various sizes exhibits certain fluctuations and remains relatively low, suggesting a minimal impact of particle size within this temperature range. Additionally, the graph indicates a downward trend in H_2 concentration as the coal particle size increases. For instance, coal samples with particle sizes ranging from 0.096 to 0.18 mm produced an H_2 concentration of 0.37% at 500 °C, whereas those between 0.38 and 0.83 mm produced a CH_4 concentration of 0.28%. Figure 2d,e display the production volumes of C_2H_4 and C_2H_6 under different particle sizes, respectively. During the oxidation phase from 30 °C to 200 °C, the coal particle size exhibits little impact on the production of C_2H_4 and C_2H_6, with gas concentrations remaining low during this stage. Peak concentrations of gases produced by spontaneous coal combustion vary with particle size; for C_2H_4, the peak concentrations for particle sizes of 0.096~0.18 mm, 0.18~0.38 mm, and 0.38~0.83 mm are 0.029%, 0.027%, and 0.023%, respectively. Similarly, for C_2H_6, the peak concentrations are 0.027%, 0.025%, and 0.021%, respectively, demonstrating a gradual decrease in peak concentrations of both gases as coal particle size increases. Moreover, the peak temperatures for C_2H_4 and C_2H_6 significantly shift towards higher values with increasing coal particle sizes. For instance, for coal samples of 0.096~0.18 mm, the peak concentrations of both C_2H_4 and C_2H_6 occur at 380 °C, whereas for samples of 0.38~0.83 mm, the corresponding peak temperatures rise to 420 °C.

The total quantity of combustible gases produced during the coal oxidation process under various particle size conditions and the proportion of the main combustible gases is illustrated in Figure 3. As shown in Figure 3a, the total amount of combustible gases generated by coal of three different particle sizes begins to rise significantly after reaching 100 °C and continues to increase as the temperature rises. Comparing the production of combustible gases among different particle sizes, it is evident that as the particle size of the coal sample increases, the total volume of combustible gases gradually decreases. It also illustrates that different types of coal exhibit peak values between 350 °C and 450 °C, primarily due to a significant reduction in CO concentration during this phase. Comparing the production of combustible gases at particle sizes of 0.096~0.18 mm, 0.18~0.38 mm, and 0.38~0.83 mm, the peak concentrations are 6.02%, 5.37%, and 4.91%, respectively, indicating a gradual decrease in the peak concentration of combustible gases as coal particle size increases. Figure 3b–d display the percentages of the three primary combustible gases—CO, CH_4, and H_2—relative to the total combustible gas volume under different particle size conditions. It is observed that the proportional concentration of CO, CH_4, and H_2 follows a similar trend across varying particle sizes, accounting for over 99% of the combustible gases in the temperature range of 30 to 500 °C. A comparison of the production patterns of these three gases across different particle sizes reveals that particle size has a minimal impact on their proportion within the total volume of combustible gases.

This section examines the impact of coal particle size on the production of combustible gases—CO, CH_4, H_2, C_2H_4, and C_2H_6—during the heating process. It shows that larger coal particles generate lower concentrations of CO, CH_4, H_2, C_2H_4, and C_2H_6 due to smaller specific surface areas and lower porosity, reducing their interaction with oxygen. The peak concentrations of these gases shift to higher temperatures as particle size increases. Smaller particles produce higher concentrations of combustible gases, especially CO, CH_4, and H_2, with the most pronounced gas production occurring between 350 °C and 450 °C. Although the total volume of combustible gases decreases with increasing particle size, CO, CH_4, and H_2 make up over 99% of the total gas volume produced, regardless of particle size. The research conclusions of Zhang et al. [28–30] on the influence of coal particle size and coal temperature on combustible gases, such as CO, are similar to those in this paper.

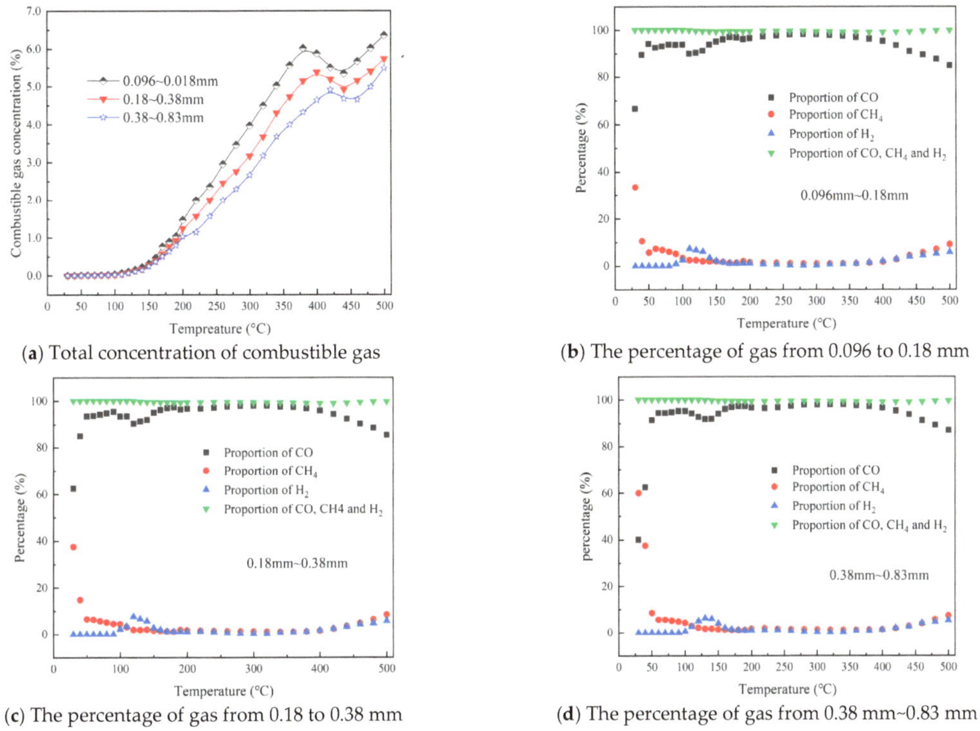

(**a**) Total concentration of combustible gas (**b**) The percentage of gas from 0.096 to 0.18 mm (**c**) The percentage of gas from 0.18 to 0.38 mm (**d**) The percentage of gas from 0.38 mm~0.83 mm

Figure 3. The percentage of combustible gas during the heating process with different particle sizes.

3.2. Segmental Analysis of Combustible Gas during the Coal Combustion Process

It is evident that the combustible gases produced during the coal self-ignition process primarily include CO, CH_4, and H_2. In the initial stages of coal self-ignition, the production of combustible gases is predominantly CO, and with the temperature of the coal rising to 300 °C, the release of CH_4 and H_2 significantly increases, with these stages primarily characterized by CH_4, CO, and H_2 as the main combustible gases. It indicates that the generation of combustible gases during coal self-ignition exhibits distinct segmental characteristics. The concentration and composition of these gases vary according to the different temperature stages of coal self-ignition. The concentration of combustible gases is segmented and fitted, as shown in Figure 4.

At temperatures below 300 °C, CO concentration rapidly increases as an exponential function; subsequently, due to a decrease in the content of carbonyl and aliphatic hydrocarbon functional groups in the coal, CO gas concentration exhibits fluctuating changes within the temperature range of 300 °C to 500 °C. For CH_4 and H_2, which are also combustible gases produced during the coal self-ignition process, their main production occurs at the high-temperature stages above 300 °C, with both gases increasing in an approximately exponential manner.

When the coal temperature is lower than 300 °C, the release law of CO gas produced by spontaneous combustion of coal with the coal temperature can be fitted as follows:

$$C_{CO} = A_0 \times \exp\left(-\frac{T}{t_0}\right) + y_0 \tag{1}$$

In the formula, A_0 and t_0 are exponential function fitting coefficients, and y_0 represents an exponential function fitting constant.

When the coal temperature is over 300 °C, the CO, CH_4, and H_2 gases generated by spontaneous combustion of coal can be fitted with the following formula, respectively:

$$C'_{CO} = B_0 + B_1 \cdot T + B_2 \cdot T^2 + B_3 \cdot T^3 + \cdots B_n \cdot T^n \tag{2}$$

$$C_{CH4} = A_1 \times \exp(-T/t_1) + y_1 \tag{3}$$

$$C_{H2} = A_2 \times \exp(-T/t_2) + y_2 \tag{4}$$

where B_0 is a polynomial function fitting constant, while B_1, B_2, B_3 ... B_n are the fitting coefficients of the polynomial function, respectively. A_1, A_2, t_1 and t_2 are exponential function fitting coefficients, while y_1 and y_2 are exponential function fitting constants.

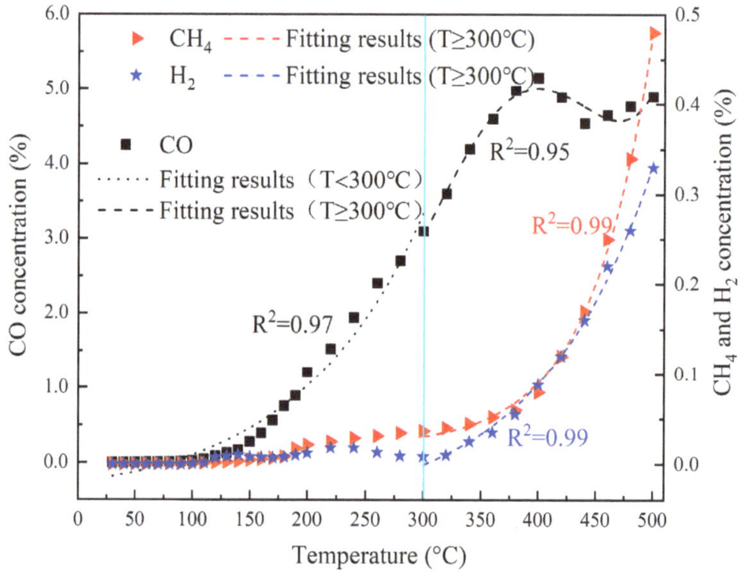

Figure 4. Fitting results of combustible gas production in the oxidation process.

The fitting results obtained for the generation of combustible gases during the coal oxidation process, as a function of temperature, are presented in Table 2. The results indicate that the correlation coefficients for the generated quantities of combustible gases and the temperature fitting equations are both above 0.95. Using these fitting equations, calculations were conducted on the generation of combustible gases during the oxidation process for different types of coals, and these were compared with experimental results, as shown in Figure 5. The established fitting equation has a high degree of fit with experimental data in different temperature ranges, indicating the reliability of the fitting equation and its ability to better capture the changes in the generation of combustible gases during coal spontaneous combustion.

Table 2. Fitting results of combustible gas generation concentration and temperature.

Coal	Temperature (T)	Fitting Equation
EMW	<300 °C	$C_{CO}(T) = 0.24475e^{(T/108.83329)} - 0.5101$, $R^2 = 0.97$
	≥300 °C	$C'_{CO}(T) = 185.87055 - 2.12727T + 0.00897T^2 - 1.62515 \times 10^{-5}T^3 + 1.62515 \times 10^{-8}T^4$, $R^2 = 0.95$ $C_{CH4}(T) = 3.77964 \times 10^{-5}e^{(T/53.10183)} + 0.01979$, $R^2 = 0.99$ $C_{H2}(T) = 0.00235e^{(T/99.44765)} - 0.05234$, $R^2 = 0.99$

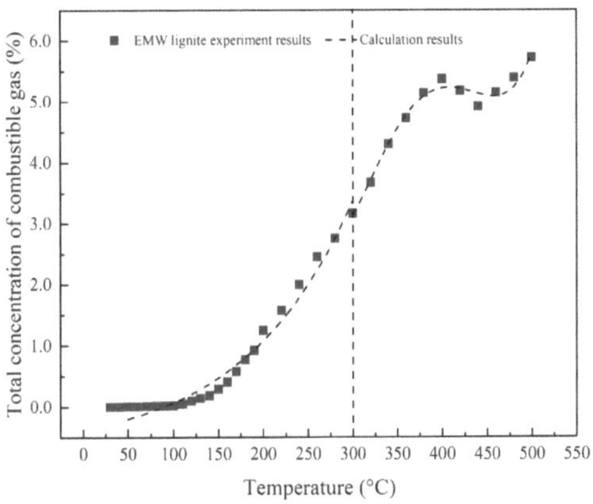

Figure 5. Comparison of experimental data and calculated results of the total concentration of combustible gas generated from EMW lignite.

According to Table 3, it can be concluded that the error between experimental data and calculated results gradually decreases with increasing temperature. When the temperature is below 100 °C, there is a significant difference between the fitted value and the experimental value. For example, at 50 °C, the experimental value is 0.0062, while the calculated value is −0.19842. This may be due to the complexity of oxidation reactions at low temperatures, as well as uncertainties such as instrument measurement errors and experimental environments. However, when the temperature reaches above 300 °C, the error between the calculated results and the experimental results gradually decreases, indicating that the fitting equation performs more accurately in the high-temperature stage.

Table 3. Comparison of experimental data and calculated results of the total concentration of combustible gas generated from EMW lignite.

Temperature/°C	Experiment Results	Calculation Results
50	0.0062	−0.19842
60	0.008	−0.15349
80	0.0137	−0.05171
100	0.023	0.06831
120	0.0953	0.20984
140	0.1826	0.37673
160	0.40532	0.57354
180	0.7709	0.80562
200	1.2442	1.0793
220	1.571	1.40203
240	1.9963	1.7826
260	2.4553	2.23137
280	2.757	2.76059
300	3.1628	3.38465 3.13495
320	3.6749	3.72034
340	4.3005	4.30055
360	4.7251	4.77585
380	5.141	5.0894
400	5.3695	5.22731
420	5.181	5.2186
440	4.913	5.13521
460	5.148	5.092
480	5.388	5.24678
500	5.723	5.79196

Although there is a large error in the low-temperature stage, in practical applications, the amount of gas generated in the low-temperature stage is relatively low, so the impact of this error on predicting the risk of coal spontaneous combustion is limited. Moreover, the overall trend of the fitted equation is highly consistent with the experimental results, verifying the feasibility of using this fitted equation for gas generation prediction.

This section analyzes the generation patterns of combustible gases (CO, CH_4, and H_2) during coal self-ignition and highlights distinct temperature-dependent phases. At temperatures below 300 °C, CO is the primary gas, increasing exponentially. As the temperature exceeds 300 °C, the production of CH_4 and H_2 rises exponentially, making them the dominant gases at higher temperatures. Fitting equations were used to simulate the release of these gases at different temperature stages, showing a strong correlation ($R^2 > 0.95$) between predicted and experimental results. The obvious segmented characteristics of combustible gases during coal spontaneous combustion can predict the danger level of coal mine spontaneous combustion based on changes in gas concentration, which is consistent with the research results of Guo et al. [31,32]. These equations provide a reliable method for predicting the concentration of combustible gases during coal oxidation, aiding in the prevention of coal combustion-related hazards.

4. Conclusions

This paper analyzes the primary sources and components of combustible gases in goaf areas where coal spontaneous combustion occurs. It focuses on the relationship between the concentration of combustible gases and temperature during coal oxidation, using a specially constructed high-temperature programmed heating experimental system for coal spontaneous combustion. This study clarifies the composition of combustible gases under different conditions, and the main conclusions are as follows:

1. To master the production patterns of combustible gases during coal oxidation, a high-temperature programmed heating test system for coal spontaneous combustion was

established. It mainly consists of a quartz reactor, a programmed heating system, an automatic gas distribution system, a data acquisition system, and a gas chromatograph, which enables the detection of gaseous products during the high-temperature stage of coal spontaneous combustion;

2. The influence of coal particle size on the production of CH_4, H_2, C_2H_4, and C_2H_6 is not significant at low temperatures but becomes considerably significant above 200 °C. As the coal particle size increases, the total amount of combustible gases gradually decreases; however, the impact of particle size on the percentage of CO, CH_4, and H_2 among the total combustible gases is minimal, with their concentration ratios reaching over 99% within the temperature range of 30 °C to 500 °C;
3. The generation of combustible gases during coal spontaneous combustion exhibits distinct phased characteristics. Initially, CO is the predominant combustible gas produced, but as the temperature rises above 300 °C, the release of CH_4 and H_2 significantly increases. During this phase, the primary combustible gases in the coal spontaneous combustion environment are CH_4, CO, and H_2. Based on these observations, functional equations relating the generation of combustible gases at different temperature stages to temperature were fitted and compared with experimental results, indicating a good agreement between the calculated results using the fitted equations and the experimental findings.

Author Contributions: D.M.: methodology, software, investigation, validation, visualization, data curation, and writing—original draft. L.Z.: supervision, conceptualization, resources, and writing—review and editing. T.Z.: supervision and resources. P.Y.: software, review, and editing. All authors have read and agreed to the published version of the manuscript.

Funding: This research was funded by National Natural Science Foundation of China grant number (52204250) and Independent Research Project of State Key Laboratory for Fine Exploration and Intelligent Development of Coal Resources, CUMT grant number (SKLCRSM23X007).

Institutional Review Board Statement: Not applicable.

Informed Consent Statement: Informed consent was obtained from all subjects involved in the study.

Data Availability Statement: The data are not publicly available due to commercial confidentiality, as they contain information that could compromise the privacy of research participants.

Conflicts of Interest: The authors declare no conflicts of interest.

References

1. Baris, K.; Kizgut, S.; Didari, V. Low-temperature oxidation of some Turkish coals. *Fuel* **2012**, *93*, 423–432. [CrossRef]
2. Wang, H.; Dlugogorski, B.Z.; Kennedy, E.M. Coal oxidation at low temperatures: Oxygen consumption, oxidation products, reaction mechanism and kinetic modelling. *Prog. Energy Combust. Sci.* **2003**, *29*, 487–513. [CrossRef]
3. Wu, Y.; Wu, S.; Gu, J.; Gao, J. Differences in physical properties and CO_2 gasification reactivity between coal char and petroleum coke. *Process Saf. Environ. Prot.* **2009**, *87*, 323–330. [CrossRef]
4. Zhang, D.; Cen, X.; Wang, W.; Deng, J.; Wen, H.; Xiao, Y.; Shu, C.M. The graded warning method of coal spontaneous combustion in Tangjiahui Mine. *Fuel* **2021**, *288*, 119635. [CrossRef]
5. Onifade, M.; Genc, B. A review of research on spontaneous combustion of coal. *Int. J. Min. Sci. Technol.* **2020**, *30*, 303–311. [CrossRef]
6. Li, Q.W.; Xiao, Y.; Zhong, K.Q.; Shu, C.M.; Lü, H.F.; Deng, J.; Wu, S. Overview of commonly used materials for coal spontaneous combustion prevention. *Fuel* **2020**, *275*, 117981. [CrossRef]
7. Kong, B.; Li, Z.; Yang, Y.; Liu, Z.; Yan, D. A review on the mechanism, risk evaluation, and prevention of coal spontaneous combustion in China. *Environ. Sci. Pollut. Res.* **2017**, *24*, 23453–23470. [CrossRef]
8. Adamus, A.; Šancer, J.; Guřanová, P.; Zubíček, V. An investigation of the factors associated with interpretation of mine atmosphere for spontaneous combustion in coal mines. *Fuel Process. Technol.* **2011**, *92*, 663–670. [CrossRef]
9. Szurgacz, D.; Tutak, M.; Brodny, J.; Sobik, L.; Zhironkina, O. The method of combating coal spontaneous combustion hazard in goafs—A case study. *Energies* **2020**, *13*, 4538. [CrossRef]
10. Zhang, M.; Wang, Z.; Wang, L.K.; Zhang, Z.; Zhang, D.; Li, C. Experimental study and thermodynamic analysis of coal spontaneous combustion characteristics. *Combust. Theory Model.* **2023**, *27*, 118–137. [CrossRef]

11. Yuan, L.; Smith, A. Experimental study on CO and CO_2 emissions from spontaneous heating of coals at varying temperatures and O_2 concentrations. *J. Loss Prev. Process Ind.* **2013**, *26*, 1321–1327. [CrossRef] [PubMed]
12. Green, U.; Aizenshtat, Z.; Hower, J.C.; Stark, S.; Weidner, C.; Cohen, H. Modes of formation of carbon oxides (CO_x (x = 1, 2)) from coals during atmospheric storage: Part 1: Effect of coal rank. *Energy Fuels* **2010**, *24*, 6366–6374. [CrossRef]
13. Green, U.; Aizenshtat, Z.; Hower, J.; Hatch, R.; Cohen, H. Modes of formation of carbon oxides (CO_x (x = 1, 2)) from coals during atmospheric storage: Part 2: Effect of coal rank on the kinetics. *Energy Fuels* **2011**, *25*, 5625–5631. [CrossRef]
14. Yang, Y.; Li, Z.; Tang, Y.; Liu, Z.; Ji, H. Fine coal covering for preventing spontaneous combustion of coal pile. *Nat. Hazards* **2014**, *74*, 603–622. [CrossRef]
15. Wen, H.; Yu, Z.; Deng, J.; Zhai, X. Spontaneous ignition characteristics of coal in a large-scale furnace: An experimental and numerical investigation. *Appl. Therm. Eng.* **2017**, *114*, 583–592. [CrossRef]
16. Wang, K.; Du, Y.; Sun, W.; Deng, J.; Zhang, Y. Coal Secondary Oxidation Thermodynamic Behaviors Affected by Pre-Heating at Different Oxygen Concentrations. *Combust. Sci. Technol.* **2023**, 1–13. [CrossRef]
17. Zhang, D.; Yang, X.; Deng, J.; Wen, H.; Xiao, Y.; Jia, H. Research on coal spontaneous combustion period based on pure oxygen adiabatic oxidation experiment. *Fuel* **2021**, *288*, 119651. [CrossRef]
18. Wen, H.; Wang, H.; Liu, W.; Cheng, X. Comparative study of experimental testing methods for characterization parameters of coal spontaneous combustion. *Fuel* **2020**, *275*, 117880. [CrossRef]
19. Dong, X.; Wen, Z.; Wang, F.; Meng, Y. Law of gas production during coal heating oxidation. *Int. J. Min. Sci. Technol.* **2019**, *29*, 617–620. [CrossRef]
20. Ren, L.; Yu, X.; Li, Q.; Tao, F.; Weng, T.F.; Zhai, X.W.; Ma, T. Thermodynamic characteristics of weakly caking coal oxidation and variation law of gaseous products in low oxygen concentration environment. *Case Stud. Therm. Eng.* **2024**, *62*, 10171. [CrossRef]
21. Sun, L.; Han, H.; Cheng, W.; Wang, H.; Shi, Q.; Qi, G.; Guo, Z. Exploring the factors influencing the low temperature CO gas production characteristics of low-metamorphic coal. *Fuel* **2024**, *359*, 13027. [CrossRef]
22. Yan, H.; Nie, B.; Liu, P.; Chen, Z.; Yin, F.; Gong, J.; Lin, S.; Wang, X.; Kong, F.; Hou, Y. Experimental assessment of multi-parameter index gas correlation and prediction system for coal spontaneous combustion. *Combust. Flame* **2023**, *247*, 112485. [CrossRef]
23. Mishra, D.P. Effects of intrinsic properties, particle size and specific surface area on WOP and spontaneous combustion susceptibility of coal. *Adv. Powder Technol.* **2022**, *33*, 103454. [CrossRef]
24. Wang, K.; Li, Y.; Zhang, Y.; Deng, J. An approach for evaluation of grading forecasting index of coal spontaneous combustion by temperature-programmed analysis. *Environ. Sci. Pollut. Res.* **2023**, *30*, 3970–3979. [CrossRef] [PubMed]
25. Küçük, A.; Kadıoğlu, Y.; Gülaboğlu, M.Ş. A study of spontaneous combustion characteristics of a Turkish lignite: Particle size, moisture of coal, humidity of air. *Combust. Flame* **2003**, *133*, 255–261. [CrossRef]
26. Singh, A.; Singh, R.; Singh, M.; Chandra, H.; Shukla, N.K. Mine fire gas indices and their application to Indian underground coal mine fires. *Int. J. Coal Geol.* **2007**, *69*, 192–204. [CrossRef]
27. Song, J.; Deng, J.; Zhao, J.; Zhang, Y.; Wang, C.; Shu, C.M. Critical particle size analysis of gas emission under high-temperature oxidation of weathered coal. *Energy* **2021**, *214*, 118995. [CrossRef]
28. Zhang, Y.; Chen, L.; Deng, J.; Zhao, J.Y.; Li, H.T.; Yang, H. Influence of granularity on thermal behaviour in the process of lignite spontaneous combustion. *J. Therm. Anal. Calorim.* **2019**, *135*, 2247–2255. [CrossRef]
29. Ren, L.; Li, Q.; Xiao, Y.; Hao, J.C.; Yi, X.; Zou, L.; Li, Z.B. Critical parameters and risk evaluation index for spontaneous combustion of coal powder in high-temperature environment. *Case Stud. Therm. Eng.* **2022**, *38*, 102331. [CrossRef]
30. Yan, H.; Nie, B.; Kong, F.; Liu, Y.; Liu, P.; Wang, Y.; Chen, Z.; Yin, F.; Gong, J.; Lin, S.; et al. Experimental investigation of coal particle size on the kinetic properties of coal oxidation and spontaneous combustion limit parameters. *Energy* **2023**, *270*, 126890. [CrossRef]
31. Guo, Q.; Ren, W.; Lu, W. Risk evaluation of coal spontaneous combustion from the statistical characteristics of index gases. *Thermochim. Acta* **2022**, *715*, 179287.
32. Chen, X.; Bi, R.; Huang, J.; Shan, W.; Xiao, J.; Wang, D. Experimental study on early prediction index gas for spontaneous combustion. *Energy Sources Part A Recovery Util. Environ. Eff.* **2020**, *46*, 7003–7017. [CrossRef]

fire

Article

Numerical Study on the Explosion Reaction Mechanism of Multicomponent Combustible Gas in Coal Mines

Dong Ma [1,*], Leilin Zhang [2], Guangyuan Han [3] and Tingfeng Zhu [4]

1 State Key Laboratory for Fine Exploration and Intelligent Development of Coal Resources, China University of Mining and Technology, Xuzhou 221008, China
2 Engineering Technology Research Centre for Safe and Efficient Coal Mining, Anhui University of Science and Technology, Huainan 232001, China; 2014017@aust.edu.cn
3 Xuzhou Coal Mining Group, Xuzhou 221004, China; hgycn@outlook.com
4 School of Safety Engineering, China University of Mining and Technology, Xuzhou 221116, China; ts23120206p31@cumt.edu.cn
* Correspondence: madongcumt@163.com

Abstract: Combustible gases, such as CO, CH_4, and H_2, are produced during spontaneous coal combustion in goaf, which may cause an explosion under the stimulation of an external fire source. It is of great significance to study the influence of combustible gases, such as CO and H_2, on the characteristics of a gas explosion. In this study, CHEMKIN software (Version 17.0) and the GRI-Mech 3.0 reaction mechanism were used to study the influences of different concentration ratios between CO and H_2 on the ignition delay time, free radical concentration, and key reaction step of a gas explosion. The results show that the increase in the initial CH_4 and CO concentrations prolonged the ignition delay time, while the increase in the H_2 concentration shortened the time and accelerated the explosion reaction. The addition of H_2 promoted the generation of free radicals (H·, O·, ·OH) and accelerated the occurrence of the gas explosion. CO generated ·OH free radicals and dominated the methane consumption through the R119 and R156 reactions. As the concentrations of CO and H_2 increased, the R38 reaction gradually became the main driving factor of the gas explosion.

Keywords: coal combustion; methane explosion; flammable gas; reaction mechanism

Citation: Ma, D.; Zhang, L.; Han, G.; Zhu, T. Numerical Study on the Explosion Reaction Mechanism of Multicomponent Combustible Gas in Coal Mines. *Fire* **2024**, *7*, 368. https://doi.org/10.3390/fire7100368

Academic Editors: Chuyuan Huang, Haipeng Jiang and Lijuan Liu

Received: 25 September 2024
Revised: 9 October 2024
Accepted: 15 October 2024
Published: 16 October 2024

1. Introduction

Gas explosions occur due to intense chemical reactions between methane and oxygen when ignited by an external source. The concentration range for methane explosions in air is generally between 5% and 16%. Sources of ignition for gas explosions in coal mines can include the spontaneous combustion of coal, electrical sparks, and heat generated by the friction of rock impacts [1–3]. The conditions conducive to gas explosions are not constant; they vary with environmental factors, such as the temperature and the addition of different gas mixtures [4,5]. Research showed that in addition to the CH_4 concentration, the presence of other combustible gases, such as CO and H_2, can also significantly alter the characteristics of explosions [6,7]. Especially hydrogen, due to its low ignition energy and high explosion rate, can significantly enhance the explosiveness of mixed gases, even in small amounts [8]. Research on the characteristics of gas explosions under various conditions not only aids in identifying the hazards associated with gas explosions but also facilitates the scientific development of explosion prevention measures [9,10]. In recent years, with the advancement of experimental techniques and numerical simulation methods, significant progress has been made in the study of gas explosion characteristics and their influencing factors. Numerical simulation methods, especially those based on chemical reaction kinetics models, such as GRI-Mech 3.0, have been able to simulate the explosion process of multi-component gases under different environmental conditions and reveal the generation and reaction pathways of free radicals. These studies provide an important theoretical basis for the development of explosion prevention and control measures, especially for the study of gas explosion characteristics in complex coal mine environments, which helps to more scientifically evaluate and prevent the occurrence of gas explosions.

In experimental studies of gas explosion characteristics, Cashdollar et al. [11] examined the impact of the reactor size, turbulence intensity, and ignition energy on the explosion limits of methane (CH_4). Cammarota et al. [12] collected data on the explosive behavior of CH_4-O_2 mixtures using photodiodes, finding that the combustion speed of the mixtures ranged from 0.7 to 10 m/s. Salzano et al. [13] conducted tests in a 5 L sealed cylindrical container to study the effect of hydrogen (H_2) on methane explosion characteristics, noting a significant impact when the H_2 content in the mixture exceeded 50%. Pekalski [14] and Gieras et al. [15] revealed through their experiments the influence of the initial temperature on the gas explosion limit parameters. Tang et al. [16] identified how methane explosion characteristic parameters change with the initial environmental pressure, demonstrating that these parameters decrease as the initial pressure is reduced. Huang et al. [17] investigated the explosion limits of methane–air mixtures under high-pressure conditions, determining that the explosion limits ranged from 2.93% to 60.75% at 30 MPa and proposed a high-pressure explosion limit model for methane–air mixtures. Takahashi et al. [18] studied the effects of different ignition materials on the explosion limits, finding that metals with higher melting points and higher voltages were more suitable for measuring explosion limits. To predict the explosion limits of combustible gas mixtures, some researchers derived formulas based on combustion kinetics or the heat and mass transfer relationships between reacting gases, which can predict the explosiveness of combustible gases to some extent [19–21].

In numerical simulation studies of gas explosion characteristics, Kobiera et al. [22] developed a model describing the explosion process in enclosed pipelines that is capable of simulating explosions under different flame front and turbulence intensity conditions. Maremonti et al. [23] simulated the explosion characteristics of gas in interconnected containers, identifying turbulence effects caused by the connection as the main factor enhancing explosions. Salnazo et al. [24] used computational fluid dynamics to simulate the flame acceleration behavior of gas in pipelines, validating the numerical model by comparing it with experimental results and finding it was well suited for slower turbulent deflagration states. Zhang et al. [25] studied the effect of ignition energy on the characteristics of gas explosions by establishing a three-dimensional geometric model, concluding that ignition energy only has a certain impact on the process of gas explosions. Yang et al. [26] used the OpenFOAM code to study the characteristics of gas explosions with non-uniform concentration gradients in mine tunnels and found that non-uniform gas explosions are more destructive than uniform gas explosions. Qiu et al. [27] studied the effects of different elbows on the propagation characteristics of methane–air premixed explosions through a combination of numerical simulation and experimental comparison and found that the attenuation of the explosion shock wave is affected by the bending angle.

In terms of the reaction mechanisms of gas combustion and explosions, researchers conducted in-depth studies and developed various mechanisms, including GRI-Mech 3.0 [28–30]. The conclusions drawn under laboratory conditions cannot fully simulate the complex situations of actual mines, and due to the limitations of experimental conditions and resources, the small sample size of data leads to low accuracy in the research results. Numerical simulation methods have revealed deeper insights into the process and intrinsic mechanisms of gas explosions, somewhat compensating for the limitations of experimental research. Although there was some research on the impact of gases, such as CO and H_2, on gas explosions, there has been a lack of systematic studies on the comprehensive effects of multi-component gas mixtures. Therefore, the research in this study is of certain significance for preventing coal mine gas explosions.

In this work, CHEMKIN and GRI-Mech 3.0 were used to analyze the impact of different concentration ratios of CO and H_2 on the ignition delay time, free radical concentration, and key reaction step of the gas explosion. The sensitivity coefficient was calculated to explain the effects of various gas concentrations on the reaction. It is of great importance to illustrate the influence of different gas concentrations on gas explosions.

2. Methodology

2.1. Selection of Reaction Mechanism

Gas explosions occur through a series of elementary reactions. In the environment of a coal mine goaf where spontaneous coal combustion occurs, combustible gases, such as CO, CH_4, and H_2, are produced, necessitating the selection of a reaction mechanism suitable for the explosion of multi-component mixed gases in the goaf. So far, the chemical mechanisms of combustion for gases like CO, CH_4, and H_2 have been extensively studied and the established reaction mechanisms are relatively mature. For this simulation, the GRI-Mech 3.0 reaction mechanism was employed, which includes 53 components and 325 elementary reactions. This mechanism is applicable to reaction systems within a temperature range of 1000–2000 K, a pressure range of 10 torr to 10 atm, and an equivalence ratio range of 0.1 to 5. This mechanism encompasses a detailed array of elementary reactions and has a broad applicability, enabling the analysis of combustion reaction kinetics for CH_4, CO, H_2, and mixed gases [31]. Here, a zero-dimensional constant-volume combustion reactor model was used for the simulation. This model represented a constant-volume adiabatic reaction system, where the combustible components and oxygen could mix completely and sufficiently, and were in a static, non-turbulent state. The technology roadmap of this study is shown in Figure 1.

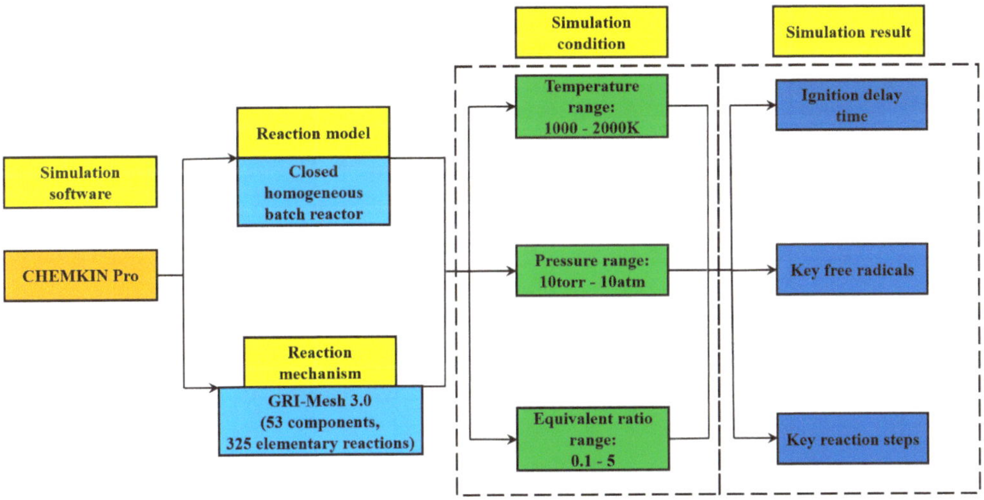

Figure 1. The technology roadmap of this study.

2.2. Initial Computational Conditions

The influence of different CO and H_2 concentration ratios on the reaction kinetics during the gas explosion process was investigated using the CHEMKIN software. The simulation assumed that the air composition was solely composed of O_2 and N_2. The gases selected for the simulation included O_2, N_2, CH_4, CO, and H_2. The initial conditions were set at a temperature of 1000 K and a pressure of 0.1 MPa, with a reaction duration of 2.0 s. To maintain consistency with experimental setups, the reactor was configured as a 20 L constant volume environment. The simulation was designed with two sets of experiments: one involved a mixture of gas and CO, with concentration ranges from 0 to 5%, and the other involved a mixture of gas, CO, and H_2, also with concentrations ranging from 0 to 5%. The specific parameters are outlined in Table 1.

Table 1. Parameters of the initial calculation conditions.

CH_4 Concentration (%)	Concentration of CO and H_2 (%)	Ratio 1 (Excluding H_2)			Ratio 2			
		CO (%)	O_2 (%)	N_2 (%)	CO (%)	H_2 (%)	O_2 (%)	N_2 (%)
7	0	0	19.53	73.47	0	0	19.53	73.47
	1	1	19.32	79.68	1	0	19.32	79.68
	2	2	19.11	78.89	1.9	0.1	19.11	78.89
	3	3	18.9	78.1	2.8	0.2	18.9	78.1
	4	4	18.69	77.31	3.5	0.5	18.69	77.31
	5	5	18.48	76.52	4	1.0	18.48	76.52
9.5	0	0	19.005	71.495	0	0	19.005	71.495
	1	1	18.795	70.705	1	0	18.795	70.705
	2	2	18.585	69.915	1.9	0.1	18.585	69.915
	3	3	18.375	69.125	2.8	0.2	18.375	69.125
	4	4	18.165	68.335	3.5	0.5	18.165	68.335
	5	5	17.955	67.545	4	1.0	17.955	67.545
11	0	0	18.69	70.31	0	0	18.69	70.31
	1	1	18.48	69.52	1	0	18.48	69.52
	2	2	18.27	68.73	1.9	0.1	18.27	68.73
	3	3	18.06	67.94	2.8	0.2	18.06	67.94
	4	4	17.85	67.15	3.5	0.5	17.85	67.15
	5	5	17.64	66.36	4	1.0	17.64	66.36

2.3. Sensitivity Analysis

Sensitivity analysis was employed to elucidate the extent to which the rate constants of the chemical reactions influenced the concentration variations of reactants within a system. This was achieved by calculating the first-order sensitivity coefficients of each elementary reaction rate constant with respect to the system parameters, such as the temperature, concentrations of the intermediate products, and concentrations of the final products. Suppose the governing equation for a variable Z is given by

$$\frac{dZ}{dt} = F(Z, t, a) \tag{1}$$

where Z (Z_1, Z_2, . . ., Z_i) is the mass fraction of each component in the system, %; a (a_1, a_2, . . ., a_i) is the pre-exponential factor of each primitive reaction step.

The first-order sensitivity coefficient can be calculated as follows:

$$W_{l,i} = \frac{\partial Z_i}{\partial a_i} \tag{2}$$

Taking the derivative of this yields

$$\frac{dw_{l,j}}{dt} = \frac{\partial F_l}{\partial Z} W_{l,j} + \frac{\partial F_l}{\partial a_i} \tag{3}$$

3. Results and Analysis

3.1. Impact of Coal Self-Combustion Environmental Mixed Gas on the Ignition Delay Time of Gas Explosions

The ignition delay time is the duration before combustion wherein a flammable gas and air mixture remains unignited under specific temperature and pressure conditions. This metric reflects the physicochemical characteristics of a mixed gas fuel. Simulations were conducted to determine the changes in the ignition delay time when CO or a CO and H_2 mixed gas was added to methane gas concentrations of 7%, 9.5%, and 11%. The results are presented in Table 2.

Table 2. Data of gas explosion ignition delay time (initial temperature 1000 K).

CH_4 Concentration (%)	Combustible Gas Concentration (%)	Ignition Delay Time (s)	
		Add CO Gas	Add CO and H_2 Gas Mixture
7	0	0.902	0.902
	1	1.106	1.106
	2	1.281	0.806
	3	1.436	0.600
	4	1.577	0.246
	5	1.707	0.092
9.5	0	1.067	1.067
	1	1.307	1.307
	2	1.527	1.060
	3	1.733	0.868
	4	1.929	0.428
	5	2.116	0.181
11	0	1.162	1.162
	1	1.418	1.418
	2	1.659	1.200
	3	1.890	1.020
	4	2.113	0.544
	5	2.328	0.246

Figure 2 illustrates the relationship between the ignition delay times and varying concentrations of added combustible gases. For pure methane gas at concentrations of 7%, 9.5%, and 11%, the ignition delay times were recorded at 0.901 s, 1.067 s, and 1.161 s, respectively. This indicates that higher methane concentrations correlated with increased ignition delay times. In the systems that contained methane mixed with CO, the addition of CO consistently resulted in a linear increase in the ignition delay times. Moreover, higher initial methane concentrations amplified the increase in the ignition delay times, as depicted by the steeper slope of the curve in Figure 1. The analysis suggests that both the increase in the initial methane concentration and the addition of the CO gas extended the ignition delay times of the methane explosions, which exerted a suppressive effect on the initial ignition phase.

For the combustible gas system comprising methane mixed with CO and H_2, the ignition delay times initially increased and then rapidly decreased as the concentration of the mixed gas rose. For example, at a mixed gas concentration of 2%, the presence of H_2 significantly reduced the ignition delay time from 1.281 s to 0.806 s for the system with a 7% methane concentration. As the H_2 content in the mixed gas concentration increased from 2% to 5%, the reductions in the ignition delay times were 37.1%, 58.2%, 84.4%, and 98.3%, respectively. Similarly, the ignition delay times for the methane explosions at concentrations of 9.5% and 11% also demonstrated consistent trends. These results indicate that the presence of H_2 in the gas mixture substantially reduced the ignition delay times, accelerating the process of the gas explosion reactions.

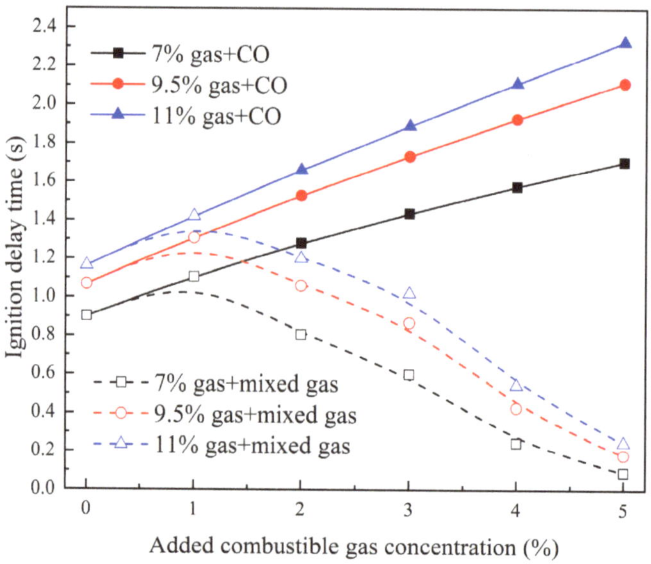

Figure 2. Relationship between ignition delay time and concentration of added combustible gas.

3.2. Effect of Gas Mixture in Coal Spontaneous Combustion Environment on Concentration of Key Free Radicals in Gas Explosion

The effects of CO on H·, O·, and ·OH in the gas explosion process of different concentrations were simulated, and the 7% concentration of gas was taken as an example for analysis, as shown in Figure 3. As can be seen from Figure 2, for the gas with a 7% concentration, the peak molar fractions of three key free radicals showed a similar change relationship with the addition amount of CO, that is, with the increase in the CO concentration, the peak molar fractions of H·, O·, and ·OH all showed an increasing trend, in which the H· radical increased the fastest, while the O· and ·OH radicals increased relatively little. By comparing the molar fractions of H·, O·, and ·OH, it can be found that the residual concentration of the ·OH radical was higher than that of H· and O· in the same reaction system. For example, in the reaction system of the 7% pure gas, the molar fractions of the H· and O· free radicals that remained in the explosion reaction were 0.0012 and 0.003, and the molar fractions of the ·OH free radicals that remained in the explosion reaction were 0.011. This was because more H· and O· free radicals that remained in the chain reaction system could combine to produce ·OH free radicals, which resulted in a higher ·OH free radical concentration after the gas explosion. In addition, it can also be seen from Figure 3 that with the increase in the CO concentration, the peak time of the three key free radicals gradually moved backward. For example, when the concentration of CO increased from 1% to 5%, the time for the H· radical to reach its peak concentration increased from 1.106 s to 1.707 s, indicating that the addition of CO delayed the reaction time of the gas explosion, which was consistent with the above simulation results on the ignition delay time of the gas explosion.

The effect of the gas mixture on the concentration of key free radicals in the gas explosion process with a 7% concentration is shown in Figure 4. Similar to the effect of CO on the gas explosion free radicals, the peak molar fractions of H·, O·, and ·OH also showed an increasing trend with the increase in the mixed gas concentration, in which the H· radical increased the fastest, while the O· and ·OH radicals increased relatively little. Different from the effect of CO on the free radicals in the gas explosion, H_2 was mixed in the experimental group when the concentration of the gas mixture was 2%, which resulted in earlier peak molar fractions of H·, O·, and ·OH free radicals. In addition, with the increase in the H_2 concentration in the mixed gas, the advance of the peak time of free radicals became more significant. This phenomenon also demonstrated that the H_2 served as an accelerator gas capable of markedly accelerating the process of the gas explosion.

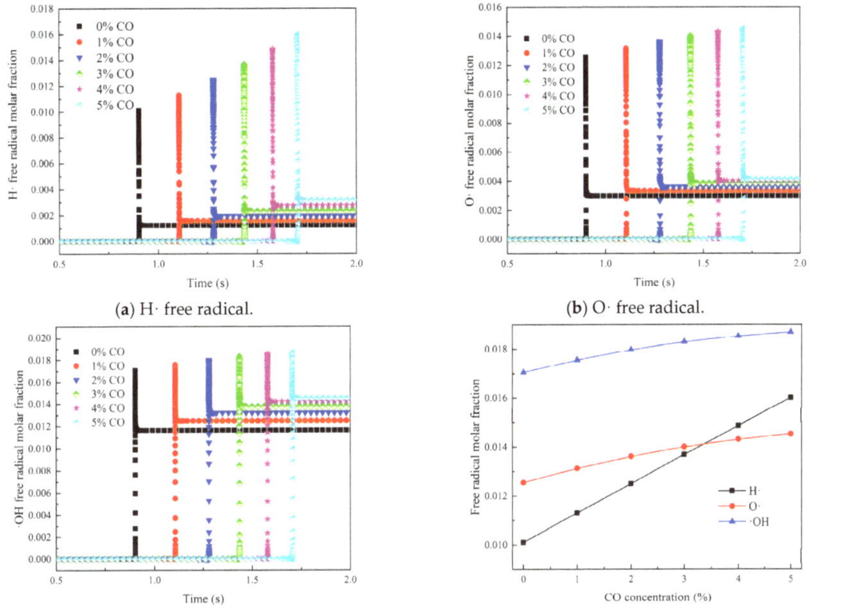

(**a**) H· free radical. (**b**) O· free radical.

(**c**) ·OH free radical. (**d**) Peak free radical concentration.

Figure 3. The influence of CO on H·, O·, and ·OH in the process of the 7% methane explosion.

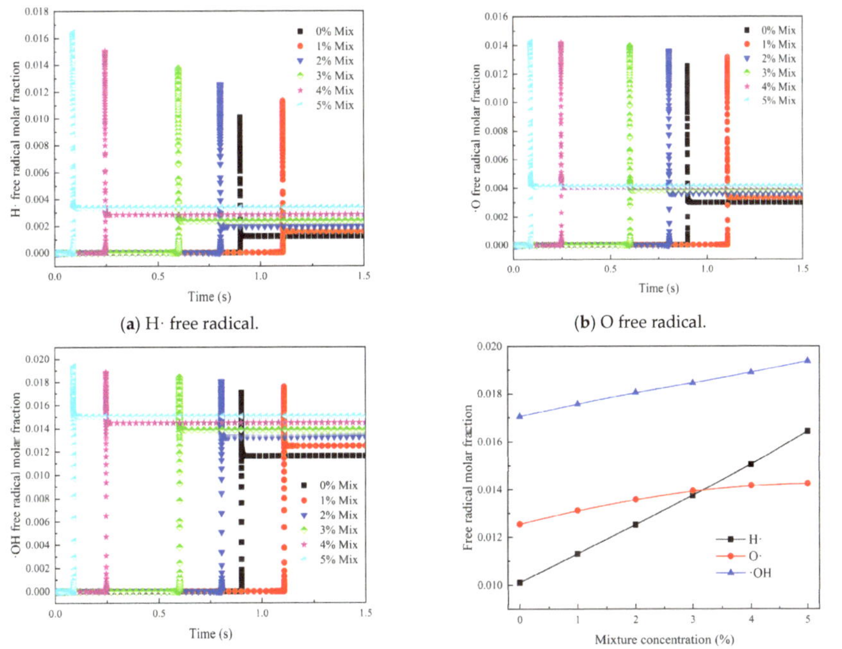

(**a**) H· free radical. (**b**) O free radical.

(**c**) ·OH free radical. (**d**) Peak free radical concentration.

Figure 4. The influence of the concentration of CO and H_2 mixed gases on H·, O·, and ·OH in the 7% methane explosion process.

Figure 5 shows the comparison of H·, O·, and ·OH free radicals in the gas explosion process by adding CO mixed with CO and H_2. As can be seen from the figure, for the 7% gas, the addition of combustible gas could promote the generation of three kinds of free radical concentrations of H·, O·, and ·OH, and their molar fractions gradually increased with the increase in the combustible gas concentration. At the same concentration, the promoting effect of mixed gas on the formation of H· and ·OH was stronger, but the promoting effect on the formation of O· was weaker than that of CO. For the 9.5% gas, the H· free radical showed an increasing trend with the increase in the combustible gas concentration, O· showed a first increasing and then decreasing trend, and ·OH showed a decreasing trend. Comparing the effect of the mixed gas and CO gas on the gas explosion, it can be found that the mixed gas had a stronger promoting effect on the generation of the H· free radical.

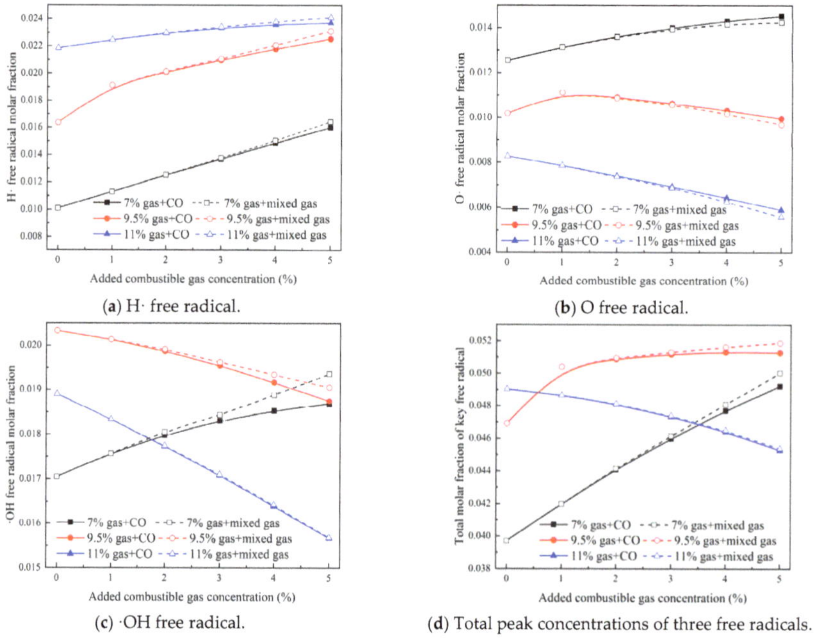

(**a**) H· free radical. (**b**) O free radical.

(**c**) ·OH free radical. (**d**) Total peak concentrations of three free radicals.

Figure 5. The influence of CO or CO and H_2 on H·, O·, and ·OH during the methane explosion.

For the 11% gas, the addition of combustible gas significantly promoted the formation of the H· free radical, but significantly inhibited the formation of the O· and ·OH free radicals. Comparing the effects of the mixed gas and CO gas on the gas explosion, it can be found that the mixed gas had a strong promotion effect on the generation of the H· free radicals, a relatively strong inhibition effect on the reduction of the O· free radicals, and the change in the mole fraction of the ·OH free radicals was similar to that of CO. In addition, it can also be found from Figure 5d that the peak total molar fraction of three key free radicals, H·, O·, and ·OH, in the gas mixture was larger than that in the explosive reaction with only CO added, indicating that the addition of H_2 in the gas mixture accelerated the generation of key free radicals in the gas explosion reaction, which promoted the gas explosion process.

3.3. *Influence of the Environmental Mixed Gas from Coal Self-Ignition on the Key Reaction Steps of Methane Consumption*

Figure 6 illustrates the impact of CO on the sensitivity of key elementary reactions in a 7% methane environment. As shown in Figure 6a, the elementary reactions that influenced the methane generation and consumption included R32, R53, R119, R156, R158, and R161.

Among these, reactions R53 and R158 exhibited positive sensitivity coefficients, indicating that they inhibited the methane consumption. Conversely, reactions R32, R119, R156, and R161 displayed negative coefficients, suggesting that they accelerated the methane consumption and facilitated the progression of the system's reactions. With the increased CO concentration in the methane reaction system, the sensitivity of these elementary reactions to methane was significantly enhanced, suggesting that the addition of CO intensified the influence of these reactions on methane. The introduction of CO in the system altered the elementary reactions that affected the methane consumption and generation. As the CO concentration increased from 1% to 5%, the elementary reactions that impacted the methane generation shift from R53 and R158 to R53, R120, and R158, and finally to R98, R120, and R158. Reaction R158, crucial for chain scission, impedes methane explosions by destroying chain carriers. Reactions R53 and R98, which consume H· and ·OH radicals, respectively, inhibit methane explosions by depleting reactive radicals. Although reaction R120 generates ·OH radicals, its overall influence on CO consumption and the interaction with various elementary reactions result in a suppressive effect on methane consumption.

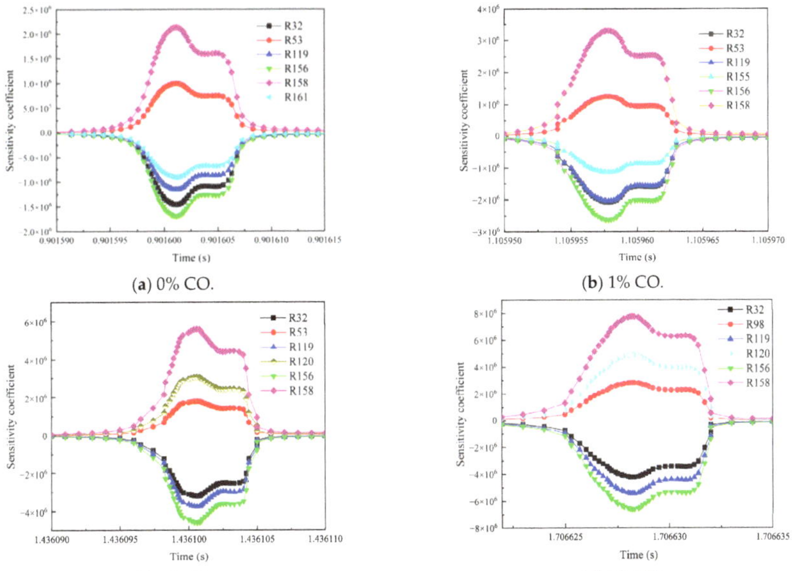

(**a**) 0% CO. (**b**) 1% CO.

(**c**) 3% CO. (**d**) 5% CO.

Figure 6. Influence of CO on the reaction sensitivity of the 7% gas key elements.

In this study of elemental reactions that affected the gas consumption, it was observed that with an increased CO concentration, the key elemental reactions influenced the gas consumption transition from R32, R119, R156, and R161 to R32, R119, R155, and R156, and eventually to R32, R119, and R156. The introduction of CO shifted the primary elemental reactions that promoted gas explosions from R32 and R156 to R119 and R156. Notably, R119, characterized by the reaction $HO_2 + CH_3$ <=> $OH + CH_3O$, facilitates the production of ·OH radicals. Similarly, R156, described by the reaction $CH_3 + O_2$ <=> $OH + CH_2O$, also generates ·OH radicals. Consequently, for the gas mixture with a 7% gas content, the addition of CO made R119 and R156 the most critical reactions that impacted the gas consumption by promoting gas explosions through the increased generation of ·OH radicals.

Figures 7 and 8 illustrate the impact of CO on the gas sensitivity of the critical elementary reactions for the gas concentrations of 9.5% and 11%. The figures indicate that for both concentrations, the elementary reactions R32, R53, R119, R156, R158, and R161 significantly influenced the gas generation and consumption. Notably, reactions R53 and

R158 had the most substantial impacts on the gas generation, while reactions R32, R119, R156, and R161 predominantly affected the gas consumption. A comparison with the primary elementary reactions in a 7% gas explosion revealed that as the concentration of gas increased, the major elementary reactions that affected the gas consumption remained unchanged; however, their sensitivity coefficients increased. For a 9.5% gas mixture, when the CO concentration was increased from 1% to 5%, the key elementary reactions that influenced the gas consumption remained consistent with those in the absence of CO, but their corresponding sensitivity coefficients increased, indicating that the higher CO concentrations progressively amplified the impacts of these reactions on the gas consumption. For an 11% gas mixture, increasing the CO concentrations in the reaction system also significantly heightened the sensitivity of the elementary reactions that affected the gas consumption, suggesting that the addition of CO intensified the influence of these reactions on the gas consumption at this concentration. Similar to the conditions at 7% and 9.5% gas concentrations, the addition of CO shifted the primary elementary reactions that facilitated the gas explosion from R32 and R156 to R119 and R156.

Simulations of the sensitivity effects of a CO and H_2 mixed gas on the gas consumption under different initial computational conditions are depicted in Figures 9–11. The sensitivity analysis results for the mixed gas concentrations of 0% and 1% were identical to those for CO's impact on the gas, hence only the results for the 3% and 5% mixed gas concentrations are presented here. This analysis particularly focused on the influence of H_2 in the mixed gas on the key reaction steps in the gas consumption.

As shown in Figure 9, the critical elementary reactions that affected the gas consumption were reactions R32, R38, R119, R120, R156, and R158, where R120 and R158 had positive sensitivity coefficients, indicating their major roles in the gas production. Conversely, reactions R38, R119, R120, and R156, with negative coefficients, significantly influenced the gas consumption. By comparing the elementary reactions influenced by the pure CO gas, it is evident that the principal elementary reactions affected by the mixed gas shifted from R156 to R38. Reaction R38, represented by $H + O_2 <=> O + OH$, produces highly reactive O· and ·OH radicals that promote the branched chain reactions leading to gas explosions, accelerating the explosion process. As the concentration of mixed gas in the reaction system increased, the sensitivity of these critical elementary reactions to gas consumption showed a general trend of weakening, suggesting that the addition of mixed gas reduced the influence of these reactions on the gas consumption.

In the suppression of the gas consumption, the elementary reactions R158, R120, and R98 dominated as the concentration of the mixed gas increased from 3% to 5%. However, the extent of their impact on the gas consumption changed with increased gas concentrations. For the reactions that promoted the gas consumption, the relevant elementary reactions shifted from R119, R38, and R156 to R38, R119, and R84 as the concentration increased from 3% to 5%. Notably, the most influential reaction changed from R119 to R38, indicating that higher concentrations of mixed gases altered the key reaction steps that affected the gas consumption, where R38 had the greatest impact. In comparison with the main reactions at the 5% CO concentration that influenced the gas consumption, the R38 reaction, which produces O· and ·OH radicals, had a more pronounced effect on enhancing the gas explosions, making it the most significant contributor to gas explosion enhancement within the mixture.

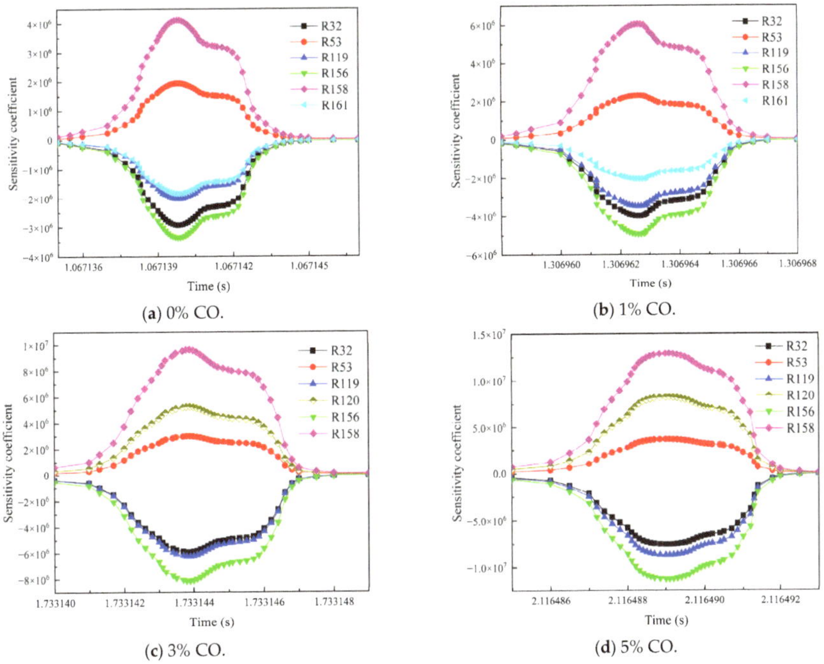

(**a**) 0% CO. (**b**) 1% CO.

(**c**) 3% CO. (**d**) 5% CO.

Figure 7. Effect of CO on the gas sensitivity of the 9.5% gas key element reaction.

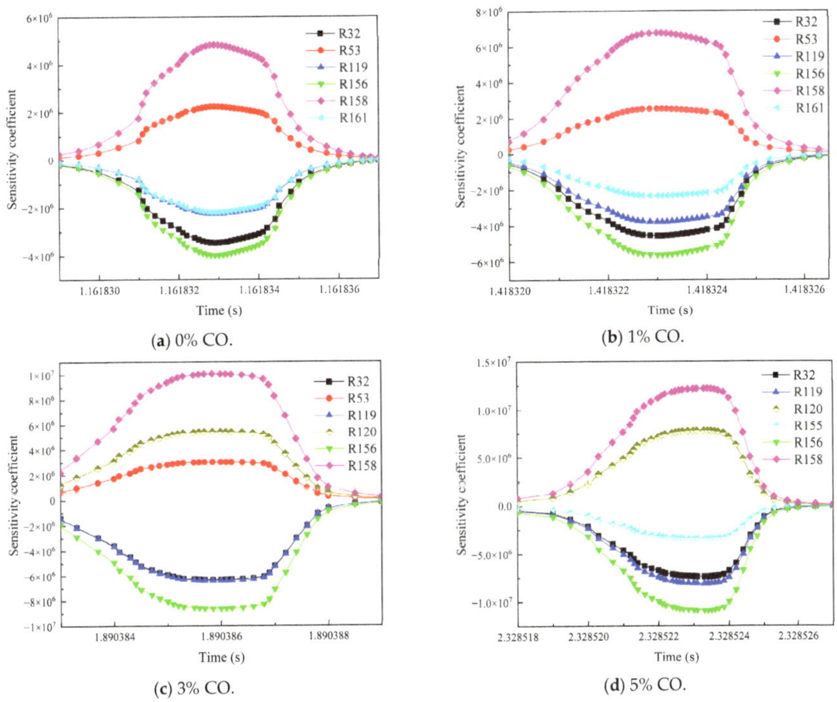

(**a**) 0% CO. (**b**) 1% CO.

(**c**) 3% CO. (**d**) 5% CO.

Figure 8. Effect of CO on gas sensitivity of 11% gas key elements.

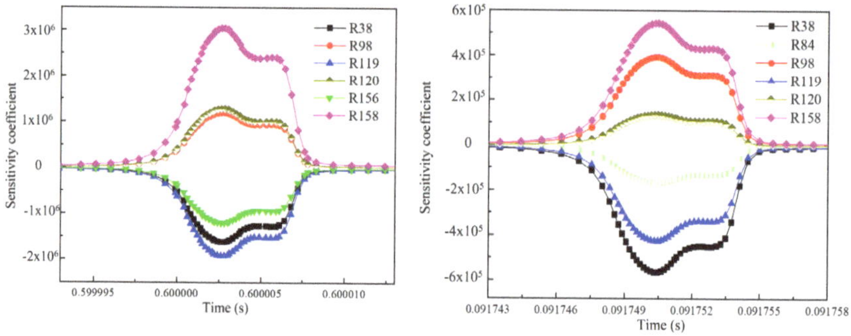

(**a**) 3% gas mixture. (**b**) 5% gas mixture.

Figure 9. Influence of the CO and H_2 mixture on the sensitivity of 7% gas key elements.

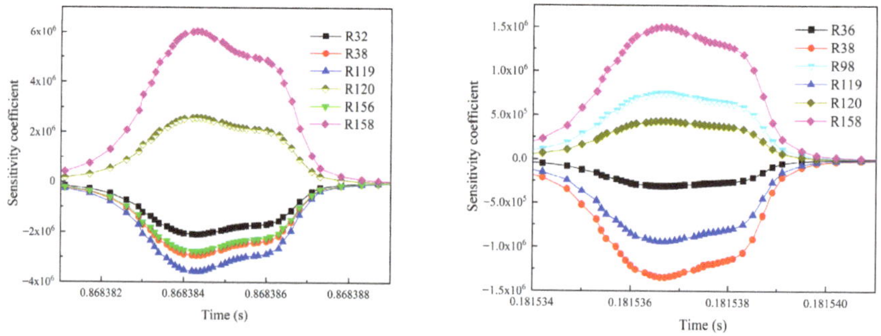

(**a**) 3% gas mixture. (**b**) 5% gas mixture.

Figure 10. Influence of the CO and H_2 mixture on the reaction sensitivity of the 9.5% gas key elements.

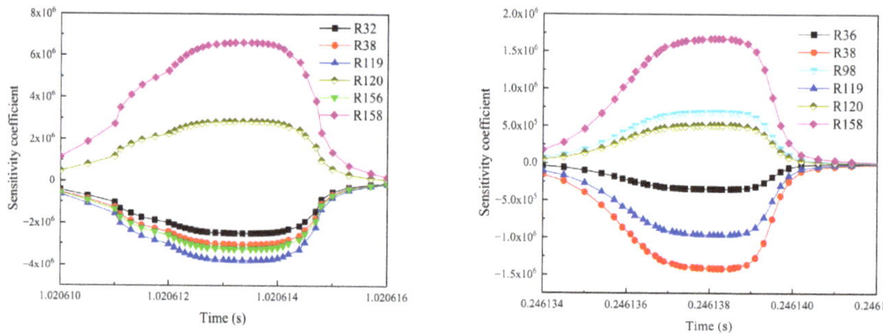

(**a**) 3% gas mixture. (**b**) 5% gas mixture.

Figure 11. Influence of CO and H_2 mixture on the sensitivity of the 11% gas key elements.

Figures 10 and 11 illustrate the impact of a mixed gas on the key elementary reactions sensitive to the 9.5% and 11% concentrations of gas, respectively. From the diagrams, it is evident that for the 9.5% gas mixture, as the concentration of the mixed gas increased, the key elementary reactions that influenced the gas formation and consumption underwent changes. Specifically, the reactions critical to gas production shifted from R120 and R158 to R98, R120, and R158. Meanwhile, those vital for gas consumption changed from R32, R156, R38, and R119 to R36, R119, and R38. Additionally, the increase in the mixed gas concentration enhanced the influence of reaction R38 on the gas consumption, indicating an elevated impact of this reaction step. For the 11% gas mixture, the key elementary reactions that affected the gas formation and consumption remained identical to those in the 9.5%

scenario; however, the degree of influence of each reaction on the gas consumption varied slightly. Moreover, as the concentration of the mixed gas rose to 5%, reaction step R38 progressively became the most influential in terms of the gas consumption.

4. Conclusions

This study used CHEMKIN software and the GRI-Mech 3.0 reaction mechanism to simulate in detail the explosion characteristics of methane–air–CO–H_2 mixed gas in a coal spontaneous combustion environment. Through simulation experiments, the specific effects of different concentrations of CO and H_2 on the ignition delay time, free radical generation, and reaction mechanism were analyzed. The following were the main conclusions:

(1) At the CH_4 concentrations of 7%, 9.5%, and 11%, the ignition delay time was significantly prolonged as the CO concentration increased from 1% to 5%. The presence of CO suppressed the initial ignition stage of the explosion and delayed the start of the combustion chain. When H_2 was added, the result was completely the opposite, with a significant reduction in the ignition delay time. H_2 had a significant acceleration effect in the mixed gases, where it greatly reduced the ignition time and increased the risk of explosion.

(2) As the concentration of the mixed gas increased, the peak molar fractions of three critical free radicals—H·, O·, and ·OH—also showed an increasing trend, with their occurrence times advancing. The total peak molar fractions of these critical free radicals—H·, O·, and ·OH—in the mixed gas were greater than in the reactions that involved only CO, indicating that the integration of H_2 into the mixed gas promoted the generation of key free radicals, which facilitated the occurrence of gas explosions.

(3) As the concentration of CO increased, R119 and R156 became the dominant reaction steps, which promoted the gas explosion by generating more ·OH radicals. The addition of H_2 changed the main reaction pathway, and the sensitivity of various elemental reactions that affected the gas consumption showed a general trend of weakening. R38 became the most critical reaction step. This reaction accelerated the explosive chain reaction by generating O· and ·OH radicals, which significantly enhanced the explosiveness of the mixed gas.

(4) The explosive behavior of the multi-component mixed gases was significantly influenced by the concentrations of CO and H_2. In actual coal spontaneous combustion zones, the presence of CO prolongs the reaction time of the explosion, and the addition of H_2 significantly shortens the ignition delay time, accelerating the generation of free radicals and the explosion chain reaction. This discovery has important guiding significance for the prevention and control of gas explosions in actual coal mines. In practical applications, special attention must be paid to the presence of H_2, as even small concentrations of H_2 may significantly increase the risk of explosion.

Author Contributions: D.M.: methodology, software, investigation, validation, visualization, data curation, and writing—original draft. L.Z.: supervision, conceptualization, resources, writing—review and editing. G.H.: supervision and resources. T.Z.: software, writing—review and editing. All authors have read and agreed to the published version of the manuscript.

Funding: This research was funded by [National Natural Science Foundation of China] grant number (52204250).

Institutional Review Board Statement: Not applicable.

Informed Consent Statement: Informed consent was obtained from all the subjects involved in this study.

Data Availability Statement: The data are not publicly available due to commercial confidentiality, as they contain information that could compromise the privacy of the research participants.

Acknowledgments: This work was supported by the National Natural Science Foundation of China (no. 52204250).

Conflicts of Interest: Guangyuan Han is employed by Xuzhou Coal Mining Group. The authors declare no conflicts of interest.

References

1. Li, L.; Qin, B.; Ma, D.; Zhuo, H.; Liang, H.; Gao, A. Unique spatial methane distribution caused by spontaneous coal combustion in coal mine goafs: An experimental study. *Process Saf. Environ. Prot.* **2018**, *116*, 199–207. [CrossRef]
2. Zheng, Y.; Li, Q.; Zhu, P.; Li, X.; Zhang, G.; Ma, X.; Zhao, Y. Study on multi-field evolution and influencing factors of coal spontaneous combustion in goaf. *Combust. Sci. Technol.* **2023**, *195*, 247–264. [CrossRef]
3. Hu, S.; Li, Y.; Zhu, C.; Lin, B.; Li, Q.; Li, B.; Huang, Z. Investigating the Influence of Flue Gas Induced by Coal Spontaneous Combustion on Methane Explosion Risk. *Fire* **2024**, *7*, 105. [CrossRef]
4. Cui, G.; Yang, C.; Li, Z.-L.; Zhou, Z.; Li, J.-L. Experimental study and theoretical calculation of flammability limits of methane/air mixture at elevated temperatures and pressures. *J. Loss Prev. Process Ind.* **2016**, *41*, 252–258. [CrossRef]
5. Wan, X.; Zhang, Q.; Shen, S. Theoretical estimation of the lower flammability limit of fuel-air mixtures at elevated temperatures and pressures. *J. Loss Prev. Process Ind.* **2015**, *36*, 13–19. [CrossRef]
6. Ma, D.; Shi, Z.; Zhu, T.; Zhang, L.; Yin, C. Effect of flammable gases produced from coal smoldering on methane explosion limits under nitrogen dilution in coal mines. *Fuel* **2025**, *380*, 133164. [CrossRef]
7. Luo, Z.; Liu, L.; Gao, S.; Wang, T.; Su, B.; Wang, L.; Yang, Y.; Li, X. Simulation studies on the influence of other combustible gases on the characteristics of methane explosions at constant volume and high temperature. *Arch. Min. Sci.* **2021**, *2021*, 279–295.
8. Wang, H.; Gu, S.; Chen, T. Experimental Investigation of the Impact of CO, C_2H_6, and H_2 on the Explosion Characteristics of CH_4. *ACS Omega* **2020**, *5*, 24684–24692. [CrossRef]
9. Zhao, Q.; Zhou, L.; Wang, H.; Zhang, Y.; Li, Y.; Qiu, D.; Chen, X. Effects of methane concentration on flame propagation mechanisms and dynamic characteristics of methane/coal dust explosions. *Powder Technol.* **2024**, *439*, 119744. [CrossRef]
10. Deng, J.; Qu, J.; Wang, Q.-H.; Xiao, Y.; Cheng, Y.-C.; Shu, C.-M. Experimental data revealing explosion characteristics of methane, air, and coal mixtures. *RSC Adv.* **2019**, *9*, 24627–24637. [CrossRef]
11. Cashdollar, K.L.; Zlochower, I.A.; Green, G.M.; Thomas, R.A.; Hertzberg, M. Flammability of methane, propane, and hydrogen gases. *J. Loss Prev. Process Ind.* **2000**, *13*, 327–340. [CrossRef]
12. Cammarota, F.; Di Sarli, V.; Salzano, E.; Di Benedetto, A. Measurements of pressure and flame speed during explosions of $CH_4/O_2/N_2/CO_2$ mixtures. *J. Loss Prev. Process Ind.* **2016**, *44*, 771–774. [CrossRef]
13. Salzano, E.; Cammarota, F.; Di Benedetto, A.; Di Sarli, V. Explosion behavior of hydrogen-methane/air mixtures. *J. Loss Prev. Process Ind.* **2012**, *25*, 443–447. [CrossRef]
14. Pekalski, A.; Schildberg, H.; Smallegange, P.; Lemkowitz, S.; Zevenbergen, J.; Braithwaite, M.; Pasman, H. Determination of the explosion behaviour of methane and propene in air or oxygen at standard and elevated conditions. *Process Saf. Environ. Prot.* **2005**, *83*, 421–429. [CrossRef]
15. Gieras, M.; Klemens, R.; Rarata, G.; Wolański, P. Determination of explosion parameters of methane-air mixtures in the chamber of 40 dm3 at normal and elevated temperature. *J. Loss Prev. Process Ind.* **2006**, *19*, 263–270. [CrossRef]
16. Tang, C.; Zhang, S.; Si, Z.; Huang, Z.; Zhang, K.; Jin, Z. High methane natural gas/air explosion characteristics in confined vessel. *J. Hazard. Mater.* **2014**, *278*, 520–528. [CrossRef]
17. Huang, L.; Wang, Y.; Pei, S.; Cui, G.; Zhang, L.; Ren, S.; Zhang, Z.; Wang, N. Effect of elevated pressure on the explosion and flammability limits of methane-air mixtures. *Energy* **2019**, *186*, 115840. [CrossRef]
18. Takahashi, A.; Urano, Y.; Tokuhashi, K.; Nagai, H.; Kaise, M.; Kondo, S. Fusing ignition of various metal wires for explosion limits measurement of methane/air mixture. *J. Loss Prev. Process Ind.* **1998**, *11*, 353–360. [CrossRef]
19. Abbas, Z.; Zinke, R.; Gabel, D.; Addai, E.K.; Darbanan, A.F.; Krause, U. Theoretical evaluation of lower explosion limit of hybrid mixtures. *J. Loss Prev. Process Ind.* **2019**, *60*, 296–302. [CrossRef]
20. Liaw, H.-J.; Chen, C.-C.; Lin, N.-K.; Shu, C.-M.; Shen, S.-Y. Flammability limits estimation for fuel-air-diluent mixtures tested in a constant volume vessel. *Process Saf. Environ. Prot.* **2016**, *100*, 150–162. [CrossRef]
21. Abdelkhalik, A.; Askar, E.; Markus, D.; Brandes, E.; El-Sayed, I.; Hassan, M.; Nour, M.; Stolz, T. Explosion regions of propane, isopropanol, acetone, and methyl acetate/inert gas/air mixtures. *J. Loss Prev. Process Ind.* **2016**, *43*, 669–675. [CrossRef]
22. Kobiera, A.; Kindracki, J.; Zydak, P.; Wolanski, P. A new phenomenological model of gas explosion based on characteristics of flame surface. *J. Loss Prev. Process Ind.* **2007**, *20*, 271–280. [CrossRef]
23. Maremonti, M.; Russo, G.; Salzano, E.; Tufano, V. Numerical simulation of gas explosions in linked vessels. *J. Loss Prev. Process Ind.* **1999**, *12*, 189–194. [CrossRef]
24. Salzano, E.; Marra, F.; Russo, G.; Lee, J. Numerical simulation of turbulent gas flames in tubes. *J. Hazard. Mater.* **2002**, *95*, 233–247. [CrossRef] [PubMed]
25. Zhang, X.-B.; Gao, J.-L.; Ren, J.-Z.; Wang, C.-X. Analysis of the Characteristics and Influencing Factors of Gas Explosion in Heading Face. *Shock. Vib.* **2020**, *2020*, 10709622. [CrossRef]
26. Yang, A.; Liu, Y.; Gao, K.; Li, R.; Li, Q.; Li, S. Numerical Simulation of Gas Explosion with Non-uniform Concentration Distribution by Using OpenFOAM. *ACS Omega* **2023**, *8*, 48789–48812. [CrossRef]
27. Qiu, J.; Jiang, B.; Tang, M.; Zhou, L.; Ren, B. Effect of Different Bend Pipes on the Propagation Characteristics of Premixed Methane-Air Explosion in Confined Spaces. *Geofluids* **2021**, *2021*, 6635156. [CrossRef]

28. Coppens, F.; Deruyck, J.; Konnov, A. The effects of composition on burning velocity and nitric oxide formation in laminar premixed flames of CH_4+H_2+O_2+N_2. *Combust. Flame* **2007**, *149*, 409–417. [CrossRef]
29. Berg, P.A.; Hill, D.A.; Noble, A.R.; Smith, G.P.; Jeffries, J.B.; Crosley, D.R. Absolute CH concentration measurements in low-pressure methane flames: Comparisons with model results. *Combust. Flame* **2000**, *121*, 223–235. [CrossRef]
30. Sung, C.; Law, C.; Chen, J.-Y. Further Validation of an Augmented Reduced Mechanism for Methane Oxidation: Comparison of Global Parameters and Detailed Structure. *Combust. Sci. Technol.* **2000**, *156*, 201–220. [CrossRef]
31. Wang, T.; Luo, Z.; Wen, H.; Cheng, F.; Liu, L.; Su, Y.; Liu, C.; Zhao, J.; Deng, J.; Yu, M. The explosion enhancement of methane-air mixtures by ethylene in a confined chamber. *Energy* **2021**, *214*, 119042. [CrossRef]

fire

Article

Study on the Control Effect of Borehole Gas Extraction in Coal Seams Based on the Stress–Seepage Coupling Field

Hongsheng Wang [1], Huaming An [2,3,*] and Bingbing Yang [2]

[1] Beijing Chemical Occupational Disease Prevention and Control Institute, Beijing 100093, China
[2] Faulty of Public Security and Emergency Management, Kunming University of Science and Technology, Kunming 650093, China
[3] Geotechnical Institute, TU Bergakademie Freiberg, 09599 Freiberg, Germany
* Correspondence: huaming.an@kust.edu.cn

Abstract: In order to determine the reasonable parameters of high-gas and extra-thick coal seam drainage, considering the factors of the coal seam metamorphic degree, stress condition, gas occurrence state, and permeability dynamic change, the gas desorption, diffusion, and transport process of coal seam gas are analyzed. A secondary distribution model of coal around the borehole, a porosity variation model of coal around the borehole, a stress–seepage coupling model, a pore flow model of the pressure-driven transition flow zone, and a free molecular flow zone are established. Taking the gas drainage of Zhangcun Coal Mine of Lu'an Group as the research object, the influence of drilling hole diameter, coal seam permeability, gas original pressure, and other factors on the control range of coal seam drainage drilling is simulated by ANSYS Fluent 6.3.26. The results show that secondary stress distribution occurs in the coal seam drill hole under the action of lead stress, which leads to the change in porosity; the seepage zone, transition zone, molecular flow zone, and original rock stress zone are presented around the drill hole; and the range of influence of the drill hole is mainly based on the seepage zone and the transition zone, supplemented by the molecular flow zone. The control range of the drill hole is in a positive proportional relationship to the diameter of the drill hole, the porosity of the coal seam, and the original pressure of the gas.

Keywords: stress field; seepage field; multi-field coupling; pumping control range; numerical simulation

Citation: Wang, H.; An, H.; Yang, B. Study on the Control Effect of Borehole Gas Extraction in Coal Seams Based on the Stress–Seepage Coupling Field. *Fire* **2024**, *7*, 410. https://doi.org/10.3390/fire7110410

Academic Editors: Chuyuan Huang, Haipeng Jiang and Lijuan Liu

Received: 3 October 2024
Revised: 21 October 2024
Accepted: 28 October 2024
Published: 8 November 2024

1. Introduction

As an important basic energy source for China's economic and social development, coal has made remarkable progress both in terms of the scope and intensity of its exploitation. Coal plays an indispensable role in national economic development and modernization. With the development of the national economy and the continuous progress of science and technology, large-scale mega coal mines have been put into operation one after another, and the coal mining method has shifted from traditional artillery mining and general mining to synthesis mining, which has a higher efficiency. The depth record of coal seams has been broken continuously, and the high pressure and content of coal seam gas has become one of the important characteristics of modern coal mining. According to the requirements of the relevant laws and regulations, high-gas mines must extract the coal seam gas before mining, so that the coal seam gas pressure and content can be reduced to the mining requirements. However, in the case of high-gas and extra-thick coal seam integrated mining, the high mining intensity, the high coal residue in the hollow mining area, and the high gas content in the neighboring layers and surrounding rocks make it very easy for the integrated working face to have the phenomenon of gas exceeding the limitation, which may lead to gas explosions, gas asphyxiation, and other disasters and bring a great safety hazard to the production of the mines.

Since the first synthesized working face was put to the test in China in 1984, the technology of synthesized-roof coal mining has been developed rapidly. At present, the working faces of integrated-roof coal mining in China are among the top in the world in terms of the number and scope of use, the advancement of technology, and the results achieved. However, with the increase in demand for coal and the improvement in production technology, coal seams mined by integrated-roof coal mining are gradually characterized by their high gas and high protrusion levels. At the same time, production practices have also gradually exposed the many hidden disasters or dangers of comprehensive roof coal mining technology, such as a high absolute gas outflow, frequent spontaneous combustion and fire, and other problems, and this approach has triggered a number of various types of major disasters and accidents. It is easy to see that gas accidents have become one of the most serious types of accidents in China's coal mines, and the number of deaths in gas accidents (such as gas explosions, coal and gas protrusion, and gas asphyxiation) has been the most common type of large accident for many years.

In recent decades, many scholars have conducted extensive scientific research and production practices on the prevention and control technologies for major gas disasters in comprehensive mining. They have proposed various gas management measures, including: pre-mining gas extraction, gas extraction during tunneling and mining, high-yield drilling for gas extraction, burying pipes in the upper corners for gas discharge, increasing air distribution, and arranging the working face along the roof plate to facilitate gas discharge towards higher-level tunnels. These measures aim to effectively prevent and control gas disasters. Wang et al. [1] determined the effective influence radius of gas extraction by numerical simulation for the reasonable arrangement of gas-extraction boreholes along a coal seam, which provided a basis for the optimization of single gas-extraction boreholes along coal seams. Liu et al. [2] theoretically studied the effective gas-extraction radius of bedding boreholes. Zhao [3] gave full consideration to the technical advantages of high-level roof drilling and improved the hole-forming rate of high-level drilling. Zhu [4] proposed an evaluation method of drilling control effects based on a physical examination method. This method divides a coal seam into several control bodies whose side lengths are equal to the effective radius of extraction and uses computer technology to evaluate the control effect of drilling. In the Hebi mining area, the layout method of using cross boreholes is used to arrange a pair of boreholes every 2.5~3.0 m, in which one hole is parallel to the working face and the other hole is oblique to the working face. The angle between the two holes is 10°~15°, and the aperture is 79 mm. This method has achieved a good drainage effect. These fruitful results provide a basis for mine gas control, reducing the harm of gas in coal mining, and provide a guarantee for safe mine production. However, due to the unpredictability of coal and rock storage conditions and the special nature of production conditions, the prevention and control measures of major disasters often fail to meet the design objectives, such as when the gas pumping effect of the mining area is generally poor, with a low pumping rate. Because serious accidents caused by gas are repeated, resulting in the majority of accidental deaths, the mortality rate per million tons of coal production remains high, a serious threat to the majority of China's coal mine workers' safety, social stability, and unity, causing an extremely negative impact. At the same time, it has also brought significant casualties and property losses to individuals, coal mining enterprises, and the State.

Solving the problem of gas disasters in high-gas synthesized working faces is an urgent task facing us. Multiple accident causative factors, a wide range of hazards, and serious consequences have become the characteristics of roof coal mining in high-gas mines. Therefore, it is necessary to summarize the valuable experience of predecessors on the basis of the current mine production situation, to crack these technical problems, and to provide theoretical support and a technological guarantee for the realization of the development of high-gas comprehensive roof coal mining mines, with the goal of a high output, high efficiency, high effectiveness, and high safety. With the increased application of seepage diffusion theory in the field of coal seam gas drainage and the continuous deepening and improvement of theoretical research, domestic and foreign experts and scholars have summarized the process of gas transport and its influencing factors through theoretical analysis and experimental research. It has been proven that the gas transportation process is a result of the joint action of multiple factors, such as the geological conditions of the coal seam, the degree of coal body metamorphism, the original pressure of the gas, and the ground stress. However, due to the special nature of its storage conditions, the storage and seepage process of gas is subject to a variety of external forces, such as the ground stress field, ground electric field, ground temperature field, etc., and each of these factors will have different degrees of influence on its flow in a porous medium. Worse still, the coupling of the various fields enhances the likelihood of composite disaster and destruction, which makes the prevention and control become more difficult. In this regard, some scholars have explored and researched the multi-field coupling law of coal seam gas and put forward many fluid–solid coupling models [5–13]. Enever (1997) [14,15] analyzed the relationship between the ground stress and the permeability of a coal seam and obtained an exponentially correlated rule of change between the two. Zhen et al. [16] established a coupled model of stress–seepage damage for hydraulic fracturing of rock bodies. In order to determine the law of temporal and spatial evolution, Jasinge et al. [17] conducted gas permeability testing experiments on coal samples under different enclosing pressures and obtained the law describing the stress–permeability change of Australian lignite coals. Cao et al. [18] established a nonlinear seepage equation of water injection in coal seams by analyzing the stress distribution of the workings and its influence on the characteristics of water injection into coal seams. Zhao et al. [19] established a multi-field (including coal stress field, damage field, gas diffusion field, and seepage field) coupling model based on the non-homogeneity of a coal body. Liu et al. [20] further discussed the characteristics of the dynamic evolution distribution of stress and seepage throughout the coal seam water injection process based on a numerical simulation and experimental results.

In recent years, with the popularization and application of simulation (CFD) software, domestic scholars have carried out simulation research on the air flow field and gas migration in goaf and achieved certain results. Based on CFD simulation software, Yuan et al. [21] systematically studied the dynamic changes in the stress field, fracture field, and gas flow field in the No.11 coal seam of the Guqiao Coal Mine. Based on Fluent simulation software, a three-dimensional heterogeneous numerical model of gas flow in goaf was established by Hu et al., Li et al., and Zhang et al. [22–24]. In a study by Wang et al. [25] based on the Fick diffusion law, Darcy seepage law, and Fourier heat conduction law, a multi-physical field coupling model of goaf considering temperature change was established by using mass conservation and energy conservation equations based on the three-dimensional porosity model. The above researchers have analyzed the mechanism and influencing factors of the multi-field coupling of coal seam gas through coal mine field monitoring, laboratory experiments, and numerical simulation. However, the multi-field coupling mechanism is not yet clear, the relevant prevention and control measures are still not effective, and the research on the stress–seepage coupling field is relatively few. Based on the secondary distribution of stress caused by coal seam drilling, combined with the change in porosity around the borehole, this paper studies the seepage process formed by the coupling of the stress field and seepage field and the void and micro-pore flow process in the pressure-driven transition flow zone and free molecular flow zone [26–29]. According to the secondary distribution characteristics of borehole stress in coal seams, the spatial variation law of permeability around boreholes, and the fluid–solid coupling model of gas, the variation law of the control range of a gas drainage borehole in a high-gas and extra-thick coal seam is simulated by ANSYS Fluent 6.3.26. The research results can provide theoretical reference for the optimal layout of drainage boreholes.

2. Theoretical Analyses and Model Establishment

In this section, the main sources of gas and its transport law in the face of comprehensive mining are studied and analyzed by reviewing relevant information; applying theoretical knowledge and technical methods of geology, mining, fluid mechanics, the gas disaster control principle, the comprehensive roof coal mining process, and other related disciplines; applying the theory of seepage in porous media combined with the actual characteristics of a mined coal seam; and establishing equations of gas desorption, diffusion, and transportation of a coal seam under porous media. The secondary distribution of the stress field in the coal body around the borehole, the change rule of the void degree under the influence of stress–strain, the dynamic model of fluid–solid coupling in the gas pumping process, and the pore flow model of the pressure-driven transition flow zone and free molecular flow zone were determined.

2.1. Theoretical Analyses of Gas Migration Law of Pre-Drainage Coal Seam in Fully Mechanized Caving Face

A coal seam is a dual medium of pore and fracture. Gas exists in the matrix in the form of adsorption. It undergoes the process of desorption from the coal body, diffusion to the pore, and fracture and transport in the coal seam fracture before production [30,31].

1. Coal seam gas desorption

There are three main types of models describing coal seam gas adsorption, namely the Langmuir isothermal adsorption model, potential difference theory model, and Gibbs model [32]. Among them, the Langmuir isothermal adsorption model is the most commonly used:

$$V_E(p) = V_L \frac{p}{p_L + p} \tag{1}$$

where $V_E(p)$ is the adsorption capacity, g/m^3; p is pressure, MPa; V_L is the Langmuir volume, g/m^3; and p_L is the Langmuir pressure, MPa. Langmuir pressure is the pressure when the adsorption capacity reaches 50% of the limit adsorption capacity.

2. Coal seam gas diffusion

Gas diffusion in a coal seam is divided into quasi-steady-state diffusion and non-steady-state diffusion.

(1) Quasi-steady-state diffusion (based on Fick's first law [33,34])

It is believed that the diffusion process of gas in a coal seam can be expressed by the change in the average concentration in each time period being proportional to the difference of time:

$$\frac{dC_m}{dt} = D_m F_s (C_2 - C_m) \tag{2}$$

where:

$$q_m = -G \frac{dC_m}{dt} \tag{3}$$

$$\tau - \frac{1}{D_m F_s} \tag{4}$$

$$F_s = 8\pi / S^2 \tag{5}$$

where C_m is the concentration in the matrix, g/m^3; D_m is the mass diffusion coefficient in porous media, m^2/s; C_2 is the concentration at the edge of the matrix block, g/m^3; F_s is the matrix block shape factor, m^{-2}; q_m is the amount of gas desorption; G is a geometric factor; τ is the adsorption time constant, s; and S is cleat spacing, m.

It can be seen from the above formula that the adsorption time is related to the diffusion distance; that is, it is proportional to the square of the fracture spacing.

(2) Unsteady diffusion (based on Fick's second law [35,36])

It is considered that the diffusion process of gas in the geometric block of the matrix can be described in the form of partial differential equations. Coal seam gas mainly includes two parts: one part is the free gas contained in pores and fractures and the other part is the gas adsorbed on the surface of the matrix. The concentration of free gas is:

$$C_1 = \rho_1 \phi_m = \frac{MP_m \phi_m}{RTZ} \tag{6}$$

The concentration of adsorbed gas is:

$$C_2 = V_L \frac{p}{p_L + p_m} \tag{7}$$

The total gas concentration in the matrix block is:

$$C_m = \frac{MP_m \phi_m}{RTZ} + V_L \frac{p}{p_L + p_m} \tag{8}$$

According to Fick's second diffusion law, the equation of pressure in pores is:

$$\frac{\partial}{\partial t}\left(\frac{MP_m \phi_m}{RTZ} + V_L \frac{p}{p_L + p_m}\right) = \nabla\left[D_m \nabla\left(\frac{MP_m \phi_m}{RTZ} + V_L \frac{p}{p_L + p_m}\right)\right] \tag{9}$$

After the gas concentration is solved by Equations (8) and (9), the amount of gas flowing out of the matrix block is:

$$q_m = -\frac{A_1}{V_1} J = -\frac{A_1}{V_1} D_m \nabla C_m \tag{10}$$

where C_1 is the concentration of free gas, g/m^3; ϕ_m is matrix porosity; C_m is the sum of the concentration of adsorbed gas C_{ma} and free gas C_{mf} in the matrix; p_m is the gas pressure in the pore, MPa; A_1 is the surface area of the matrix block unit, m^2; J is the diffusion flux, $g/m^2{\cdot}s$; and V_1 is the volume of matrix block unit, m^3.

3. Coal seam gas transport

Due to the continuous gas diffusion into the cracks in the coal matrix block, it should be regarded as a continuous source distribution in the continuity equation. The mass conservation equation of gas transport in cracks is [37]:

$$\frac{\partial}{\partial t}\left(\phi_f s_{fg} \rho_{fg}\right) = -\nabla \cdot \left(\rho_{fg} V_{fg}\right) + q_m \tag{11}$$

Among them is the quality of the gas that diffuses into the crack. The velocity consists of two parts: the macroscopic seepage velocity following Darcy's law and the diffusion velocity following Fick's law, that is:

$$V_{fg} = -\left(\frac{K_g}{\mu_g} \nabla P_{fg} + \frac{D_f}{C_f} \nabla C_f\right) \tag{12}$$

The density and concentration in Equations (11) and (12) are expressed by pressure, respectively:

$$\rho_{fg} = \frac{M}{RT}\left(\frac{P_{fg}}{Z}\right) \tag{13}$$

When isothermal,

$$\frac{\nabla C_f}{C_f} = \nabla\left(s_{fg} \frac{P_{fg}}{Z}\right) / \left(s_{fg} \frac{P_{fg}}{Z}\right) \tag{14}$$

Equation (11) becomes:

$$\frac{\partial}{\partial t}\left(\frac{\phi_f s_g p_g}{Z}\right) = \nabla \cdot \left[\frac{p_g}{Z}\frac{K_g}{\mu_g}\nabla P_g + \frac{D_f}{s_g}\nabla\left(\frac{s_g p_g}{Z}\right)\right] + \frac{RT}{M}q_m \tag{15}$$

where D_f is the gas diffusion coefficient in the fracture; C_f is the gas concentration in the fracture, $C_f = \phi_f \rho_g$; ϕ_f is the fracture porosity; s_g is the gas saturation; K is the coal seam permeability, m^2; μ is gas viscosity, Pa·s; and m, f, and g represent the matrix, fracture, and gas, respectively.

In summary, a complete description of the coal seam gas output process was described above.

2.2. Establishment of Secondary Distribution Model of Stress Field of Coal Around Borehole

The occurrence of gas disasters is ultimately a process of energy transfer and release. Fully mechanized caving mining destroys the original stress equilibrium state of mining coal seams and breaks the dynamic equilibrium condition of gas adsorption–desorption in a coal seam, resulting in a large amount of adsorbed gas released from mining layer, adjacent layer, and surrounding rock, forming gas accumulation in local space. Due to the drilling construction, the original stress state of the coal seam changes, and the stress of the coal body around the drilling site is redistributed. After the hole is formed, the coal body at the hole wall produces large secondary stress and shear failure. The damaged coal body still bears part of the overburden pressure; however, most of the pressure is transmitted to the external coal body unit and generates concentrated stress, which further damages the coal body. After the stress is balanced, the pressure relief zone, the stress concentration zone, and the original stress zone are formed along the radial direction of the borehole [38].

According to the static equilibrium equation, it can be seen that the equilibrium conditions along the tangential direction of the arc are naturally satisfied for the axisymmetric, volume, and shape-invariant unit body, resulting in shear stress. The components of tangential stress and radial stress in the radius direction are:

$$(\sum \sigma_\theta)_r = 2\sigma_\theta dr sin\,\frac{d\theta}{2} = \sigma_\theta dr d\theta \tag{16}$$

$$(\sum \sigma_r)_r = \sigma_r r d\theta - (\sigma_r - d\sigma_r)(r + dr)d\theta \tag{17}$$

From $(\sum \sigma_\theta)_r + (\sum \sigma_r)_r = 0$, and by eliminating $d\theta$ and omitting the second-order trace term $drd\sigma_r$, the static equilibrium differential equation is obtained:

$$r\frac{d\sigma_r}{dr} + \sigma_r - \sigma_\theta = 0 \tag{18}$$

Using geometric equations and physical equations to solve Equation (18):

$$\begin{cases} r\frac{d\sigma_r}{dr} + \sigma_r - \sigma_\theta = 0 \\ \varepsilon_r = \frac{du}{dr} \\ \varepsilon_\theta = \frac{u}{r} \\ \varepsilon_z = 0 \\ \sigma_z = \mu(\sigma_\theta + \sigma_r) \\ \varepsilon_r = \frac{1-\mu^2}{E}(\sigma_r - \frac{\mu}{1-\mu}\sigma_\theta) \\ \varepsilon_\theta = \frac{1-\mu^2}{E}(\sigma_\theta - \frac{\mu}{1-\mu}\sigma_r) \end{cases} \tag{19}$$

There are:

$$\frac{d^2\sigma_r}{dr^2} + \frac{3}{r}\frac{d\sigma_r}{dr} = 0 \tag{20}$$

Combined with boundary conditions:

$$\begin{cases} \sigma_r = p & r \to \infty \\ \sigma_r = 0 & r = a \end{cases} \tag{21}$$

The solution of Equation (20) is:

$$\begin{cases} \sigma_r = p(1 - \frac{a^2}{r^2}) \\ \sigma_\theta = p(1 + \frac{a^2}{r^2}) \end{cases} \tag{22}$$

where σ_θ is the tangential stress, MPa; σ_r is the radial stress, MPa; r is the stress radius, m; a is the diameter of the borehole, m; and p is the original rock stress, MPa.

According to Equation (22), the range of secondary stress distribution around boreholes with different apertures can be obtained, and the change curve is shown in Figure 1a–d.

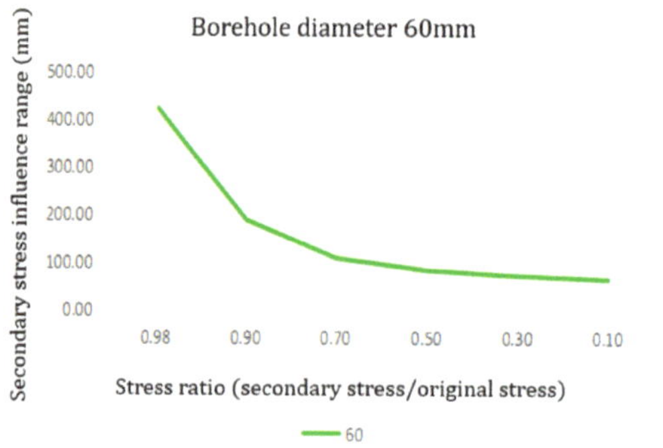

(**a**)

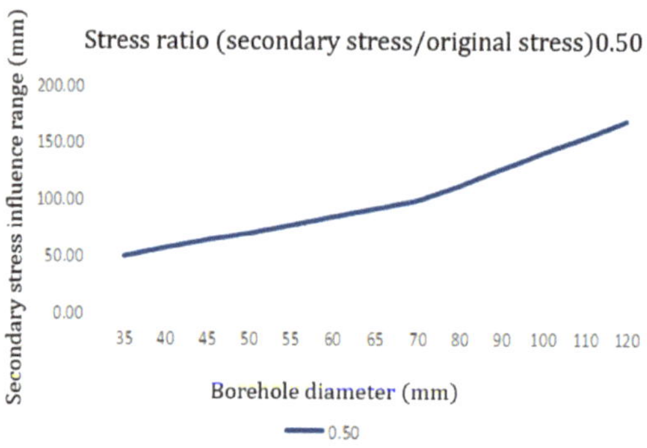

(**b**)

Figure 1. *Cont.*

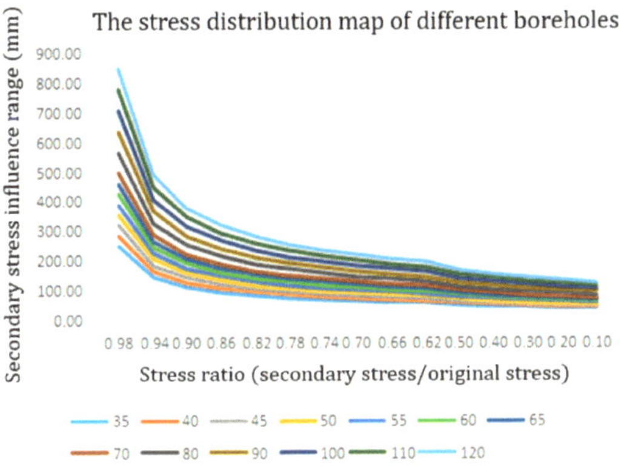

(c)

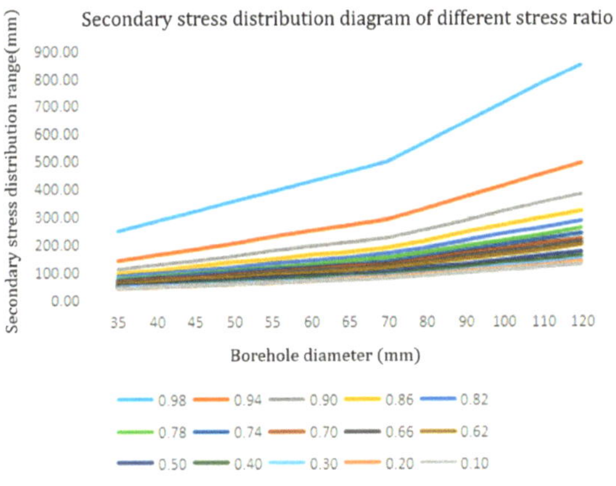

(d)

Figure 1. (**a**) The secondary stress distribution around the borehole with a diameter of 60 mm. (**b**) Secondary stress distribution map around the borehole at a stress ratio of 0.5. (**c**) Secondary stress distribution around different boreholes. (**d**) Secondary stress distribution around boreholes with different stress ratios.

Figure 1a shows that when the diameter of the borehole is constant, the distribution range of the secondary stress around the borehole decreases with the decrease in the stress ratio, but the variation law is nonlinear. When the stress ratio is less than 0.5, the secondary stress distribution line around the borehole is similar to a straight line, indicating that the secondary stress in this area decreases very fast, so its influence range is small. When the stress ratio is in the range of 0.5~0.9, the secondary stress distribution line around the borehole is similar to the circular arc line, indicating that the secondary stress decline rate in this area gradually slows down, and the influence range gradually increases. When the stress ratio is greater than 0.9, the secondary stress around the borehole is affected by the borehole very little, so the area is very wide.

Taking the stress ratio of 0.5 as an example, Figure 1b shows that the distribution range of secondary stress around boreholes with different diameters is quite different. When the borehole diameter is less than 70 mm, the increment value of secondary stress distribution around boreholes is basically the same. With the increase in borehole diameter, the distribution range of secondary stress around boreholes also increases obviously. For example, when the borehole diameter is 35 mm, the distribution range of secondary stress is about 50 mm. When the borehole diameter is 70 mm, the distribution range of secondary stress is about 100 mm, the borehole diameter increases by 35 mm, and the influence range of secondary stress increases by 50 mm. When the borehole diameter is greater than 70 mm, the secondary stress distribution around the borehole changes significantly, indicating that the stress increment increases significantly with the increase in the borehole diameter and also that the larger the borehole is, the wider the secondary stress distribution range is.

Figure 1c shows that the distribution of secondary stress around the borehole varies obviously with different borehole diameters. When the stress ratio is less than 0.5, the increase in the borehole diameter has little effect on the secondary stress distribution. With the increase in the stress ratio, the influence range of the increase in the borehole diameter on the secondary stress distribution increases gradually, and the increment of the shear influence range increases obviously. When the strain is relatively large, the increase in secondary stress around the borehole is more obvious. When the stress ratio is 0.3, the influence range of the borehole diameter of 35 mm is about 41 mm, the influence range of the borehole diameter of 70 mm is about 83 mm, and the influence range of the borehole diameter of 110 mm is about 131 mm. When the stress ratio is 0.78, the influence range of the borehole diameter of 35 mm is about 75 mm, the influence range of the borehole diameter of 70 mm is about 150 mm, and the influence range of the borehole diameter of 110 mm is about 234 mm. When the stress ratio is 0.98, the influence range of the borehole diameter 35 mm is about 245 mm, the influence range of the borehole diameter 70 mm is about 500 mm, and the influence range of the borehole diameter 110 mm is about 780 mm. It can be seen that different stress ratios have a significant effect on the secondary stress distribution of different diameter boreholes.

Figure 1d shows that different stress ratios have significantly different distribution areas around the borehole. When the stress ratio is below 0.75, the area of secondary stress distribution is about twice the diameter of the borehole. When the stress ratio increases to 0.9, the influence area of the secondary stress increases to about three times the diameter of the borehole. When the stress ratio increases to 0.98, the influence area of the secondary stress increases to about seven times the borehole diameter. The area of secondary stress distribution around the borehole is mainly concentrated in the range of twice the diameter of the borehole.

2.3. Establishment of Porosity Variation Model Under the Influence of Stress–Strain

Due to the secondary stress distribution around the coal seam borehole, the coal seam void volume will change [39].

The change in coal seam porosity around the borehole is as follows:

$$d\phi = d\left(\frac{V_p}{V_b}\right) = \frac{dV_p}{V_b} - \phi\frac{dV_b}{V_b} \tag{23}$$

where:

$$dV_p = -C_{pc}V_p d\sigma + C_{pp}V_p dp \tag{24}$$

$$dV_b = -C_{bc}V_b d\sigma + C_{bp}V_b dp \tag{25}$$

This is arranged as:

$$d\phi = -\phi C_{pc}d\sigma + \phi C_{pp}dp + \phi C_{bc}d\sigma - \phi C_{bp}V_b dp \tag{26}$$

Using the relationship between the compression coefficients, the relationship between porosity and stress is:

$$d\phi = -[C_{bc}(1-\phi) - C_s](d\sigma - dp) \tag{27}$$

Similarly, the expression of the volume strain is:

$$d\varepsilon_V = -\frac{dV_b}{V_b} = C_{bc}d\sigma - C_{bp}dp = C_{bc}(d\sigma - dp) + C_s dp \tag{28}$$

Therefore, the relationship between porosity and volume strain is:

$$d\varepsilon_V = -\left(1 - \frac{C_s}{C_{bc}} - \phi\right)(d\varepsilon_V - C_s dp) \tag{29}$$

This is arranged as:

$$\phi = 1 - \frac{C_s}{C_{bc}} + \frac{d\varepsilon_V}{d\varepsilon_V - C_s dp} \tag{30}$$

where ϕ is porosity; V_p is the physical examination of coal pores with a volume of V_b, m^3; V_b is the volume of coal, m^3; C is the compression coefficient; and p is the original rock stress, Mpa.

Equation (30) shows that when the compression coefficient of coal and gas and the original rock pressure of the coal seam are known, the change in porosity should become a similar inverse proportional function relationship with the body. When the coal body is in a compressed state, that is, greater than zero, the function is a decreasing function. When the coal body is in the expansion state, that is, less than zero, the function is an increasing function. Therefore, the pressure relief zone along the borehole is the coal body expansion zone, and the porosity increases with the increase in the strain in the pressure relief zone. The stress concentration zone is the coal body compression zone, and the porosity decreases with the increase in the strain in the stress concentration zone. In the original stress zone, the porosity of the region is unchanged because the coal body is not strained.

2.4. Establishment of Gas Migration Law Model Under the Condition of Stress–Seepage Coupling Field

Under the action of an in situ stress field, the secondary stress distribution will occur around the borehole. Affected by the secondary stress, the pressure relief zone, stress concentration zone, and original stress zone will be formed along the radial direction of the borehole. The porosity of each area around the borehole has also changed greatly. Due to the difference in porosity, the gas migration law around the coal seam gas pre-drainage borehole is divided into three parts. In the pressure relief area near the borehole center, the gas will undergo a fluid–solid coupling seepage process [40]. In the excessive flow area of stress concentration, the gas will undergo pipeline flow and molecular free flow, and the gas movement will be stagnant when it is close to the original rock stress area. Therefore, the stress–seepage coupling field model of the coal seam gas pre-drainage process includes the dynamic model of the fluid–solid coupling field and the pressure-driven pore flow model.

2.4.1. Establishment of Fluid–Solid Coupling Dynamic Model of Pressure Relief Zone Around Gas Drainage Boreholes

In the process of coal seam gas drainage, its motion process satisfies the gas continuity equation, gas motion equation, and gas state equation [1,2,41,42]. The above equations are combined to obtain:

$$\begin{cases} -\left[\frac{\partial(\rho_g v_x)}{\partial x} + \frac{\partial(\rho_g v_y)}{\partial y} + \frac{\partial(\rho_g v_z)}{\partial z}\right] = \frac{\partial(\phi\rho_g)}{\partial t} \\ v_x = -\frac{K}{\mu}\frac{\partial p}{\partial x} \\ v_y = -\frac{K}{\mu}\frac{\partial p}{\partial y} \\ v_z = -\frac{K}{\mu}\frac{\partial p}{\partial z} \\ pV = nZRT \end{cases} \tag{31}$$

Considering the viscosity and compression factor of real gas, the above equation is solved, and the seepage differential equation of gas can be obtained as follows:

$$\frac{\partial}{\partial x}\left[\frac{p}{\mu(p)Z(p)}\frac{\partial p}{\partial x}\right]+\frac{\partial}{\partial y}\left[\frac{p}{\mu(p)Z(p)}\frac{\partial p}{\partial y}\right]+\frac{\partial}{\partial z}\left[\frac{p}{\mu(p)Z(p)}\frac{\partial p}{\partial z}\right]=\frac{\phi}{K}\frac{\partial}{\partial t}\left[\frac{p}{Z(p)}\right] \tag{32}$$

The upper right side is:

$$\frac{\partial}{\partial t}\left[\frac{p}{Z(p)}\right]=\left[\frac{1}{p}-\frac{1}{Z(p)}\frac{\partial p}{\partial Z(p)}\right]\cdot\frac{p}{Z(p)}\frac{\partial p}{\partial t}=C(p)\frac{p}{Z(p)}\frac{\partial p}{\partial t} \tag{33}$$

This is arranged as:

$$\nabla\left[\frac{p}{\mu(p)Z(p)}\nabla p\right]=\frac{\phi\mu(p)C(p)}{K}\left[\frac{p}{\mu(p)Z(p)}\frac{\partial p}{\partial t}\right] \tag{34}$$

2.4.2. Establishment of Channel Flow Model in Stress Concentration Area Around Gas Drainage Borehole

The stress concentration zone around the borehole can be subdivided into the over-flow zone and the free-flow zone. It is assumed that the velocity distribution of the pressure-driven over-flow zone and the free-flow zone is parabolic, and the slip velocity can be expressed as follows:

$$U_s-U_w=\frac{2-\sigma_r}{\sigma_r}\left[\frac{K_n}{1-bK_n}\left(\frac{\partial U}{\partial n}\right)_s\right] \tag{35}$$

The dimensionless velocity scaling equation of micro-porous flow can be obtained:

$$U^*(r,K_n)=\frac{U(x,r)}{\overline{U(x)}}=\left[\frac{-\left(\frac{r}{d}\right)^2+1+2\times\frac{K_n}{1-bK_n}}{\frac{1}{2}+2\times\frac{K_n}{1-bK_n}}\right] \tag{36}$$

where d is the aperture, and then the volume flow rate in the hole is:

$$\dot{Q}=-\frac{\pi d^4}{8\mu_0}\times\frac{dp}{dx}(1-dK_n)\left(1+\frac{4K_n}{1-bK_n}\right) \tag{37}$$

Since b = −1 can be determined from the linearized Boltzmann solution, the only parameter to be determined in the model is d. However, d should change from the zero value in the slip flow region to the constant asymptotic value d_0 in the molecular flow region. The asymptotic value of the mass flow rate at that time can be obtained by Equation (38).

$$d_{K_n\to\infty}\equiv d_0=\left[\frac{64}{3\pi\left(1-\frac{4}{b}\right)}\right] \tag{38}$$

The variation of d as a K_n function obtained from the numerical and experimental data is as follows:

$$d=d_0\frac{2}{\pi}tan^{-1}\left(d_1\overline{K_n^{\beta}}\right) \tag{39}$$

The volume flow rate of the channel flow in the excessive flow area is:

$$\dot{Q}=C(AR)\frac{wh^3}{12\mu}\left(-\frac{dp}{dx}\right) \tag{40}$$

The volume flow rate of the rarefied gas flow in the molecular flow region is:

$$\dot{Q} = C(AR)\frac{wh^3}{12\mu_0} \times \left(-\frac{dp}{dx}\right)(1+\alpha K_n)\left(1+\frac{6K_n}{1-6K_n}\right) \tag{41}$$

where $C(AR)$ is the correction factor, generally taking 1 in the micro-pore flow.

So far, the mathematical model of the gas migration law under the condition of the stress–seepage coupling field in the coal seam gas pre-drainage process has been established.

In this model, ρ is the gas density, Kg/m^3; v is speed, m/s; K is permeability, m^2; μ is the viscosity coefficient, Pa·s; K_n is the number of effort; p is pressure, Pa; Vis volume, m^3; n is the amount of gas substance, mol; R is a gas constant, $Pa \cdot m^3/(mol \cdot K)$; T is temperature, K; Z is the gas compression factor.

3. Numerical Simulation and Result Analysis

The difference of the drainage technology in the gas control of a high-gas and extra-thick coal seam is directly related to the safety and overall progress requirements of the mining work. In view of the complexity and non-repeatability of the coal seam gas drainage system, the CFD commercial software Fluent is used to numerically simulate the gas migration law of a pre-drainage coal seam in the fully mechanized caving face. The CFD simulation study is used to obtain a numerical solution of the fluid flow control equation. It obtains the mathematical description of the overall flow field concerned by solving the space–time solution. The basis of CFD is to establish the Navier–Stokes equation. It is composed of a series of partial differential equations describing the conservation law of fluid flow, and the conservation equations of mass and momentum are solved. The software has the ability to simulate complex flow fields from incompressible to highly compressible. ANSYS Fluent 6.3.26 contains a very rich physical model that can simulate complex multiphase flow. The model of the gas migration law based on a stress coupling field is also solved by the mass conservation equation of fluid flow. ANSYS Fluent 6.3.26 has powerful operation energy to simulate the law of fluid flow in porous media and is widely used in simulating the law of gas migration in goaf. In this section, the numerical simulation method is used to simulate the gas drainage control range and drainage effect by using computer technology. The core is to study the migration law of gas under different coal seam geologies and occurrence conditions, study the distribution law of gas movement under different boundary conditions, and study the main factors affecting the gas accumulation in fully mechanized caving mining so as to provide a basis and guidance for the application of gas disaster control technology in fully mechanized caving mining.

3.1. The Establishment of a Geometric Model and the Determination of Model Parameters

Taking a mine in Shanxi as an example, the coal seam occurrence conditions and gas parameters of the mine are shown in Table 1. Under the condition of ignoring the borehole pressure loss, the gas migration process and pressure change law of any section perpendicular to the borehole axis are the same. Combined with the secondary stress distribution law of the coal seam after drilling, the length is 150 m, the width is 50 m, and the height is 6.31 m. For a cuboid geometric model with a borehole diameter of 0.12 m, the schematic diagram of the coal seam drainage borehole is shown in Figure 2, and the geometric simulation is shown in Figure 3.

Table 1. Coal seam occurrence conditions and gas parameters.

Parameter	Value
Coal seam thickness	6.36 m
Gas emission quantity	67.62 m^3/min
Initial stress σ/MPa	13.5 MPa
Radius of drill hole r_0/m	0.06 m
Coal seam permeability coefficient	0.0605~0.1209 m^2/MPa^2d
Coefficient of kinetic viscosity μ/Pa·s	11.2×10^{-6}
Initial void ψ_0	0.09
Initial gas pressure p_0/MPa	0.17
Adsorption constants $a/m^3 \cdot t^{-1}$	23.925
Adsorption constants $b/m^3 \cdot t^{-1}$	0.319
Coal density $\rho s/t \cdot m^{-3}$	1.39
Universal gas constant $R/(J \cdot (mol \cdot K)^{-1})$	8.3143
Absolute temperature T/K	300
Gas molar volume $n/m^3 \cdot mol^{-1}$	0.0224

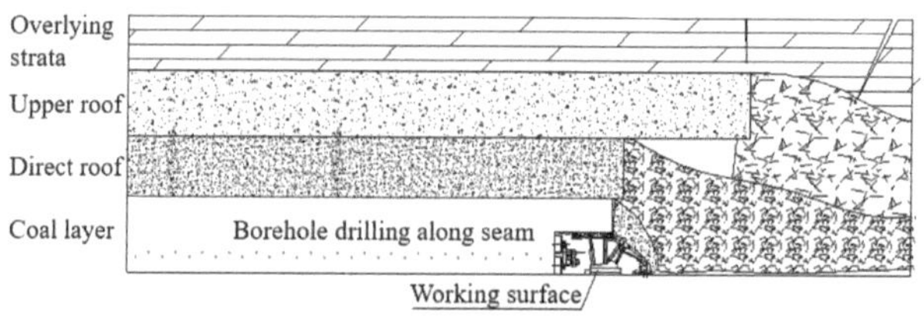

Figure 2. Pre-drainage coal seam gas drainage schematic diagram.

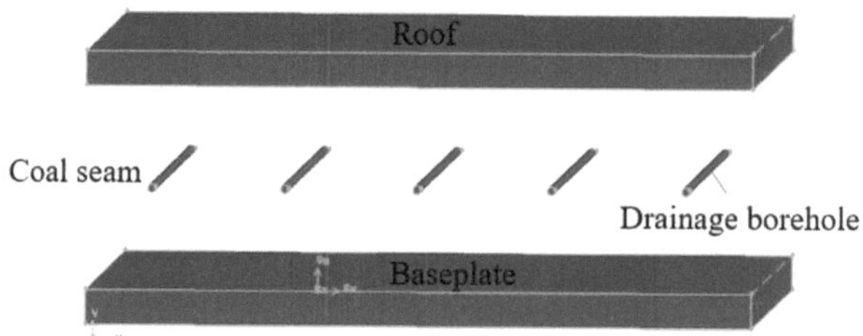

Figure 3. Geometric model of gas drainage in pre-drainage coal seam.

The setting of the boundary conditions mainly considers the porosity of the pre-drainage coal seam, the amount of gas analysis, the pre-drainage negative pressure, the pre-drainage borehole diameter, etc. The gas drilling in the coal seam is a kind of internal boundary condition for the simulation of gas drainage in the coal seam. When the gas drainage amount of the pre-drainage borehole is given, the known source and sink terms need to be supplemented in the gas motion equation. When the gas flow pressure of the pre-drainage borehole is given, it can be treated as a stable flow. The borehole flow pressure is usually known in the study of coal seam drainage. According to the relevant parameters of Table 1, the geometric area grid of the pre-drainage model is adjusted adaptively, and the numerical simulation parameters of the gas migration law of the pre-drainage coal seam are shown in Table 2.

Table 2. Simulation parameters of gas drainage in pre-drainage coal seam.

Simulated Condition	Simulation Parameters
Equation solver	Separation solver
Turbulent flow model	κ-ε two equations model
Component model	Gas
Energy model	Open
Convergence criteria	10^{-3}
Porous media model	Open
Percentage of void	0.09
Inertial resistance coefficient ($1/m^2$)	1.0×10^6
Pressure inlet (MPa)	0.17
Pressure outlet (KPa)	5

3.2. Simulation Study on Variation Law of Gas Pressure Under Different Aperture Conditions

Taking the relevant parameters of Tables 1 and 2 as the boundary conditions, the gas drainage process of 10 groups of boreholes, φ35, φ40, φ47, φ50, φ60, φ70, φ80, φ94, φ100, and φ120, is simulated. The gas pressure changes around the boreholes with diameters of φ70, φ80, φ94, φ100, and φ120 are shown in Figure 4, and the gas pressure changes around the boreholes with diameters of φ35, φ40, φ47, φ50, and φ60 are shown in Figure 5.

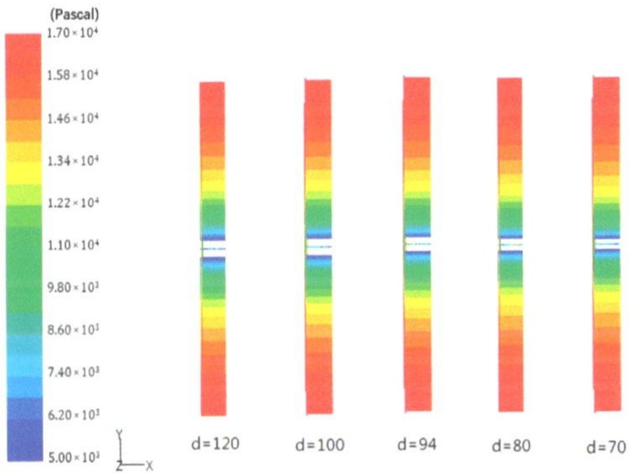

Figure 4. Gas pressure change cloud around different borehole diameters.

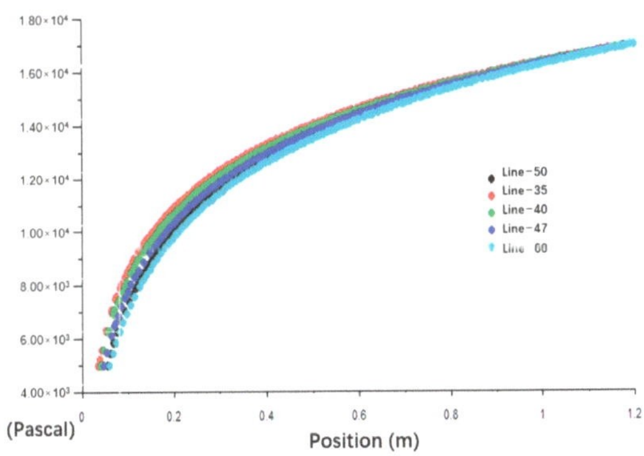

Figure 5. Gas pressure curves around different borehole diameters.

Figure 4 shows that due to the large borehole diameters, which are 70 mm, 80 mm, 94 mm, 100 mm, and 120 mm, the secondary distribution of stress around the borehole is very obvious. From Figure 4, the area range of the static pressure distribution can be clearly distinguished. Through the gas pressure distribution diagram around different boreholes, it can be seen that the size of the aperture has a different influence range on the surrounding gas pressure, and the influence range gradually increases with the increase in the borehole. The variation law of the borehole and gas pressure shown in this diagram is basically consistent with that in Figure 1, which also indirectly proves the reliability of the secondary distribution model of the stress field of the coal body around the borehole. Figure 5 shows that under the influence of drilling and drainage negative pressure, when the borehole aperture is small, as shown in Figure 5, where the aperture is 35 mm, 40 mm, 47 mm, 50 mm, and 60 mm, the gas pressure around the borehole also changes, but its influence range is small, and its influence range has little to do with the size of the borehole aperture, generally not more than 1–2 times the diameter of the borehole. The simulation results also show that the borehole must not be too small during the pre-drainage of the coal seam gas; otherwise, it will affect the drainage effect of the coal seam gas.

3.3. Simulation Study on Variation Law of Gas Pressure Under Different Permeability Conditions

Due to the different degrees of metamorphism, the permeability of the coal body also shows great differences. Under the condition that other conditions remain unchanged, a gas drainage simulation study was carried out on the three groups of coal seams with resistance viscosity coefficients of 3.54×10^{17}, 3.54×10^{13}, and 3.54×10^{9}. The change law of gas pressure around it is shown in Figures 6 and 7.

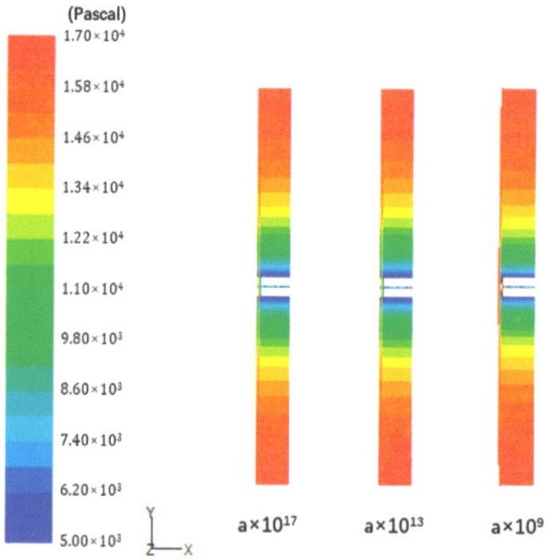

Figure 6. The pressure change cloud diagram around the aperture under different viscous resistance coefficients.

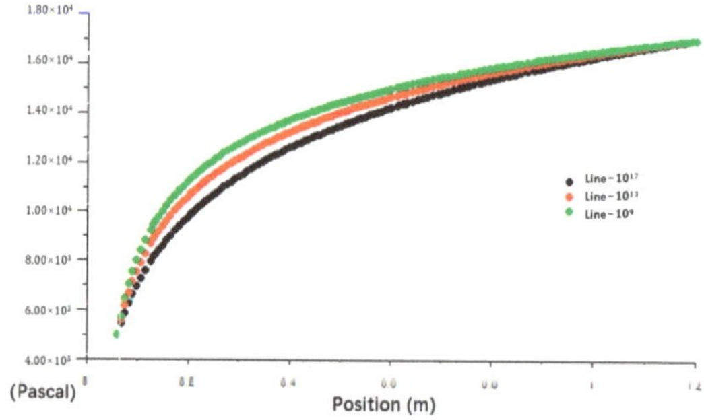

Figure 7. Pressure curves around the aperture under different viscous resistance coefficients.

Figures 6 and 7 show that the variation range of gas pressure around the borehole in the coal seam with good permeability is obviously larger than that in the coal seam with poor permeability. The variation range of gas pressure in the coal seam with good permeability is 1.7 m, while that in the coal seam with poor permeability is only 1.0 m. In the coal seam with good permeability, the change rate of gas pressure around the borehole is significantly higher than that in the coal seam with poor permeability. However, when the gas pressure drops to a certain range, the change rate of gas pressure becomes significantly weaker and eventually tends to be stable.

3.4. Simulation Study on Variation Law of Borehole Gas Pressure Under Different Initial Pressure Conditions

Due to the different occurrence conditions of coal seams, the original pressure of gas varies greatly. Under the condition that other parameters remain unchanged, the inlet pressure, P = 0.17 MPa, P = 0.5 MPa, P = 1.0 MPa, P = 1.5 MPa, and P = 2.0 MPa, is changed to simulate the pressure change law around the borehole under different original gas pressures. The change cloud diagram and curve diagram are shown in Figures 8 and 9.

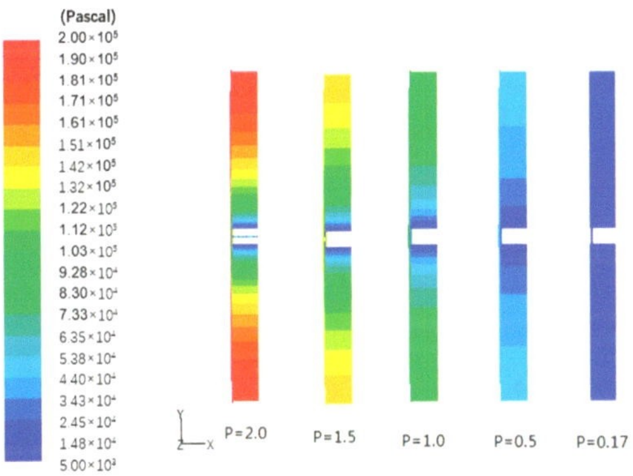

Figure 8. Pressure cloud diagram around the aperture under different original gas pressure.

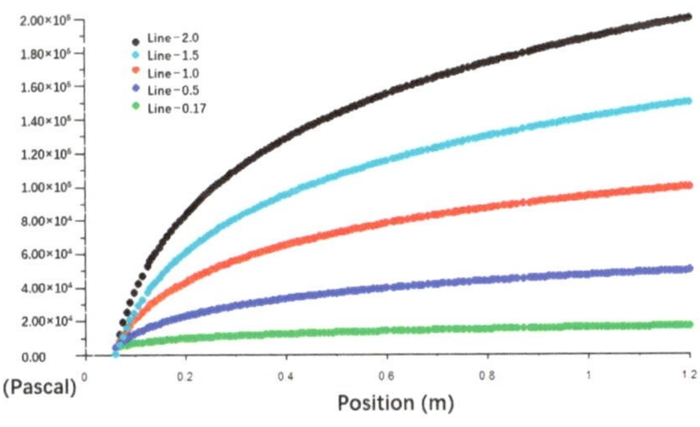

Figure 9. Pressure curve around the aperture under different original gas pressure.

Figures 8 and 9 show that the greater the original gas pressure, the greater the influence range of the gas pressure around the borehole, and the smaller the original gas pressure, the smaller the influence range of the gas pressure around the borehole; in the range of 0.4 m around the borehole, the pressure change rate gradually increases with the increase in the original gas pressure, indicating that the gas pressure attenuates rapidly under the same conditions. The influence range of gas pressure around all boreholes does not change linearly with the increase in the original pressure but changes in a curve law. Outside the 0.6 m around the borehole, the difference in the change rate of each curve gradually decreases, indicating that outside a certain range around the borehole, the original gas pressure has little effect on the control range of the borehole, and its influence range is still limited.

Figures 10–13 show that when the pressure ratio is 0.98, the gas around the pre-drainage coal seam is affected by the secondary stress, and the gas migration in the coal seam of the porous medium is quite different. When the borehole is small, such as when the diameter of the borehole is 35 mm and 45 mm, the influence range of the secondary stress around the borehole is small, and the influence areas are not evenly distributed, so the gas migration speed changes significantly. With the increase in the diameter of the borehole, the influence range of the secondary stress around the borehole gradually increases, and the range of each partition also increases, while the gas migration speed in each region is more regionally stable.

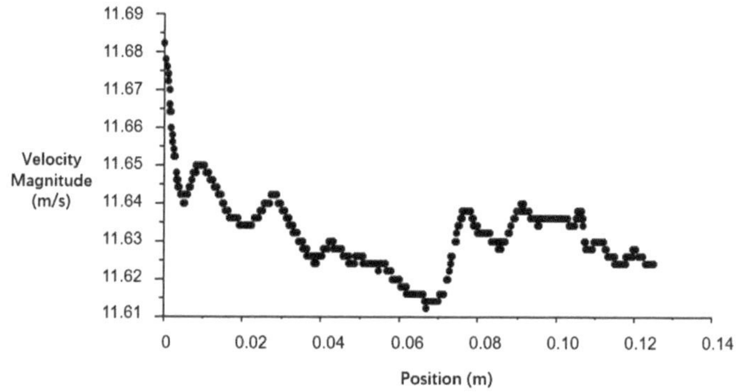

Figure 10. Gas migration velocity curve when the aperture is 35 mm and the pressure ratio is in the range of 0.98.

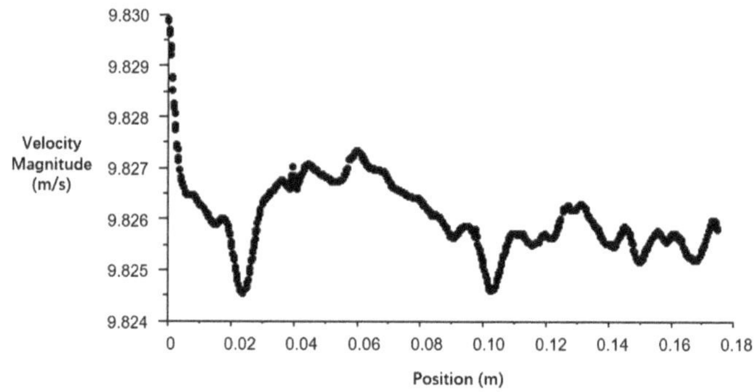

Figure 11. Gas migration velocity curve when the aperture is 50 mm and the pressure ratio is in the range of 0.98.

Figures 12 and 14–16 show that when the diameter of the borehole is the same, the pressure ratio around the borehole is different, so that the pressure of each pressure ratio area is different, and the deformation degree of the coal seam is also different. Therefore, the closer to the center of the borehole, the better the permeability of the coal seam, and the influence of the coal seam on the gas is gradually weakened. As shown in Figures 12 and 14–16, the smaller the pressure ratio, the faster the gas migration, and the more stable the operation. For example, when the pressure ratio drops to 0.3, the gas migration in the coal seam remains basically unchanged.

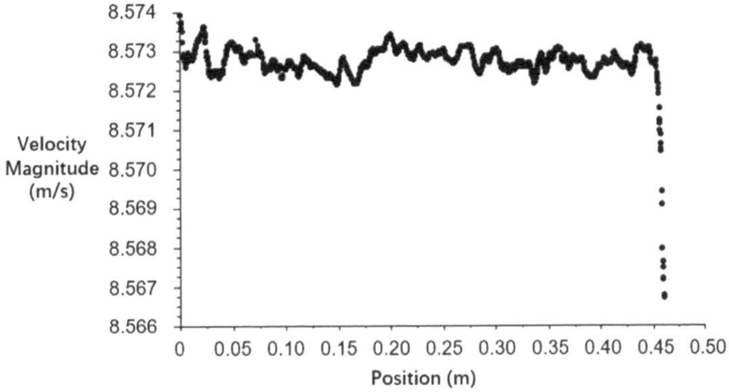

Figure 12. Gas migration velocity curve when the aperture is 65 mm and the pressure ratio is in the range of 0.98.

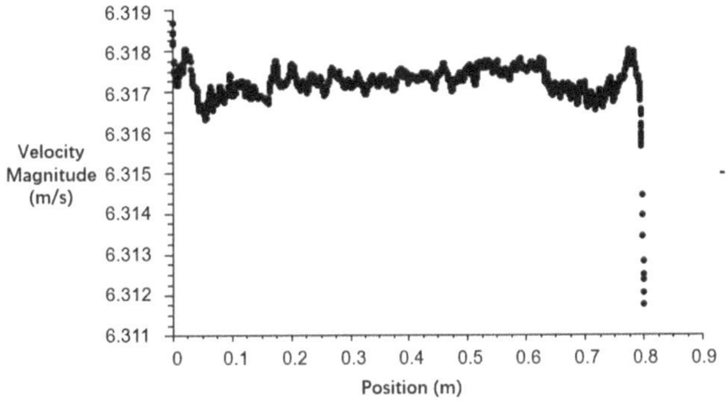

Figure 13. Gas migration velocity curve when the aperture is 80 mm and the pressure ratio is in the range of 0.98.

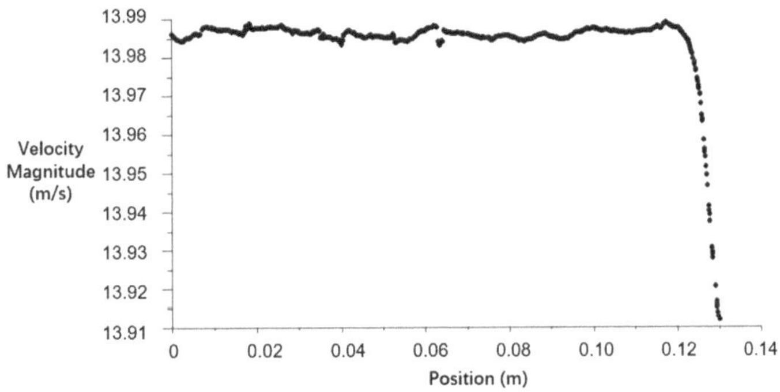

Figure 14. Gas migration velocity curve when the aperture is 65 mm and the pressure ratio is in the range of 0.75.

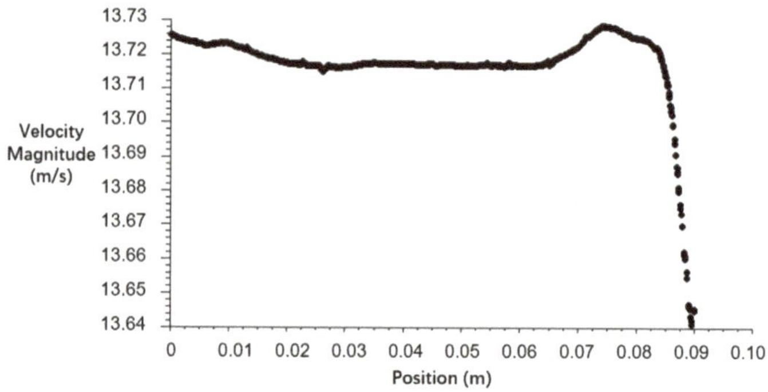

Figure 15. Gas migration velocity curve when the aperture is 65 mm and the pressure ratio is in the range of 0.5.

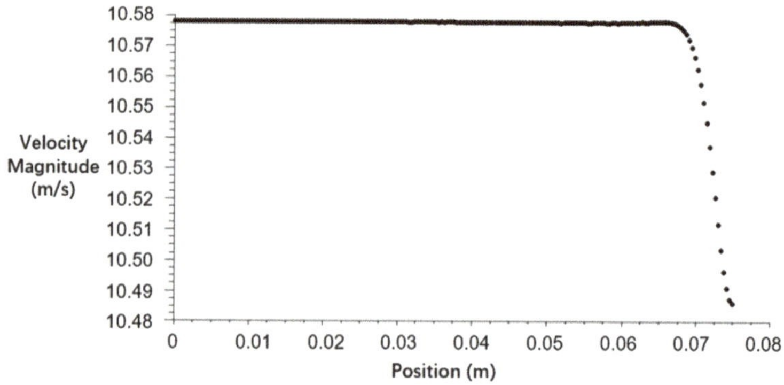

Figure 16. Gas migration velocity curve when the aperture is 65 mm and the pressure ratio is in the range of 0.3.

3.5. Analysis of Influencing Factors of Pre-Drainage Borehole Control Range

Generally speaking, the physical properties of coal, the borehole diameter, and the environment of the coal seam have a significant influence on the evolution of the flow field in the process of gas drainage. This section discusses the influence of borehole diameter, coal seam permeability, and gas original pressure on the control range, which will help to arrange gas drainage boreholes accurately and reasonably on site.

Borehole diameter. The influence range of the borehole is generally not more than 6~7 times the aperture. When the borehole aperture is 70 mm, the pressure influence range is 0.45 m, and the pressure influence range is 0.6 m when the aperture is 120 mm. The closer to the center of the borehole, the greater the change rate of the gas pressure, and the farther away from the center of the borehole, the smaller the change rate of the gas pressure, gradually approaching 0.

Coal seam permeability. The permeability of the coal seam is an important index in measuring the difficulty of gas drainage, and its control range is about 1.0~1.7 m. When the permeability of the coal seam is good, the gas pressure drop rate is small, and the drilling control range is large. When the permeability of the coal seam is poor, the gas pressure drop rate is large, and the drilling control range is small.

Gas original pressure. The control range of the gas drainage borehole and the original pressure of the gas change is a curve law. The higher the original pressure of the gas is, the larger the radius of the gas drainage is. A different original gas pressure will cause different gas pressure drop rates. In the range of 0.4 m around the borehole, the change rate of the pressure drop curve is larger. Outside 0.6 m around the borehole, the change rate of each curve tends to be consistent.

4. Field Practice and Effect Verification

In the study of the gas control effect in high-gas and extra-thick coal seams, field practice and effect verification are important links to ensure coal mine safety and efficient mining. First of all, according to the geological conditions of an extra-thick coal seam, a detailed mine gas composition analysis and pressure tests should be carried out to evaluate the gas content and its release characteristics. In the design of the drainage system, the combination of multi-point drainage and centralized drainage is usually used to achieve efficient gas removal. In the actual operation, by setting up multiple drainage holes, selecting the appropriate hole spacing and depth, and combined with the real-time monitoring system, the drainage parameters are dynamically adjusted. In this process, the gas drainage rate and the gas emission effect of the coal seam are the main assessment indicators.

The average thickness of the xxxx working face of a mine is 6.36 m. The initial stage of extraction is designed as a single row of gas extraction along the layer, and the diameter of the borehole is 70 mm. After a period of drainage, it is found that the gas drainage rate is low and the pre-drainage time is long, which cannot meet the production needs. After that, the gas drainage borehole is changed to 94 mm, and the upward and downward boreholes are arranged along the inclined direction of the working face in the belt crossheading and the auxiliary transportation crossheading, respectively. That is, half of the boreholes are parallel to the working face, and the other half of the boreholes have a certain angle with the working face and the horizontal plane. The two boreholes are alternately arranged to form a cross between the boreholes, as shown in Figure 17. According to the simulation results, the drilling control range is 0.7 m. When the cross hole is used, 70% of the coal seam can be within the drilling control range, so the drainage effect is obvious. Taking xxxx working face drainage in March 2013 as an example, the monthly gas drainage volume is 122,457 m^3, the daily maximum drainage volume is 6295 m^3, the minimum drainage volume is 2664 m^3, and the average drainage volume is 4479.5 m^3. The gas drainage rate reached 51.7%, and the pre-drainage time was greatly shortened. The drainage volume statistics are shown in Table 3.

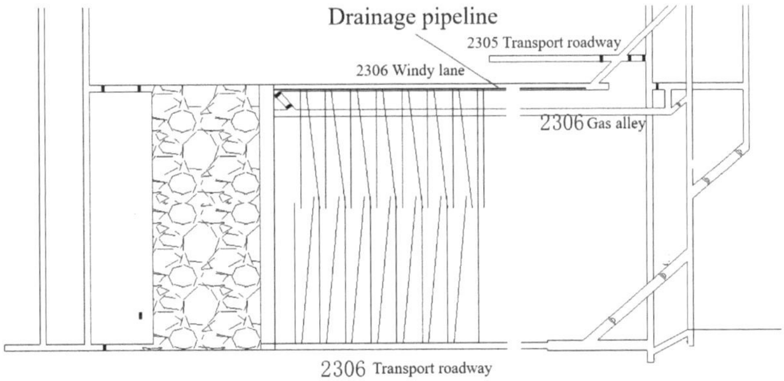

Figure 17. Working face drilling layout diagram.

The field data show that the optimized drainage design can effectively reduce the gas concentration in the coal seam and improve the safety factor. In the aspect of effect verification, the influence of gas drainage on mine safety and operation efficiency is evaluated by a comparison with the traditional drainage method. The measured results show that the newly designed drainage system not only effectively improves the gas drainage efficiency but also significantly reduces the risk of gas accidents, ensuring the safety of workers and the normal production of coal mines. In addition, the on-site feedback and continuous monitoring data provide a valuable basis for the subsequent optimization of the drainage scheme, forming a benign cycle improvement mechanism.

Table 3. Gas drainage statistics of xxxx working face in a mine.

Working Surface	Return Laneway		Tile Discharge Roadway		Upper Corner	Gas Drainage Volume (m^3/month)	Primitive Gas Content (m^3/t)	Residual Gas Content (m^3/t)	Drainage Rate (%)
	Wind Rate (m^3/min)	CH4 Average Concentration (%)	Wind Rate (m^3/min)	CH4 Average Concentration (%)	CH4 Average Concentration (%)				
xxxx	1120	0.39	310	0.54	0.45	122,457	5.53	2.67	51.7

Through the above practice and verification, the gas drainage design of a high-gas, extra-thick coal seam has achieved remarkable results in practical application, which provides a solid guarantee for safe coal mining. The results also further verify the reliability and accuracy of the numerical simulation results, which also show that the numerical simulation results can be used as a theoretical basis and reference for the gas drainage design of high-gas and extra-thick coal seams.

5. Discussion

This paper is based on the secondary distribution of coal seam stress around the borehole due to the influence of the borehole on the pre-drainage coal seam, which leads to the change in the porosity of the coal body around the borehole. Through theoretical analysis, the effect of borehole pre-drainage of coal seam gas control in the stress–seepage coupling field is studied in combination with the change in porosity around the borehole. The reliability and accuracy of the numerical simulation results are verified by the field-measured data. However, in view of the characteristics of complex coal seam occurrence conditions, large mining intensity, and many gas sources in fully mechanized top-coal caving mining, combined with the arduousness and urgency of gas disaster prevention and control technology development and research, there are still some key problems to be explored and discussed:

(1) In the process of high-gas coal seam mining, borehole drainage as an effective means of gas control has attracted extensive research attention. Based on the theoretical framework of the stress–seepage coupling field, this study deeply discusses the gas control effect of a borehole pre-drainage coal seam. The coupling relationship between stress and seepage has an important influence on the gas flow characteristics of the coal seam. Borehole drainage not only changes the pressure state in the coal seam but also significantly affects the release and migration mechanism of gas.

(2) Through the combination of numerical simulation and field practice data, we found that under different borehole apertures, coal seam permeabilities, and gas original pressures, the pre-drainage control range of boreholes shows obvious differences. In addition, the application of the stress–seepage coupling model enables us to predict the gas control effect under different drainage schemes more accurately, which provides a scientific basis for the safe mining of coal mines.

(3) In the practical application, the research results of the coupling model provide theoretical support for optimizing borehole drainage parameters and promote the development of gas control technology. However, it is still necessary to further study the drainage effect under different geological conditions and coal seam characteristics so as to provide more detailed guidance and reference for the production practice of coal mines. Future research should combine more field experiments and data analysis to verify the reliability of the theoretical model and explore more efficient gas control strategies.

6. Conclusions

Based on the theoretical analysis, this paper establishes a secondary distribution model of the stress field of a coal body around a borehole and a porosity change law model under the influence of stress–strain, The numerical simulation method is used to analyze the influence of different borehole diameters, different permeabilities, and different initial pressures on the gas pressure distribution around the coal seam borehole. According to the field-measured data, the mathematical model of the gas migration law in fully mechanized top-coal mining is verified. The simulation results have guiding significance for the study of gas control technology. The main conclusions are as follows:

(1) The law of secondary stress redistribution around the borehole and its influence range are given. The size of the coal seam stress, to a certain extent, reflects the depth of the coal seam. When the stress is large, the gas content and initial pressure of the coal seam are generally high, and the gas pressure around the borehole changes in a large range, but the larger ground stress will also affect the permeability of the coal seam. When the permeability decreases, the range of gas pressure around the borehole will also decrease.

(2) Under the premise of considering the influence of stress and strain around the borehole, a fluid–solid coupling model of the seepage zone based on different porosities around the borehole, a pressure-driven transition flow zone, and a free molecular flow zone were established. The model combines the in situ stress and the change in porosity around the borehole, which makes the simulation more objective and specific. The permeability of a coal seam is an important index in measuring the difficulty of gas drainage. When the permeability of a coal seam is good, the gas pressure drop rate is small and the drilling control range is large. When the permeability of a coal seam is poor, the gas pressure drop rate is large and the drilling control range is small.

(3) The simulation results of the influence of the borehole diameter, the permeability of the coal seam, and the original gas pressure on the control range of the gas drainage borehole in the pre-drainage coal seam after the secondary stress distribution around the borehole are compared and analyzed, which provides a theoretical basis for the selection of the gas process parameters of the pre-drainage coal seam in the coal mine.

(4) The effective control range of drilling is that the gas pressure of the coal seam is reduced to 30~50% of the original pressure. It is considered that the better the permeability of the coal seam is, the higher the original pressure of the gas is, the larger the aperture of the gas drainage borehole is, the larger the control range of the gas drainage borehole is, the lower the density of the drainage borehole is, and the more obvious the drainage effect is.

Author Contributions: Formal analysis, writing—original draft preparation, H.W.; supervision, conceptualization, methodology, resources, H.A.; validation, formal analysis, investigation, B.Y. All authors have read and agreed to the published version of the manuscript.

Funding: This research was supported in part by the Yunnan Province "Caiyun" Postdoctoral Innovative Project Plan (No. CG24056E004A).

Institutional Review Board Statement: Not applicable.

Informed Consent Statement: Not applicable.

Data Availability Statement: Data are available from the corresponding author upon reasonable request.

Acknowledgments: The authors would like to express their sincere gratitude to the Kunming University of Science and Technology 2023 Undergraduate Education Teaching Reform Research Project (No. 075, KUST Education Office [2023]) for their support. Moreover, the authors would like to thank the anonymous reviewers for their valuable comments and constructive suggestions.

Conflicts of Interest: The authors declare no conflicts of interest.

References

1. Wang, H.; Jiang, J.; Wang, Z.Q.; Hu, G.Z.; Yuan, Z.G. Numerical simulation of seepage field of gas extraction drilling of single bedding of mining-coal bed. *Chongqing Daxue Xuebao/J. Chongqing Univ.* **2011**, *34*, 24–29.
2. Liu, Q.Q.; Cheng, Y.P.; Wang, H.F.; Liu, H.Y.; Liu, J.J. Numerical Resolving of Effective Methane Drainage Radius in Drill Hole along Seam. *Coal Min. Technol.* **2012**, *17*, 5–7.

3. Zhao, J.G. Construction technology and development tendency of high level directional driling in seam roof. *Coal Sci. Technol.* **2017**, *45*, 137–141.
4. Zhu, M. 3D simulation and control effect evaluation of gas extraction drilling in coal mine. *Min. Saf. Environ. Prot.* **2024**, *51*, 50–55. [CrossRef]
5. Terzaghi, K. *Theoretical Soil Mechanics*; John Wiley & Sons: Hoboken, NJ, USA, 1943; Volume 2, pp. 180–189. [CrossRef]
6. Biot, M.A. General theory of three-dimensional consolidation. *J. Appl. Phys.* **1941**, *12*, 155–164. [CrossRef]
7. Biot, M.A. Theory of elasticity and consolidation for a porous anisotropic solid. *J. Appl. Phys.* **1955**, *26*, 182–185. [CrossRef]
8. Biot, M.A. General Solutions of the Equations of Elasticity and Consolidation for a Porous Material. *J. Appl. Mech.* **1956**, *78*, 91–96. [CrossRef]
9. Lewis, R.W.; Roberts, P.J.; Schrefler, B.A. Finite element modelling of two-phase heat and fluid flow in deforming porous media. *Transp. Porous Media* **1989**, *4*, 319–334. [CrossRef]
10. Wang, H.; Dlugogorski, B.Z.; Kennedy, E.M. Pathways for Production of CO_2 and CO in Low-Temperature Oxidation of Coal. *Energy Fuels* **2003**, *17*, 150–158. [CrossRef]
11. Beamish, B.; Lau, A.; Moodie, A.; Vallance, T.A. Assessing the self-heating behaviour of Callide coal using a 2-metre column. *J. Loss Prev. Process Ind.* **2002**, *15*, 385–390. [CrossRef]
12. Harpalani, S.; Chen, G. Estimation of changes in fracture porosity of coal with gas emission. *Fuel* **1995**, *74*, 1491–1498. [CrossRef]
13. Lewis, R.; Sukirman, Y. Finite element modelling of three-phase flow in deforming saturated oil reservoirs. *Int. J. Numer. Anal. Methods Geomech.* **1993**, *17*, 577–598. [CrossRef]
14. Zhang, Z.G. Prediction of multi-borehole undermine coalbed gas drainage. *J. Coal Sci. Eng. (China)* **2009**, *15*, 295–298. [CrossRef]
15. Li, Z.L. New definition of correlation function with interval number and its application in extenics. *Math. Pract. Theory* **2006**, *36*, 207–210.
16. Shen, Z.; Zhang, X.; Sun, Y. Research on stress-seepage-damage coupling model of hydraulic fracturing for rock mass. *Jisuan Lixue Xuebao/Chin. J. Comput. Mech.* **2009**, *26*, 523–528.
17. Jasinge, D.; Ranjith, P.; Choi, S. Effects of effective stress changes on permeability of latrobe valley brown coal. *Fuel* **2011**, *90*, 1292–1300. [CrossRef]
18. Cao, J.; Dong, L.; Ai, D.; Wang, F.; Cui, Y.G. Nonlinear seepage law of deep hole water-injection in working face under the influence of mining. *J. China Coal Soc.* **2017**, *42*, 225–232.
19. Zhao, Y.; Lin, B.; Liu, T.; Zheng, Y.; Kong, J.; Li, Q.; Song, H. Mechanism of multifield coupling-induced outburst in mining-disturbed coal seam. *Fuel* **2020**, *272*, 117716. [CrossRef]
20. Liu, Z.; Hu, P.; Yang, H.; Yang, W.; Gu, Q. Coupling Mechanism of Coal Body Stress–Seepage around a Water Injection Borehole. *Sustainability* **2022**, *14*, 9599. [CrossRef]
21. Yuan, L.; Guo, H.; Shen, B.; Qu, Q.; Xue, J.-H. Circular overlying zone at longwall panel for efficient methane capture of mutiple coal seams with low permeability. *J. China Coal Soc.* **2011**, *36*, 365.
22. Hu, Q.T.; Liang, Y.P.; Liu, J.Z. CFD simulation of goaf gas flow patterns. *Meitan Xuebao/J. China Coal Soc.* **2007**, *32*, 719–723.
23. Li, S.-G.; Zhang, W.; Lin, H.-F. Numerical simulation on the delivery law of gob gas of fully mechanized caving face. *J. Coal Sci. Eng.* **2008**, *14*, 403–406. [CrossRef]
24. Zhang, T.; Li, W.; Li, S.; Song, S.; Tang, F. Gas emission law in high gas goaf of fully mechanized caving face. *Sci. Technol. Rev.* **2015**, *33*, 58–62. [CrossRef]
25. Wang, T.; Nie, C.; Yang, X.; Wang, P. Numerical simulation of gas drainage in gob considering temperature change. *Coal Sci. Technol.* **2021**, *49*, 85–94.
26. Hao, F.; Liu, M.; Sun, L.J. Determination method of gas drainage radius based on multi physics coupling. *J. China Coal Soc.* **2013**, *38* (Suppl. S1), 106–111.
27. Yang, T.H.; Chen, S.K.; Zhu, W.C.; Liu, H.L. Coupled Model of Gas-Solid in Coal Seams Based on Dynamic Process of Pressure Relief and Gas Drainage. *Rock Soil Mech.* **2010**, *31*, 2247–2252.
28. Ding, H.; Jiang, Z.; Han, Y. Numerical simulation and application of boreholes along coal seam for methane drainage. *Chin. J. Eng.* **2008**, *30*, 1205–1210.
29. Wang, Y.G.; Li, H.Y.; Qi, Q.X.; Peng, Y.W.; Li, C.R.; Deng, Z.G. The evolution of permeability and gas extraction technology in mining coal seam. *Meitan Xuebao/J. China Coal Soc.* **2010**, *35*, 406–410.
30. Furati, K.M. Effects of Relative Permeability History Dependence on Two-Phase Flow in Porous Media. *Transp. Porous Media* **1997**, *28*, 181–203. [CrossRef]
31. Bentsen, R.G. The Physical Origin of Interfacial Coupling in Two-Phase Flow through Porous Media. *Transp. Porous Media* **2001**, *44*, 109–122. [CrossRef]
32. Zhao, Y.-J.; Guo, H.-D.; Yuan, S.-J. The Study on the Key Gas Drainage Technique by Long & large Diameter Roof Boreholes in Fully Mechanized Face. *J. Taiyuan Univ. Technol.* **2009**, *40*, 74–77.
33. Ertekin, T.; King, G.A.; Schwerer, F.C. Dynamic Gas Slippage: A Unique Dual-Mechanism Approach to the Flow of Gas in Tight Formations. *Spe Form. Eval.* **1986**, *1*, 43–52. [CrossRef]
34. Ertekin, T.; Sung, W.; Schwerer, F.C. Production Performance Analysis of Horizontal Drainage Wells for the Degasification of Coal Seams. *J. Pet. Technol.* **1988**, *40*, 625–632. [CrossRef]

35. Smith, D.M.; Williams, F.L. Diffusional effects in the recovery of methane from coalbeds. *Soc. Pet. Eng. J.* **1984**, *24*, 529–535. [CrossRef]
36. Ancell, K.; Lambert, S.; Johnson, F.S. Analysis of the coalbed degasification process at a seventeen well pattern in the Warrior Basin of Alabama. In *SPE Unconventional Gas Recovery Symposium*; SPE: Pittsburgh, PA, USA, 1980. [CrossRef]
37. Guo, Y.S.; Lin, B.Q.; Wu, C.S. Coupling relation of crack evolvement in surrounding rocks with the storage and migration of mining induced releasing gas. *J. Min. Saf. Eng.* **2007**, *24*, 4.
38. Liu, Z.; Li, Z.; Yang, Y.; Ji, H. Experimental study of effect of water on sorption and radial gas seepage of coal. *Yanshilixue Yu Gongcheng Xuebao/Chin. J. Rock Mech. Eng.* **2014**, *33*, 586–593.
39. Tao, Y.; Xu, J.; Cheng, M.; Li, S.; Peng, S. Theoretical analysis and experimental study on permeability of gas-bearing coal. *Yanshilixue Yu Gongcheng Xuebao/Chin. J. Rock Mech. Eng.* **2009**, *28*, 3363–3370.
40. Yin, G.-Z.; Li, M.; Li, S.-Z.; Li, W.; Yao, J.-W.; Zhang, Q. 3D numerical simulation of gas drainage from boreholes based on solid gas coupling model of coal containing gas. *J. China Coal Soc.* **2013**, *38*, 535–541.
41. Ju, L. Solid-Gas Coupling Dynamics Model and It's Application Research of Coal Containing Gas. *China Min. Mag.* **2013**, *22*, 126–131.
42. Si, H.; Guo, T.; Li, X.H. Analysis and numerical simulation of fluid-structure coupling of gas drainage from boreholes. *J. Chongqing Univ. (Nat. Sci.)* **2011**, *34*, 105–110.

fire

Article

Simulation Study of Gas Seepage in Goaf Based on Fracture–Seepage Coupling Field

Hongsheng Wang [1], Huaming An [2,3,*] and Xin Zhang [2]

1 Beijing Chemical Occupational Disease Prevention and Control Institute, Beijing 100093, China; cumming.an@gmail.com
2 Faculty of Public Security and Emergency Management, Kunming University of Science and Technology, Kunming 650093, China; zhangxin1219@stu.kust.edu.cn
3 Geotechnical Institute, TU Bergakademie Freiberg, 09599 Freiberg, Germany
* Correspondence: huaming.an@kust.edu.cn

Abstract: In order to solve the problem of gas overrun in the fully mechanized caving face and the upper corner of high gas and extra-thick coal seam, the fracture and caving process of the roof in the goaf is analyzed and studied by using the relevant theories of fracture mechanics and seepage mechanics. The mathematical model of fracture and caving of the immediate roof and main roof in the goaf is established. Combined with ANSYS Fluent 6.3.26, the seepage process of gas in coal and rock accumulation in the goaf under different ventilation modes is simulated. The distribution law of gas concentration in the goaf is obtained, and the application scope of different ventilation modes is determined. In addition, the influence of the tail roadway application and the wind speed size on the gas concentration in the goaf and the upper corner of the fully mechanized caving face is also explored. The results show that, affected by wind speed and rock porosity, along the strike of the goaf, about 30 m near the working face, the gas concentration is low and growth is slow. In the range of 30~160 m, the gas concentration increases rapidly and reaches a higher value. After 160 m, the gas concentration tends to be stable. Along with the tendency of the working face, the gas concentration in the goaf increases gradually from the inlet side to the return side, and the gas concentration increases noticeably near the return air roadway. Along the vertical direction of the goaf, the gas concentration gradually increases, and the concentration of the fracture zone basically reaches 100%. Different ventilation modes have different application scopes. The U-type ventilation mode is suitable for the scenario of less desorption gas in the coal seam, while U + I and U + L-type ventilation modes are suitable for the scenario of more desorption gas in coal seam or higher mining intensity. The application of the tail roadway can reduce the gas concentration in the upper corner to a certain extent, but it has limited influence on the overall gas concentration distribution in the goaf. In addition, when the wind speed of the working face should be controlled at 2.0~3.5 m/s, it is more conducive to the discharge of gas, the method of reducing the gas concentration in the upper corner by increasing the wind speed of the working face is more suitable for the case where the absolute gas emission of the fully mechanized caving face is low, and the effect is limited when the absolute gas emission is high. The above conclusions provide a reference for solving the problem of gas overrun in the goaf and the upper corner of a fully mechanized caving face.

Keywords: high gas and extra-thick coal seams; goaf; gas seepage; liquid-solid coupling; numerical simulation

Citation: Wang, H.; An, H.; Zhang, X. Simulation Study of Gas Seepage in Goaf Based on Fracture–Seepage Coupling Field. *Fire* **2024**, 7, 414. https://doi.org/10.3390/fire7110414

Academic Editors: Chuyuan Huang, Haipeng Jiang and Lijuan Liu

Received: 3 October 2024
Revised: 31 October 2024
Accepted: 7 November 2024
Published: 13 November 2024

1. Introduction

As an important basic energy source for economic and social development, coal has experienced great progress in mining scope and intensity. It plays a vital role in national economic development and modernization.

With the development of the economy and the progress of science and technology, large and super-large coal mines have been put into operation one after another. Coal mining has also changed from traditional blasting and general mining to fully mechanized mining with a higher efficiency. The record of mining coal seam depth has been consecutively broken. The high gas pressure and high content of coal seam have become important characteristics of modern coal mining. With the increase in coal mining depth, the intensity of gas explosions increases yearly [1,2]. Accompanied by gas explosions, especially serious gas explosion accidents still occur frequently. For example, from 2003 to 2013, there were 16 gas explosion accidents with more than 10 victims in Anhui Shengluling Coal Mine, Henan Daping Coal Mine, Shanxi Chenjiashan Coal Mine, Heilongjiang Dongfeng Coal Mine, Jilin Babao Coal Industry Company, Jilin Qingxing Coal Mine, Guizhou Dashan Coal Mine, Sichuan Taozigou Coal Mine, Hunan Simachong Coal Mine, and so on. There were seven gas explosion accidents with more than 100 deaths in each accident. And with the increase in coal mining intensity, gas disaster accidents are more prominent. Therefore, the prevention and control of gas disasters in the fully mechanized mining face of the high gas and extra-thick coal seam have become the focus of coal mine safety management. In order to explore the correlation between gas pressure and disaster prevention and control, researchers have carried out many related experiments, indicating that gas seepage and desorption in coal seams are the key causes of gas outbursts and explosions [3–16]. However, due to the particularity of its occurrence conditions, the seepage process of gas is affected by various external forces such as the in situ stress field, geoelectric field, and geothermal field, and various factors will have different degrees of influence on its flow in porous media. In this regard, experts and scholars have made active attempts to summarize the migration law of the coal seam and goaf gas under the action of a multi-physical field.

As early as 1943, Terzaghi K [17] studied the coupling of rock–soil–fluid. He first treated the flow of saturated fluid in one-dimensional elastic porous media as a flow–deformation coupling problem and proposed the famous effective stress formula. A one-dimensional consolidation model was established. Since then, many experts and scholars at home and abroad have made more valuable attempts to summarize the migration law of gas in the coal seam and goaf under the action of a multi-physical field [18–25]. In recent decades, Zhao et al. [26] proposed a nonlinear coupling mathematical model of solid deformation and gas seepage and established a finite element model for the numerical analysis to explain the migration and exchange of methane in the rock matrix and fractures. In 2006, Yang et al. [27] discussed the influence of fracture pore pressure on methane and carbon dioxide gas seepage in coal seams. It was found that the gas permeability coefficient in rock fractures changed parabolically with fracture pore pressure, and the gas permeability was negatively exponential with normal deformation and tangential deformation. Based on the influence mechanism of permeability, Connell (2009) [28] established a coupling model of gas flow and geomechanics and explored the applicability of the rock mechanics hypothesis to coal seam gas extraction. In 2010, Zhao et al. [29] studied the seepage law of gas in the goaf and the overlying strata. By establishing the mathematical model of the seepage field in the goaf, the seepage field equation was solved by the LBM digital method. Ronny et al. (2011) [30] established a one-dimensional mathematical coupling model that includes gas flow, adsorption, and geomechanical properties. In 2012, Li et al. [31] studied the influence of deep coal seam mining on gas seepage. In 2015, Wang et al. [32] studied the gas seepage characteristics of the coal seam by using the self-developed gas seepage triaxial experimental system, analyzed the influence of the Klinkenberg effect on gas seepage, and proposed a method for calculating the permeability of the coal seam considering the influence of gas dynamic viscosity, the compression coefficient, and the Klinkenberg effect. Danesh et al. (2016, 2017) [33,34] established a multi-field coupling model considering creep, simulated the creep-seepage evolution of the coal seam during gas recovery, and discussed the influence of creep on coal seam permeability under different conditions. Bertrand et al. (2017) [35] developed a hydraulic model for CBM production and analyzed a range of parameters that can highlight the impact of production scenarios or key reservoir-related parameters. In 2017, Zhang et al. [36] studied the influence of loading rate on gas seepage and temperature in coal and found that the seepage velocity of gas had critical deceleration characteristics under load. In 2019, Qiu et al. [37] established a microscopic pore structure model of coal and simulated the seepage behavior of gas in coal under 3 MPa pressure. It was found that the heterogeneity of the coal structure would lead to irregular changes in the cross-sectional area of pore channels along the seepage direction, which made the average seepage velocity of gas fluctuate greatly. With the increase in seepage length, the average seepage pressure and gas mass flow gradually decreased. In the same year, Xue et al. [38] established a hydraulic–mechanicaldamage coupling model considering the coupling relationship between coal damage, gas seepage, and coal deformation. They obtained the migration law of gas with coal seam failure. In 2021, Yan et al. [39] used the lattice Boltzmann method (LBM) at the representative elementary volume (REV) scale to simulate the gas seepage of the reconstructed pore structure and obtain the gas velocity distribution. The effects of porosity and pore structure on gas seepage were studied. It was found that the gas seepage velocity and apparent permeability increased with the increase in porosity. In 2024, based on the smooth particle hydrodynamics (SPH) method, Mu et al. [40] proposed a meshless numerical model of gas seepage in fractured coal bodies, studied the seepage law of gas in homogeneous and heterogeneous coal seams and coal bodies with single prefabricated defects, and revealed the seepage and stress coupling mechanism of fractured coal bodies during gas extraction.

Reviewing the existing research shows that the influence of various factors, including in situ stress, pore pressure, heterogeneity of the coal seam, and so on, on gas seepage has been discussed, and the law of gas seepage under different factors has been obtained. However, in the existing research field, few studies consider the influence of the ventilation mode and wind speed on gas seepage. In addition, due to the low permeability of the coal seam and the short time of gas extraction, the gas concentration in the working face, especially in the upper corner, is still high in the mining process of high gas and extra-thick coal seam. At present, the commonly used treatment methods in mines are buried pipe gas drainage in the goaf [41], using high drainage roadways to extract gas in the fracture zone, excavating directional long drilling groups [42], and excavating tail roadways to control gas seepage in the goaf. Among them, tail roadway extraction technology is one of the most important methods of controlling gas in the upper corner. It is important to facilitate the extraction of gas sources such as the goaf, adjacent layer, and surrounding rock in the working face by excavating the special gas discharge roadway so as to reduce the gas concentration in the upper corner of the working face, thus eliminating the gas overrun in the upper corner. This technology has been widely used, but there are few studies on the mechanisms of gas overrun in the upper corner of the tail roadway, especially for analyzing the gas seepage law. Starting from the theoretical basis, this paper fully analyzes the formation and evolution law of the goaf in the process of fully mechanized top coal caving mining, and it establishes the mathematical model of the initial fracture of the coal seam roof and the separation and caving model of the immediate roof and main roof, which lays a foundation for the study of the gas migration law in the goaf by a numerical simulation method. The influence of the working face wind speed and working face ventilation mode on gas concentration in the working face, goaf, and upper corner is emphatically analyzed. It also makes up for the shortcomings of other experts and scholars in this field.

In summary, this paper takes the caving time effect of the roof surrounding rock as the access point and uses the actual situation of gas control in the goaf of the N2306 working face of a mine in Shanxi Province. By studying the fracture and caving process of the roof, the ANSYS Fluent 6.3.26 is used to simulate the fluid–solid coupling process of gas in the collapsed coal and rock mass under different ventilation modes to analyze the distribution law of gas concentration in the goaf and to explore the influence of the ventilation mode and wind speed on gas seepage in the goaf. In addition to them, by focusing on the analysis of the influence of different ventilation methods (i.e., whether to set up tail roadways and different types of tail roadways) on the law of gas seepage, the application of outburst tail roadways has a positive effect on solving the problem of gas accumulation in the upper corner.

2. Materials and Methods

When the fully mechanized caving mining is adopted, the goaf is formed when the upper coal body is released. There are the pressure arch hypothesis, cantilever beam hypothesis, hinged rock block hypothesis, pre-formed fracture hypothesis, and masonry beam hypothesis in the failure process of the goaf roof. Although the contents and forms expressed by each hypothesis differ, they all show a delay in roof fracture and caving in the goaf. That is, with the advancement of the working face, the upper rock will form a certain length of probe in the goaf under the influence of its strength. The formation and existence of this probe create favorable conditions for the tail roadway to extract gas from the goaf and reduce gas concentration in the upper corner.

2.1. Initial Fracture Step Distance of Coal Seam Roof

Under the condition that the coal seam roof is simply supported on one side of the goaf and fixed on three sides, when it is in the limit suspension state, the three fixed edges form a negative bending moment area and the maximum main bending moment value, M_a, is in the middle of the long fixed edge. The positive bending moment area is formed in the center of the goaf, and the maximum main bending moment, $M_c = \sqrt{M_x^2 + M_y^2}$. According to the modified Marcus solution, the following can be obtained:

$$|M_a| = \frac{(1-\mu^2)(1+\mu\lambda_1^2)}{12(1+\lambda_1^4)} \cdot qa_1^2 \tag{1}$$

$$|M_a| = \frac{(1-\mu^2)\lambda_1^2(\mu+\mu_1^2)}{12(1+\lambda_1^4)} \cdot qb_1^2 \tag{2}$$

According to the following relationship between bending moment and stress:

$$M_a = \frac{h^2\sigma_s}{6} \tag{3}$$

By substituting Equation (3) into Equation (2), the relationship between the first fracture step distance of the coal seam roof under the condition of simple support and three-sided fixed support on one side of the goaf can be obtained as follows:

$$a = \frac{2h}{1-\mu^2}\sqrt{\frac{\sigma_s}{q} \cdot \frac{2+\lambda^4}{4+3\mu\lambda^2}} \tag{4}$$

Let $l_m = \frac{h}{1-\mu^2}\sqrt{\frac{\sigma_s}{q}}$,
Then, Equation (4) is modified as follows:

$$a = \begin{cases} b^4\sqrt{\frac{l_m^2}{2(b^2-l_m^2)}} \\ \frac{b}{l_m}\sqrt{b^2 - \sqrt{b^4 - 2l_m^4}} \end{cases} \tag{5}$$

where μ is the Poisson's ratio of the rock stratum; q is the weight of the rock stratum and its load; and λ_1 is the geometric shape coefficient of the goaf, $\lambda_1 = a_1/b$, a_1 it is the advancing distance of the working face, and b is the length of the working face. Moreover, h is the thickness of the main roof strata, and σ_s is the tensile strength of the main roof strata.

2.2. The Separation Layer Collapse of the Immediate and Main Roofs

Whether the immediate roof and the main roof fall off from the layer can be judged by the maximum disturbance.

The maximum disturbance of the main roof is given as follows:

$$y_{max} = \frac{(\gamma h_1 + q_1)L_1^4}{384E_1J_1} \tag{6}$$

where q_1 is the load applied to the main roof, γ is the load per unit length of the main roof itself, L_1 is the initial caving step distance, h_1 is the thickness of the main roof, E_1 is the elastic modulus of the main roof, and J_1 is the section inertia moment of the main roof.

The maximum disturbance of the immediate roof is given as follows:

$$y_{max} = \frac{\sum h\gamma L_1^4}{384E_2J_2} \tag{7}$$

where $\sum h$ is the thickness of the immediate roof, E_2 is the elastic modulus of the immediate roof, and J_2 is the section moment of inertia of the immediate roof.

According to the conditions of the separation phenomenon between the immediate roof and the main roof, that the following can be observed:

When $\frac{(\gamma h_1+q_1)L_1^4}{384E_1J_1} < \frac{\sum h\gamma L_1^4}{384E_2J_2}$, the immediate roof and the main roof are prone to separation and collapse.

By simplifying Equation (7), it can be approximated that when the thickness of the immediate roof is less than or equal to the thickness of the main roof, the immediate roof separation layer is prone to collapse.

3. Numerical Simulation

3.1. *Overview of Fully Mechanized Caving Face*

The N2306 working face of a mine in Shanxi has a length of 275 m, an average height of 6.31 m, an inclination angle of 2~6°, a mining height of 3.5 m, a caving height of about 2.8 m, and a mining/caving ratio of 1:1.04. The cycle occurs six times a day, and the cycle progress is 0.8 m. The gas content of the coal seam is 1.81~9.68 m^3/t, with an average of 5.49 m^3/t. On the top of the coal seam is a false roof, the lithology is generally black carbonaceous mudstone and mudstone; the thickness is not stable, generally 0.25~0.44 m, only in local development; the immediate roof is composed of gray-black mudstone and sandy mudstone, and it is sometimes interbedded with sand and mud. The thickness is generally 1.0~7.0 m, with an average thickness of 5.35 m. The lithology of the main roof is medium-fine gray-white quartz feldspar sandstone, locally coarse sandstone, with an average thickness of 5.8 m, and the height of the caving zone in the goaf is about 30 m. The relevant parameters of the working face are shown in Table 1. Combined with Equation (6) and Equation (7), it can be seen that in advancing the working face, the immediate and main roofs of the coal seam will separate and collapse. According to Equation (5), the length of the immediate roof suspension beam is 0.91 m, and the accumulation height of the immediate roof caving rock layer is 6.4 m. The length of the main roof suspension beam is 6.35 m, the accumulation height of the direct roof and the main roof caving rock mass is 13.4 m, and the height of the goaf fracture zone is about 30 m.

Table 1. Parameters of N2306 working face coal seam roof.

Strata Name	Strata Thickness (m)	Bulking Coefficient	Poisson Ratio	Elastic Modulus (MPa)	Tensile Strength (MPa)	Rock Load (MPa)	Coal Seam Thickness (m)	Fracture Interval (m)	Falling Body Height (m)
Immediate roof	5.35	1.2	0.25	1.5×10^4	2.2	13.5	6.31	0.91	6.4
Main roof	5.8	1.2	0.25	3.5×10^4	7	13.7	6.31	6.35	13.4

3.2. *The Establishment of the Geometric Model and the Determination of Simulation Parameters*

3.2.1. Establishment of the Geometric Model of Goaf

According to the theory of spatial zone division of rock mass deformation around the stope, the goaf can be divided into 'three zones' in both vertical and horizontal directions, shown in Figure 1.

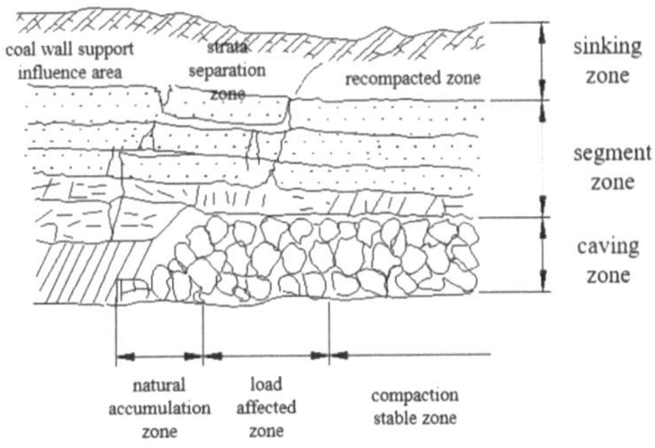

Figure 1. Division of the three horizontal and vertical zones in goaf.

Due to the limitations of the site, it is not suitable to carry out the actual measurement, so the height of the caving zone and the fracture zone are estimated by the empirical formula. The height of the caving zone in the goaf is affected by the thickness of the mining coal seam and the occurrence conditions and bulkiness of the roof rock, and it is usually three to five times the thickness of the coal seam. In the actual mining process, the height of the air-conducting zone and the caving zone can be expressed by Equation (8) and Equation (9), as follows:

$$h = \frac{100m}{1.6m + 3.6} \pm 5.6 \quad (8)$$

where m is the thickness of the mining coal seam.

$$h = \frac{m}{(k_p - 1)cos\alpha} \quad (9)$$

where α is the dip angle of coal seam and k_p represents the bulkiness of the rock.

In the N2306 fully mechanized caving face, the thickness of the coal seam is 6.31 m, and the dip angle is 2~6°. According to Equation (9), the height of the caving zone is given as follows:

$$h = \frac{6.11}{(1.4 - 1)cos(4/180)} = 15.28 \text{ m} \quad (10)$$

In the N2306 working face, the roof and floor are based on mudstone and contain a small amount of fine-grained sandstone. The characteristics of the rock strata are treated according to the medium-hard rock strata, and the height of the air-conducting zone (the sum of the caving zone and the fractured zone) can be obtained as follows:

$$h = \frac{100 \times 6.11}{1.6 \times 6.11 + 3.6} \pm 5.6 = 45.68 \pm 5.6 \text{ m} \quad (11)$$

Therefore, the height of the caving zone in the N2306 working face is 15.28 m, the fracture zone is 30.4 m, and the air conduction zone is 45.68 m.

According to the actual situation of the site, combined with the need for numerical simulation, the geometric model of the goaf in the N2306 working face will be simplified. The details are as follows:

- The working face, goaf, intake and return air roadways, and tail roadway are all regarded as cuboids without considering the equipment of the working face;
- The goaf is divided into caving zone 1, caving zone 2, caving zone 3, fracture zone 1, and fracture zone 2 to distinguish different porosities;
- The length of the working face is 275 m, the width is 5.5 m, and the height is 3.5 m. The strike length of the goaf is 200 m, the dip length is 285 m (including the intake and return air roadways), and the vertical height is 40 m;

- The length, width, and height of the intake and return air roadways are 15 m, 5 m, and 3.5 m, respectively, and the length, width, and height of the tail roadway are 20 m, 4.4 m, and 3.5 m, respectively.

The geometric model of the working face is shown in Figure 2.

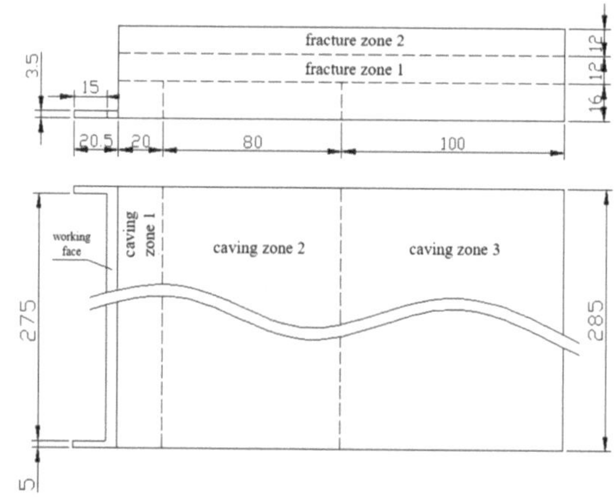

Figure 2. Geometric model of working face.

3.2.2. Determination of Simulation Parameters

Through the detailed investigation of the layout of the fully mechanized caving face, ventilation, gas occurrence, and the operation of the main equipment on the spot, based on reality, the relevant simulation parameters, including the rock porosity, the gas emission of each part of the goaf, the viscous resistance coefficient, and the boundary conditions are set as follows:

(1) Determination of porosity

As the fully mechanized caving face advances, the roof of the goaf begins to collapse, the rock falling behind the goaf is gradually compacted, and the expansion coefficient of the goaf from shallow to deep gradually decreases. According to the empirical formula, the bulking coefficient of the three parts of the caving zone can be taken as 1.5, 1.3, and 1.2. The porosity of the caving zone can be obtained from the bulking coefficient of 0.333, 0.231, and 0.167. The porosity of fracture zones 1 and 2 can be 0.05 and 0.02.

(2) Determination of gas emission (source term) in each part of the goaf

Using the source prediction method, it is predicted that the gas emission in the goaf of the N2306 working face is about 44.89 m^3/min, the gas emission in the fracture zone is about 18 m^3/min, and the gas emission in the caving zone is about 27 m^3/min. Among them, gas emissions of fracture zones 1 and 2 are 13.54 m^3/min and 4.45 m^3/min, respectively. The gas source term of each part of the caving zone is calculated according to the broken expansion coefficient, in which the gas density is 0.716 kg/m^3.

The total gas source term of the caving zone is given as follows:

$$\frac{2.78 \div 60 \times 0.716}{200 \times 285 \times 16} = 3.51 \times 10^{-7}\ \mathrm{kg}/\left(\mathrm{m}^3{\cdot}\mathrm{s}\right) \tag{12}$$

According to the length and bulking coefficient of each part of the caving zone, the source terms of the three parts of the caving zone can be calculated according to the following weighted average method:

The gas source term of caving zone 1 is as follows:

$$\frac{1.5 \times 20}{1.27 \times 200} \times 3.51 \times 10^{-7} = 0.419 \times 10^{-7}\ \mathrm{kg}/\left(\mathrm{m}^3{\cdot}\mathrm{s}\right) \tag{13}$$

The gas source term of caving zone 2 is as follows:

$$\frac{1.3 \times 80}{1.27 \times 200} \times 3.51 \times 10^{-7} = 1.433 \times 10^{-7}\ \mathrm{kg}/\left(\mathrm{m}^3 \cdot \mathrm{s}\right) \tag{14}$$

The gas source term of caving zone 3 is as follows:

$$\frac{1.2 \times 100}{1.27 \times 200} \times 3.51 \times 10^{-7} = 1.658 \times 10^{-7}\ \mathrm{kg}/\left(\mathrm{m}^3 \cdot \mathrm{s}\right) \tag{15}$$

The gas source term of fracture zone 1 is as follows:

$$\frac{13.54 \div 60 \times 0.716}{200 \times 275 \times 12} = 1.75 \times 10^{-7}\ \mathrm{kg}/\left(\mathrm{m}^3 \cdot \mathrm{s}\right) \tag{16}$$

The gas source term of fracture zone 2 is as follows:

$$\frac{4.45 \div 60 \times 0.716}{200 \times 275 \times 12} = 0.8 \times 10^{-7}\ \mathrm{kg}/\left(\mathrm{m}^3 \cdot \mathrm{s}\right) \tag{17}$$

(3) Determination of viscous resistance coefficient

Due to the different degrees of rock collapse in the goaf, the permeability is also quite different. However, since measuring the expansion coefficient everywhere in the goaf is difficult, it is regarded as heterogeneous isotropic in the numerical simulation process and calculated according to the empirical formula.

The empirical formula for the permeability of each partition in the goaf is as follows:

$$k = \frac{D^2 n^3}{180(1-n)^2} \tag{18}$$

where D represents the average particle size of the rock and n is the porosity.

According to the empirical formula, the permeability of the natural accumulation zone, the load affected zone, and the compaction stable zone can be set to 1.5×10^{-7} m^2, 0.7×10^{-7} m^2, and 0.2×10^{-7} m^2. The permeability of fracture zone 1 and fracture zone 2 can be considered as 1.0×10^{-5} m^2 and 0.5×10^{-5} m^2, respectively. The viscous resistance coefficient is the reciprocal of the permeability of the goaf.

(4) Setting boundary conditions

The setting of boundary conditions mainly considers the intake wind speed and outlet pressure. The intake wind speeds were 1.0, 1.5, 2.0, 2.5, and 3.0 m/s, respectively. The turbulent k-epsilon and species transport models are selected to open the energy equation. The numerical simulation parameters of the goaf gas seepage law are shown in Table 2.

Table 2. Simulation parameters of gas seepage law in goaf.

Name	Caving Zone 1	Caving Zone 2	Caving Zone 3	Fracture Zone 1	Fracture Zone 2
Porosity	0.333	0.231	0.167	0.05	0.02
Coefficient of viscous resistance (10^7 m^2)	0.7	1.4	5	100	200
Methane source term (10^{-7} Kg/m^3·s)	0.419	1.433	1.658	1.75	0.8

4. Numerical Results and Analysis

4.1. Numerical Simulation of Gas Seepage Law in Goaf Under U-Type Ventilation Mode

4.1.1. Before the Initial Weighting

From the beginning of mining to the formation of the initial weighting before the collapse of the main roof of the fully mechanized caving face, the main roof of the coal seam does not collapsed under the influence of its strength. The cracks in the roof and floor of the goaf are not developed, and effective communication with the adjacent coal seam is not formed. The mining face is less affected by the gas of the adjacent coal seam. At this time, the main source of gas in the working face is the gas emitted from the coal wall of the fully mechanized caving face, the gas desorbed from the coal falling in the working face and the goaf, and a small amount of gas from the seepage of the surrounding rock. Before the initial weighting, only the gas emitted from the coal wall and desorbed from the falling coal is considered the gas source.

According to the technical parameters of the N2306 working face coal cutter, the roof control distance of the working face is 5.16~5.96 m, the model is 5.5 m, the mining height of the working face is 3.5 m, and the tendency length is 285 m. In the model, the intake and return air roadway length is 15 m, and the section is 5×3.5 m^2. The first caving step distance of the main roof is 25~40 m, and the goaf is 30 m along the strike. The height of the goaf is the sum of the mining coal seam's thickness and the immediate roof's thickness, which is 8.5 m. The three-dimensional geometric model of the goaf in the fully mechanized caving face is shown in Figure 3.

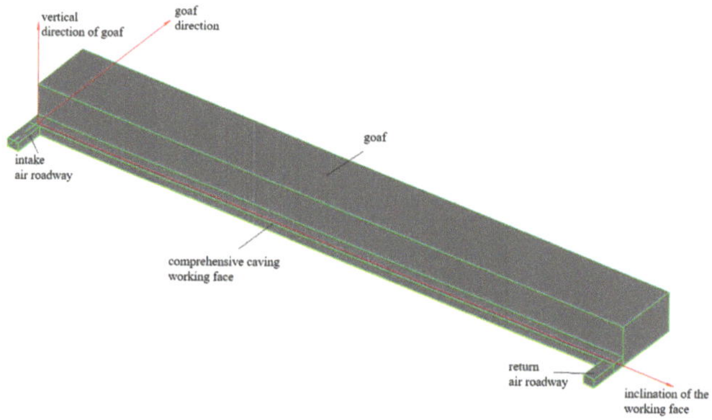

Figure 3. Geometric model of goaf before initial weighting under U-type ventilation mode.

The numerical simulation is carried out according to the parameters and boundary conditions of the goaf before the first weighting of the fully mechanized caving face. The velocity cloud map of the goaf is shown in Figure 4, the velocity vector map in Figure 5, and the gas concentration distribution map in Figure 6.

From Figures 4 and 5, it can be seen that under the conditions of U-type ventilation, the ventilation airflow mainly flows through the working face from the return air roadway, and a very small part of the airflow bypasses the support between the working face and the goaf into the goaf. Affected by the support and the caving coal and rock mass, the airflow velocity entering the goaf decreases rapidly and tends to stop as the wind speeds away from the working face. When the airflow moves in the working face, the wind speed near the roof and floor is low, which is also one of the reasons for the high gas concentration near the roof of the working face.

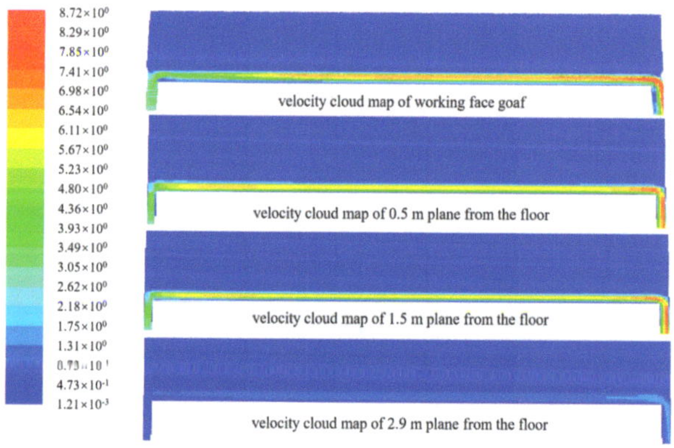

Figure 4. Velocity cloud map of roof before initial weighting.

According to Figure 6, before the initial weighting, the maximum gas concentration in the working face appears near the coal wall near the return airway. The maximum value is about 0.85% (blue circle area in Figure 6), the gas concentration in the upper corner of the working face is about 0.70% (orange circle area in Figure 6), and the gas concentration in the return airway is 0.80%. The gas concentration is higher in the range of about 130 m from the return air roadway along the tendency of the goaf, and the farther away from the return air roadway, the lower the gas concentration; the gas concentration near the roof and floor is high, and the gas concentration in the middle of the working face is low.

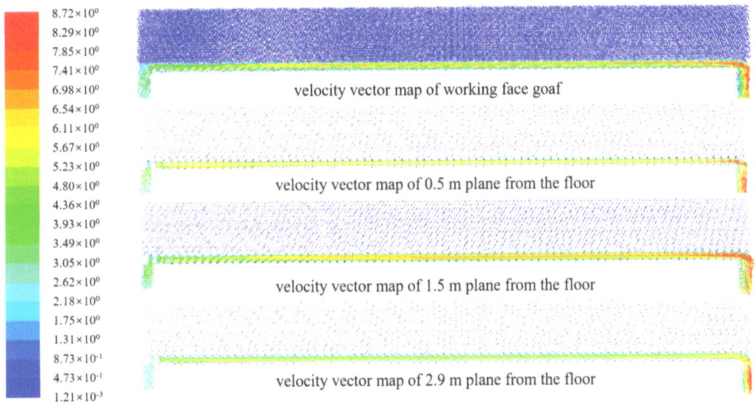

Figure 5. Velocity vector map of roof before initial weighting.

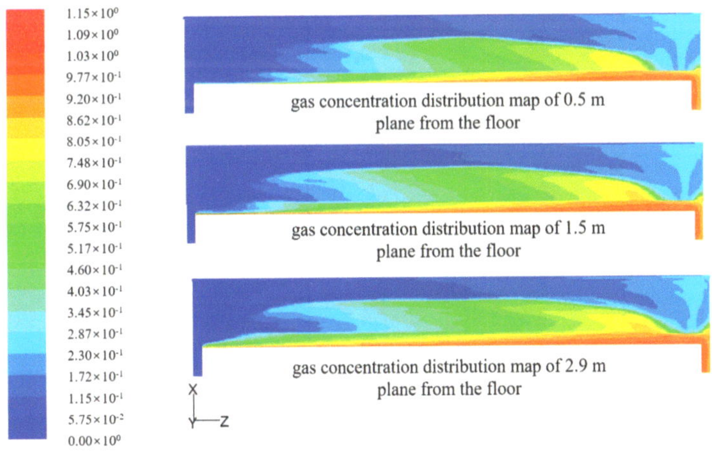

Figure 6. Gas concentration distribution map of roof before initial weighting.

4.1.2. After the Initial Weighting

In the N2306 fully mechanized caving face, after the initial weighting, the coal seam roof falls under gravity, and the bending subsidence zone produces many cracks. The caving coal and rock mass is under the influence of fragmentation, self-weight, and external force, so the goaf of the working face presents a strip distribution. According to the distribution law of each strip, a three-dimensional geometric model of the goaf of the working face with a tendency of 285 m, a strike of 200 m, and a height of 40 m is established, as shown in Figure 7.

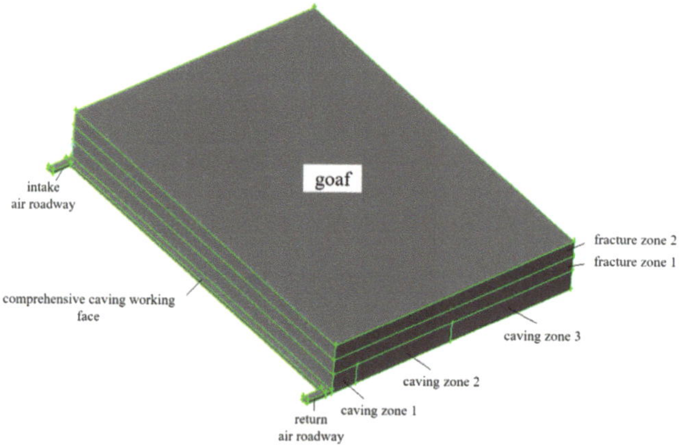

Figure 7. Geometric model of goaf after initial weighting under U-type ventilation mode.

(1) Simulation of air leakage in goaf after initial weighting under U-type ventilation mode

The intake velocity is set to 2.85 m/s, the outlet is the free outlet, and the goaf is the porous medium. The multi-component transport model is used to simulate the goaf after the first roof pressure, and the velocity cloud map of the goaf is shown in Figure 8.

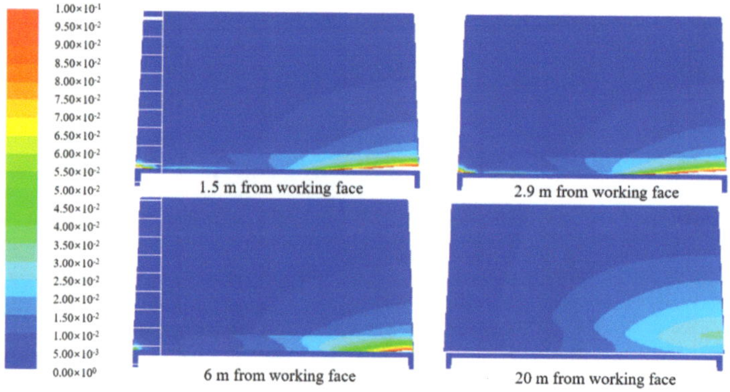

Figure 8. Distribution cloud map of air leakage velocity in goaf under U-type ventilation mode.

It can be seen from the figure that under the ventilation airflow in the working face, there is also a weak airflow in the goaf. Its influence range is concentrated in the lower and upper corners, and the influence range is greater in the upper corner. When the wind speed of the working face is 2.85 m/s, the wind speed of the goaf at the upper corner is 0.1 m/s in the area of 5~10 m from the working face, and the wind speed is 0.05 m/s in the area of 15~18 m from the working face. The farther away from the working face, the lower the wind speed. The wind speed also shows a decreasing trend in the vertical height. At 20 m from the floor, the wind speed has been below 0.01 m/s.

(2) Numerical simulation of gas concentration distribution law in goaf under different wind speed conditions of working face

By adjusting the boundary conditions, the distribution law of gas concentration in the goaf is simulated under wind speed conditions of 1.0 m/s, 1.5 m/s, 2.0 m/s, 2.5 m/s, and 3.0 m/s in the fully mechanized caving face. Figures 9–11 depict 1.5 m/s, 2.0 m/s, 2.5 m/s of wind speed, respectively, from the floor z = 3 m, and the 10 m goaf gas concentration distribution cloud.

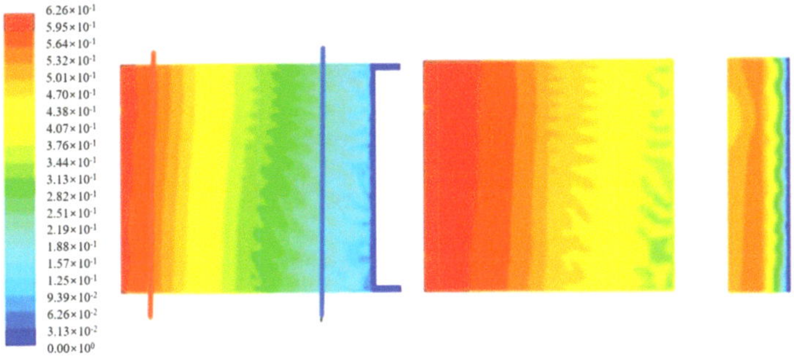

(a) (b) (c)

Figure 9. Gas concentration cloud map when v = 1.5 m/s under U-type ventilation mode: (**a**) Gas concentration cloud map in goaf from floor z = 3 m; (**b**) Gas concentration cloud map in goaf from floor z = 10 m; (**c**) Gas concentration cloud map in vertical direction of goaf.

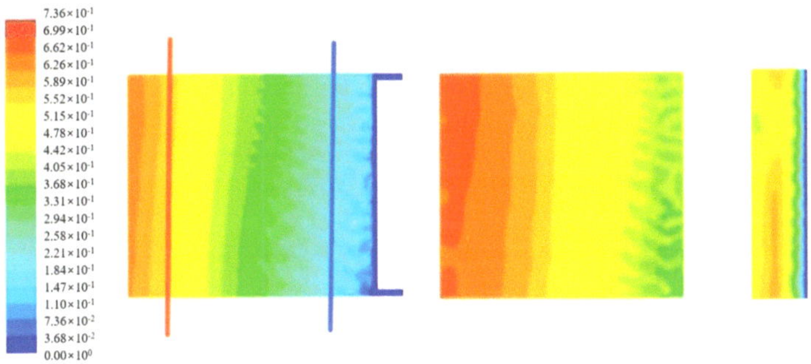

(a) (b) (c)

Figure 10. Gas concentration cloud map when v = 2.0 m/s under U-type ventilation mode: (**a**) Gas concentration cloud map in goaf from floor z = 3 m; (**b**) Gas concentration cloud map in goaf from floor z = 10 m; (**c**) Gas concentration cloud map in vertical direction of goaf.

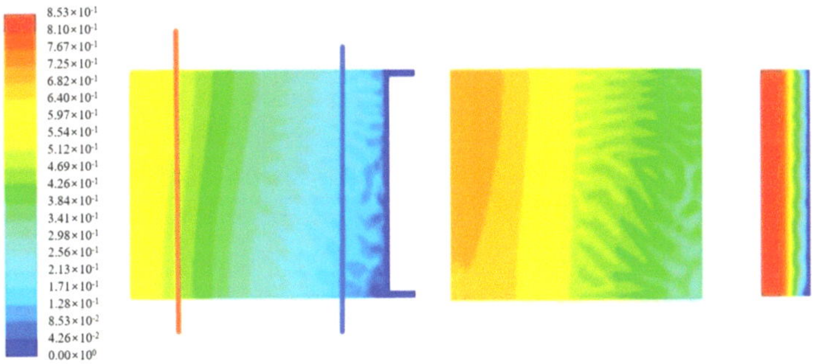

(a) (b) (c)

Figure 11. Gas concentration cloud map when v = 2.5 m/s under U-type ventilation mode: (**a**) Gas concentration cloud map in goaf from floor z = 3 m; (**b**) Gas concentration cloud map in goaf from floor z = 10 m; (**c**) Gas concentration cloud map in vertical direction of goaf.

It can be seen from the figures that the gas concentration can be divided into three stages in the whole goaf trend as follows: the gas concentration increases slowly, and the gas concentration is low in the first 30 m near the working face (location of the blue line in the picture); in the range of 30 m~160 m (blue line and orange line area in the picture), the gas concentration increases rapidly, and there is obvious stratification; after 160 m (orange line position in the picture), the growth of gas concentration gradually slows down and eventually stabilizes. The main reason for this phenomenon is the effect of the wind speed of the working face. The air leakage in the goaf near the working face is high, and the air leakage gradually decreases with the depth increase. On the other hand, because the porosity in the goaf gradually decreases from shallow to deep, the bulking coefficient in the re-compacted area is close to 1, which leads to its accumulation in the deep part of the goaf, and the concentration can reach more than 90%.

Along the dip direction of the working face, the gas concentration distribution in the goaf shows a trend of increasing gas concentration from the intake side to the return side, especially in the gas concentration on the return side of the working face. The reason for this phenomenon is that in the early stage of coal release, the immediate roof and the main roof are prone to the cantilever beam phenomenon under influence of their strength, and the caving rock in the goaf accumulates freely, the porosity is great, and the air leakage wind speed is high. Part of the gas accumulates on the return side driven by the airflow, and with the increase in the depth of the goaf, the air leakage airflow gradually disappears, and its influence on the gas concentration gradually disappears.

In the vertical direction of the goaf, the gas concentration gradually increases from the bottom plate, and the gas concentration in the fracture zone can even reach 100%. The main reason for this phenomenon is that the density of gas is low relative to the air, and it migrates to the upper space of the goaf under gravity and buoyancy. At the same time, due to the large difference in air leakage flow between the caving zone and the fracture zone in the goaf, the gas in the caving zone is easily taken away by the airflow, while the accumulated gas in the fracture zone cannot be taken away so that the gas forms a gradient distribution in the vertical direction of the goaf.

Figure 12 is the change curve of gas concentration in the upper corner of the working face when the wind speed is 1.0 m/s, 1.5 m/s, 2.0 m/s, 2.5 m/s, and 3.0 m/s, respectively, under the U-type ventilation mode of the working face.

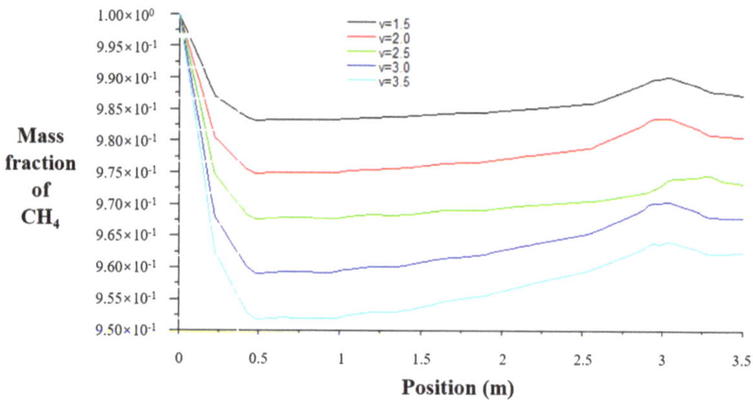

Figure 12. Curve of gas concentration change in upper corner under U-type ventilation mode.

It can be seen from Figure 12 that the change in wind speed in the working face greatly influences the gas concentration in the upper corner. However, in practice, this should be determined by the amount of gas emission from the working face. For the fully mechanized caving face with a high absolute gas emission, reducing the gas concentration in the upper corner and the return air side is limited to only increasing the wind speed (pressure difference). As the pressure difference increases, the air leakage volume also increases, removing more gas from the goaf so that the gas concentration in the return air side and the upper corner does not decrease significantly or even increase.

4.2. Numerical Simulation of Gas Seepage Law in Goaf Under U + I-Type Ventilation Mode

Based on the process parameters of the N2306 working face, the working face is designed in accordance with a U + I-type ventilation mode. A three-dimensional geometric model of the goaf in a fully mechanized caving face is established, which is 15 m in length, 5 m in width, and 3.5 m in the height for the intake and return air roadway; 15 m in length, 4.4 m in width, and 3 m in height for the tail roadway; 10 m from the coal seam floor to the bottom of the tail roadway and 20 m from the horizontal distance for the return air roadway; 275 m in length, 5 m in width, and 3.5 m in height for the working face; and 285 m in length, 200 m in width, and 40 m in height for the goaf, as shown in Figure 13.

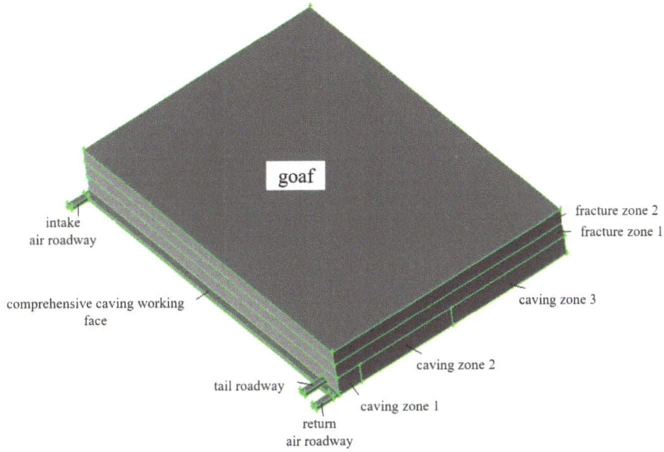

Figure 13. Geometric model of goaf under U + I-type ventilation mode.

Under the condition that other parameters remain unchanged, the intake roadway is selected as the speed intake, the wind speed is 2.5 m/s, and the return air roadway and the tail roadway are free exits. The absolute gas emission in the goaf is 44.89 m^3/min, respectively, to simulate the distribution and migration law of gas concentration in the goaf when the U + I-type ventilation mode is simulated.

Figures 14 and 15 show the gas concentration distribution cloud maps in the horizontal and vertical directions of the goaf under the U + I-type ventilation mode.

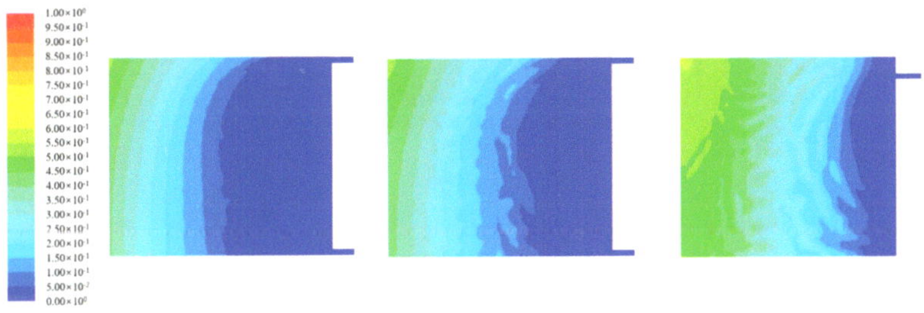

Figure 14. Gas concentration in different horizontal sections of goaf under U + I-type ventilation mode: (**a**) Gas concentration distribution at horizontal section z = 0 m; (**b**) Gas concentration distribution at horizontal section z = 3 m; (**c**) Gas concentration distribution at horizontal section z = 8 m.

It can be seen from the figure that compared with the U-type ventilation mode, some gas is discharged with the 'I' roadway in the U + I-type, and the gas concentration in the upper corner and the support influence area can be effectively reduced in the horizontal direction. In the vertical direction, the stratification of gas concentration distribution in the deep goaf is more obvious. The reason for this phenomenon is that the gas in the shallow caving zone of the goaf is eliminated with the airflow of the tail roadway. The deep wind speed in the goaf is very low, and the gas mainly depends on gravity and free diffusion. In addition, the viscous resistance coefficient of the fracture zone is high, the porosity is low, and the gas accumulation gradually increases the concentration.

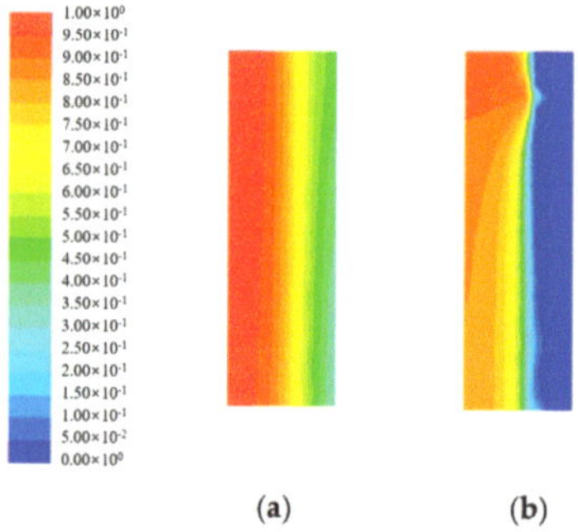

Figure 15. Gas concentration in different vertical sections of goaf under U + I-type ventilation mode: (**a**) Gas concentration distribution at vertical section z = 20 m; (**b**) Gas concentration distribution at vertical section z = 265 m.

4.3. Numerical Simulation of Gas Seepage Law in Goaf Under U + L-Type Ventilation Mode

Based on the process parameters of the N2306 working face, the working face is designed as a U + L-type ventilation mode. A three-dimensional geometric model of the goaf in a fully mechanized caving face with the same size of intake and return air roadways and tail roadway is established, with a length of 15 m, a width of 5 m, and a height of 3.5 m. The tail roadway is parallel to the return air roadway, and the horizontal spacing is 15 m. The working face has a length of 275 m, a width of 5 m, and a height of 3.5 m. The goaf has a length of 285 m, a width of 200 m, and a height of 40 m, as shown in Figure 16.

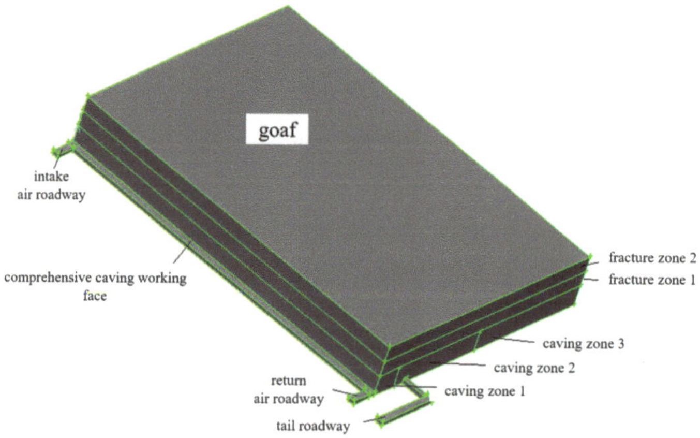

Figure 16. Geometric model of goaf under U + L-type ventilation mode.

Under the condition that other parameters remain unchanged, by changing the ratio of the air volume of the return air roadway and the 'L' roadway, the distribution and migration law of the gas concentration in the goaf are simulated when the air volume ratio is 2:1, 3:1, and 4:1 in the U + L-type ventilation mode.

When the air volume ratio of the return air roadway and the 'L' roadway is 2:1 (the wind speed of the return air roadway is 1.33 m/s, and the wind speed of the 'L' roadway is 0.67 m/s), 3:1 (the wind speed of the return air roadway is 1.5 m/s, and the wind speed of the 'L' roadway is 0.5 m/s), and 4:1 (the wind speed of the return air roadway is 1.6 m/s, and the wind speed of the 'L' roadway is 0.4 m/s), the gas concentration distribution cloud map of the goaf is shown in Figures 17–19.

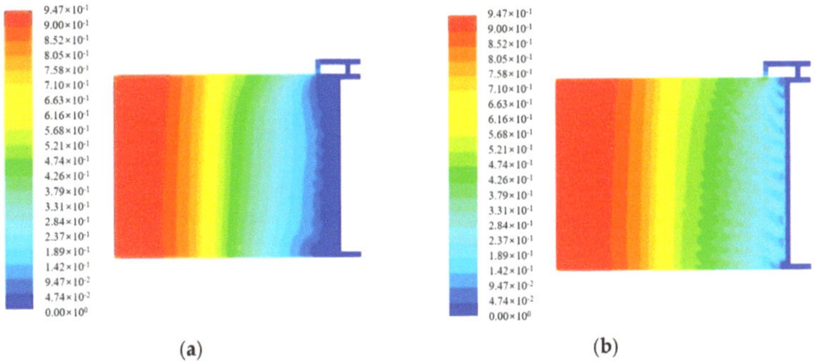

Figure 17. Gas concentration of different horizontal sections in goaf under U + L-type ventilation mode when air distribution volume is 2:1: (**a**) Gas concentration distribution at horizontal section z = 0 m; (**b**) Gas concentration distribution at horizontal section z = 3 m.

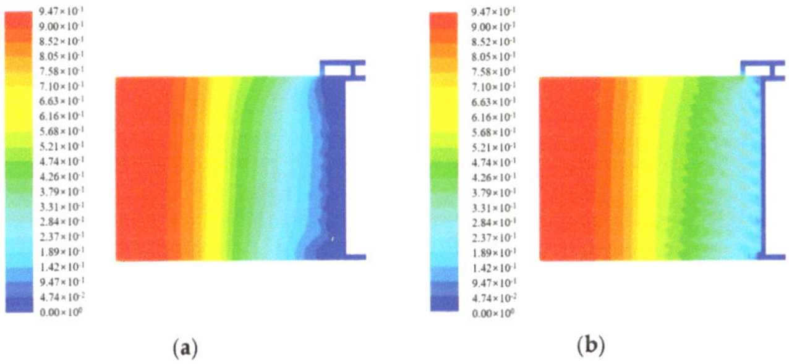

Figure 18. Gas concentration of different horizontal sections in goaf under U + L-type ventilation mode when air distribution volume is 3:1: (**a**) Gas concentration distribution at horizontal section z = 0 m; (**b**) Gas concentration distribution at horizontal section z = 3 m.

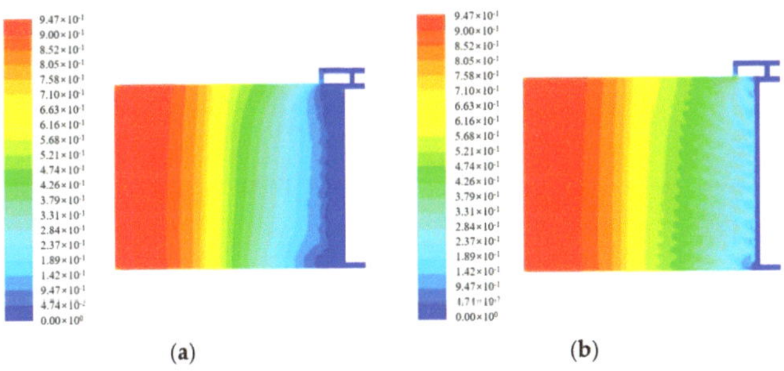

Figure 19. Gas concentration of different horizontal sections in goaf under U + L-type ventilation mode when air distribution volume is 4:1: (**a**) Gas concentration distribution at horizontal section z = 0 m; (**b**) Gas concentration distribution at horizontal section z = 3 m.

It can be seen from the figure that due to the existence of the 'L' roadway, the gas in the goaf flows to the 'L' roadway with the airflow, so the concentration of the coal mining face and the return air roadway is low, and the gas concentration of the outer tail roadway is high. The influence of the tail roadway on the gas flow field in the goaf of the working face is mainly reflected near the upper corner. The farther away from the tail roadway, the smaller the gas concentration change. Therefore, the outer tail roadway can effectively reduce the gas concentration in the upper corner and alleviate the gas overrun problem. By comparing the gas concentration distribution in the goaf with the air volume ratios of 2:1, 3:1, and 4:1 in the return air roadway and the outer staggered tail roadway, it can be seen that the concentration in the upper corner is significantly lower than that in the U-type ventilation. However, the change in air volume ratio has a limited impact on the gas concentration in the upper corner. If conditions permit, increasing the air volume of the outer staggered tail roadway will help reduce the gas concentration in the gas tail roadway and ensure safety.

5. Discussion

Using the three ventilation modes of U, U + I, and U + L, and the three wind speed conditions of 1.5 m/s, 2.0 m/s, and 2.5 m/s in the intake air roadway, the gas seepage law in the goaf was simulated. Figures 20–22 depict the curves of gas concentration change along the goaf strike at 0.5 m from the working face under the three ventilation modes.

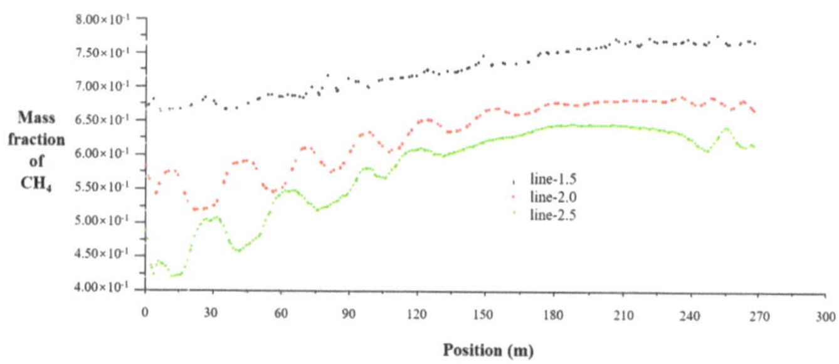

Figure 20. The curve of gas concentration at 0.5 m away from the working face under U-type ventilation mode.

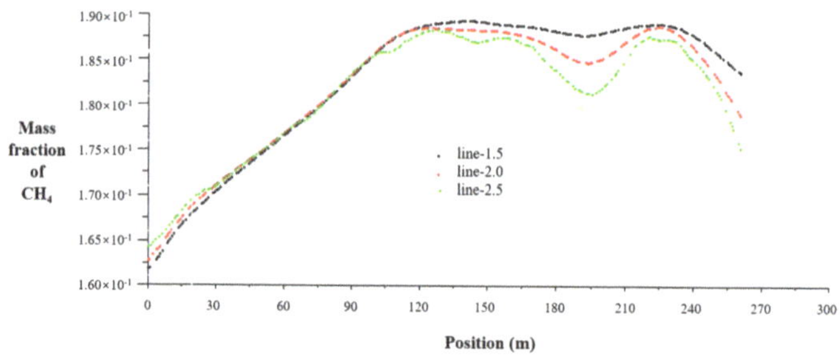

Figure 21. The curve of gas concentration at 0.5 m away from the working face under U + I-type ventilation mode.

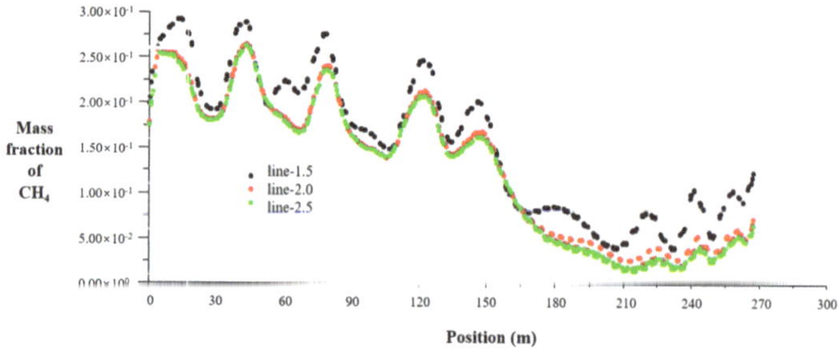

Figure 22. The curve of gas concentration at 0.5 m away from the working face under U + L-type ventilation mode.

It can be seen from the figure that compared with the U-type mode, the gas concentration in the upper corner of the goaf is significantly reduced under the U + I and U + L ventilation modes. The existence of the tail roadway is beneficial in reducing the gas concentration in the upper corner of the goaf in the fully mechanized caving face. However, the influence of the tail roadway on the overall gas concentration distribution in the goaf is limited. Therefore, for the hidden danger control of gas disasters in high gas fully mechanized caving faces, the ventilation mode should be determined according to the specific situation of the mine to reduce the occurrence of gas accumulation accidents in the upper corner of the working face.

Figure 23 represents the U + L ventilation mode. Under the three wind speed conditions, the gas concentration change curve is 80 m from the return air roadway along the goaf.

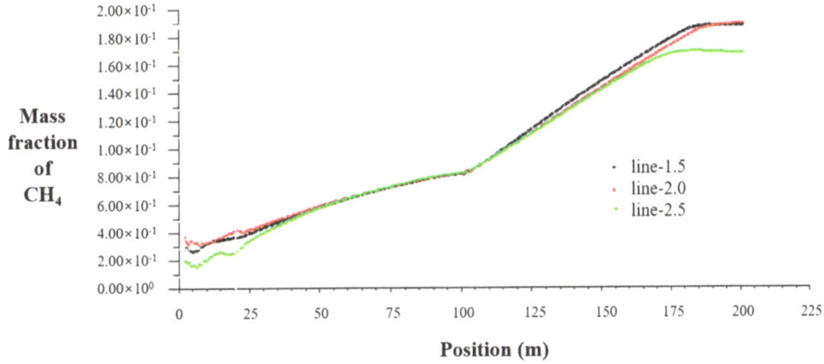

Figure 23. Gas concentration curve of goaf cross-section under U + L-type ventilation mode.

It can be seen from Figure 23 that the gas concentration gradually increases along the goaf, while the change rate gradually decreases, and the change tends to be gentle. When it is close to the 'asphyxiation zone', the gas concentration tends to be stable. This trend is mainly because when the airflow in the working face enters the working face from the air intake, part of the airflow penetrates the goaf, and the gas in the goaf continuously oozes from the goaf under the influence of the airflow, thus reducing the gas concentration. As the goaf is away from the working face, the influence of the working face airflow on the goaf gradually weakens until it disappears. Therefore, the influence of the airflow on the goaf is also weakened or may even disappear. As the goaf is away from the working face, the gas concentration gradually increases and fills the entire 'asphyxiation zone'.

6. Conclusions

According to the technological characteristics of the fully mechanized caving face in a mine in Shanxi Province, the numerical simulation of gas seepage law in the goaf of a fully mechanized caving mining is carried out. The mathematical model of the gas seepage law in a fully mechanized top coal caving mining is verified, and the main factors affecting gas accumulation in the working face are determined. The simulation results are significant in guiding future studies of gas control technology. The main conclusions are as follows:

1. The gas concentration distribution in the goaf shows a similar law for different ventilation modes (including U-type, U + I-type, and U + L-type), affected by wind speed and rock porosity. The gas concentration in the whole goaf can be divided into the following three stages: in the first 30 m near the working face, the gas concentration increases slowly, and the gas concentration is low; in the range of 30 m~160 m, the gas concentration increases rapidly, and there is obvious stratification; and after 160 m, the growth of gas concentration gradually slowed down and eventually stabilized. Along the dip direction of the working face, the gas concentration distribution in the goaf shows a trend of an increasing gas concentration from the intake side to

the return side, especially in the gas concentration on the return side of the working face. In the vertical direction of the goaf, the gas is distributed in a gradient. The gas concentration gradually increases from the bottom plate, and the gas concentration in the fracture zone can even reach 100%;

2. Different ventilation methods have different scopes of application. When there us less desorption gas in the coal seam, it is easy to adopt the U-type ventilation mode and the U + I or U + L ventilation modes when there is more desorbable gas, or the mining intensity is high. Compared with the U-type ventilation mode, the U + I and U + L-type ventilation modes are more conducive to reducing gas seepage from the goaf to the upper corner of the working face. That is, applying the tail roadway can solve the problem of gas accumulation in the upper corner caused by eddy currents. To a certain extent, this can reduce the gas concentration in the upper corner of the goaf of the fully mechanized caving face. However, in general, the influence of the tail roadway on the overall gas concentration distribution in the goaf is limited;
3. The wind speed at the working face is one of the most important factors affecting the gas concentration in the upper corner of the working face's mining area. In the case of U-type ventilation, i.e., when the absolute gas outflow from the workface is low, the change in the wind speed at the workface has a greater impact on the gas concentration in the upper corner, and by increasing the wind speed appropriately, the gas concentration in the upper corner can be significantly reduced. In the case of U + I- and U + L-type ventilation, i.e., when the absolute gas outflow from the workface is high, there are great limitations in reducing the gas concentration in the upper corner and the return side by only increasing the wind speed (pressure difference), because when the pressure difference is increased, the amount of air leakage increases, which will take away more gas from the mining area, thus making the reduction in the gas concentration in the return side and the upper corner insignificant or even increasing.

Author Contributions: Formal analysis, Writing—original draft preparation, H.W.; Supervision, Conceptualization, Methodology, Resources, H.A.; Validation, Formal analysis, Investigation, X.Z. All authors have read and agreed to the published version of the manuscript.

Funding: This research was supported in part by the Yunnan Province "Caiyun" Postdoctoral Innovative Project Plan (No. CG24056E004A).

Institutional Review Board Statement: Not applicable.

Informed Consent Statement: Not applicable.

Data Availability Statement: The original contributions presented in the study are included in the article, further inquiries can be directed to the corresponding authors.

Acknowledgments: The authors would like to express their sincere gratitude to the Kunming University of Science and Technology 2023 Undergraduate Education Teaching Reform Research Project (No. 075, KUST Education Office [2023]) for their support. Moreover, the authors would like to thank the anonymous reviewers for their valuable comments and constructive suggestions.

Conflicts of Interest: The authors declare no conflicts of interest.

References

1. Lu, S.; Li, L.; Cheng, Y.; Sa, Z.; Zhang, Y.; Yang, N. Mechanical failure mechanisms and forms of normal and deformed coal combination containing gas: Model development and analysis. *Eng. Fail. Anal.* **2017**, *80*, 241–252. [CrossRef]
2. Guo, H.; Cheng, Z.; Wang, K.; Qu, B.; Yuan, L.; Xu, C. Coal permeability evolution characteristics: Analysis under different loading conditions. *Greenh. Gases Sci. Technol.* **2020**, *10*, 347–363. [CrossRef]
3. Guan, P.; Wang, H.; Zhang, Y. Mechanism of instantaneous coal outbursts. *Geology* **2009**, *37*, 915–918. [CrossRef]
4. Baisheng, N.; Xiangchun, L. Mechanism research on coal and gas outburst during vibration blasting. *Saf. Sci.* **2012**, *50*, 741–744. [CrossRef]
5. Skoczylas, N. Laboratory study of the phenomenon of methane and coal outburst. *Int. J. Rock Mech. Min. Sci. Geomech. Abstr.* **2012**, *55*, 102–107. [CrossRef]

6. Wang, S.; Elsworth, D.; Liu, J. Rapid decompression and desorption induced energetic failure in coal. *J. Rock Mech. Geotech. Eng.* **2015**, *7*, 345–350. [CrossRef]
7. Tu, Q.; Cheng, Y.; Guo, P.; Jiang, J.; Wang, L.; Zhang, R. Experimental study of coal and gas outbursts related to gas-enriched areas. *Rock Mech. Rock Eng.* **2016**, *49*, 3769–3781. [CrossRef]
8. Yin, G.; Jiang, C.; Wang, J.G.; Xu, J.; Zhang, D.; Huang, G. A new experimental apparatus for coal and gas outburst simulation. *Rock Mech. Rock Eng.* **2016**, *49*, 2005–2013. [CrossRef]
9. Xu, L.; Jiang, C. Initial desorption characterization of methane and carbon dioxide in coal and its influence on coal and gas outburst risk. *Fuel* **2017**, *203*, 700–706. [CrossRef]
10. Zhang, C.; Peng, S.; Xu, J.; Geng, J.; Yang, H.; Luo, X. Temporospatial evolution of gas pressure during coal and gas outburst. *Rock Soil Mech.* **2017**, *38*, 81–90.
11. Ding, Y.; Yue, Z.Q. An experimental investigation of the roles of water content and gas decompression rate for outburst in coal briquettes. *Fuel* **2018**, *234*, 1221–1228. [CrossRef]
12. Guo, B.; Li, Y.; Jiao, F.; Luo, T.; Ma, Q. Experimental study on coal and gas outburst and the variation characteristics of gas pressure. *Geomech. Geophys. Geo-Energy Geo-Resour.* **2018**, *4*, 355–368. [CrossRef]
13. Lu, S.; Zhang, Y.; Sa, Z.; Si, S. Evaluation of the effect of adsorbed gas and free gas on mechanical properties of coal. *Environ. Earth Sci.* **2019**, *78*, 218. [CrossRef]
14. Sa, Z.; Liu, J.; Li, J.; Zhang, Y. Research on effect of gas pressure in the development process of gassy coal extrusion. *Saf. Sci.* **2019**, *115*, 28–35. [CrossRef]
15. Guo, H.; Tang, H.; Wu, Y.; Wang, K.; Xu, C. Gas seepage in underground coal seams: Application of the equivalent scale of coal matrix-fracture structures in coal permeability measurements. *Fuel* **2021**, *288*, 119641. [CrossRef]
16. Guo, H.; Wang, K.; Wu, Y.; Tang, H.; Wu, J.; Guan, L.; Chang, C.; Xu, C. Evaluation of the weakening behavior of gas on the coal strength and its quantitative influence on the coal deformation. *Int. J. Min. Sci. Technol.* **2021**, *31*, 451–462. [CrossRef]
17. Terzaghi, K. *Theoretical Soil Mechanics;* Wiley: New York, NY, USA, 1943.
18. Biot, M.A. General theory of three-dimensional consolidation. *J. Appl. Phys.* **1941**, *12*, 155–164. [CrossRef]
19. Biot, M.A. Theory of elasticity and consolidation for a porous anisotropic solid. *J. Appl. Phys.* **1955**, *26*, 182–185. [CrossRef]
20. Biot, M.A. General solutions of the equations of elasticity and consolidation for a porous material. *J. Appl. Mech.* **1956**, *23*, 91–96. [CrossRef]
21. Lewis, R.W.; Roberts, P.J.; Schrefler, B.A. Finite element modelling of two-phase heat and fluid flow in deforming porous media. *Transp. Porous Media* **1989**, *4*, 319–334. [CrossRef]
22. Wang, H.; Dlugogorski, B.Z.; Kennedy, E.M. Pathways for production of CO_2 and CO in low-temperature oxidation of coal. *Energy Fuels* **2003**, *17*, 150–158. [CrossRef]
23. Beamish, B.; Lau, A.; Moodie, A.; Vallance, T. Assessing the self-heating behaviour of Callide coal using a 2-metre column. *J. Loss Prev. Process Ind.* **2002**, *15*, 385–390. [CrossRef]
24. Harpalani, S.; Chen, G. Estimation of changes in fracture porosity of coal with gas emission. *Fuel* **1995**, *74*, 1491–1498. [CrossRef]
25. Lewis, R.W.; Sukirman, Y. Finite element modelling of three-phase flow in deforming saturated oil reservoirs. *Int. J. Numer. Anal. Methods Geomech.* **1993**, *17*, 577–598. [CrossRef]
26. Zhao, Y.; Hu, Y.; Zhao, B.; Yang, D. Nonlinear coupled mathematical model for solid deformation and gas seepage in fractured media. *Transp. Porous Media* **2004**, *55*, 119–136. [CrossRef]
27. Yang, D.; Zhao, Y.; Hu, Y. The constitute law of gas seepage in rock fractures undergoing three-dimensional stress. *Transp. Porous Media* **2006**, *63*, 463–472. [CrossRef]
28. Connell, L. Coupled flow and geomechanical processes during gas production from coal seams. *Int. J. Coal Geol.* **2009**, *79*, 18–28. [CrossRef]
29. Zhigang; Guirong, T. Gas seepage field of god and the LBM simulation. *Disaster Adv.* **2010**, *3*, 421–423.
30. Pini, R.; Storti, G.; Mazzotti, M. A model for enhanced coal bed methane recovery aimed at carbon dioxide storage. *Adsorption* **2011**, *17*, 889–900. [CrossRef]
31. Li, S.; Xiao, P.; Pan, H.; Lin, H.; Cheng, L. Experimental investigation on the seepage law of pressure-relieved gas under the influence of mining. *Saf. Sci.* **2011**, *50*, 614–617. [CrossRef]
32. Wang, D.-K.; Wei, J.-P.; Fu, Q.-C.; Liu, Y.; Xia, Y.-L. Seepage law and permeability calculation of coal gas based on Klinkenberg effect. *J. Cent. South Univ.* **2015**, *22*, 1973–1978. [CrossRef]
33. Danesh, N.N.; Chen, Z.; Connell, L.D.; Kizil, M.S.; Pan, Z.; Aminossadati, S.M. Characterisation of creep in coal and its impact on permeability: An experimental study. *Int. J. Coal Geol.* **2017**, *173*, 200–211. [CrossRef]
34. Danesh, N.N.; Chen, Z.; Aminossadati, S.M.; Kizil, M.S.; Pan, Z.; Connell, L.D. Impact of creep on the evolution of coal permeability and gas drainage performance. *J. Nat. Gas Sci. Eng.* **2016**, *33*, 469–482. [CrossRef]
35. Bertrand, F.; Cerfontaine, B.; Collin, F. A fully coupled hydro-mechanical model for the modeling of coalbed methane recovery. *J. Nat. Gas Sci. Eng.* **2017**, *46*, 307–325. [CrossRef]
36. Zhang, C.; Liu, X.; Xu, G.; Wang, X. Effects of loading rate on gas seepage and temperature in coal and its potential for coal-gas disaster early-warning. *Energies* **2017**, *10*, 1246. [CrossRef]
37. Qiu, L.; Zhou, G.; Zhang, W.; Han, W. Simulations on the micro-seepage rules of gas and water based on micro-CT/CFD and the related contrastive analysis. *Arab. J. Geosci.* **2019**, *12*, 549. [CrossRef]

38. Xue, Y.; Cao, Z.; Dang, F.; Wang, S.; He, M.; Du, F. Effect of damage on gas seepage mechanism in coal seam based on a coupled model. *Therm. Sci.* **2019**, *23 Part A*, 1323–1328. [CrossRef]
39. Yan, M.; Zhou, M.; Li, S.; Lin, H.; Zhang, K.; Zhang, B.; Shu, C.-M. Numerical investigation on the influence of micropore structure characteristics on gas seepage in coal with lattice Boltzmann method. *Energy* **2021**, *230*, 120773. [CrossRef]
40. Mu, D.; Ma, Q.; Zhang, K.; Zhao, J. A gas-mechanical-damage coupling model based on the TLF-SPH method and its application to gas seepage in fractured coal. *Comput. Geotech.* **2024**, *171*, 106352. [CrossRef]
41. Zhang, L.; Yang, M.; Luo, H. Parameter optimization and application of gas extraction technology by buried pipes in goaf of multi-roadway U-type working face. *Combust. Sci. Technol.* **2024**, *196*, 820–837. [CrossRef]
42. Xie, S.-R.; Zhao, Y.-J.; Zhang, S.-B.; Yang, H.-Z.; Xiao, D.-C.; Tian, C.-Y. Mechanism and experiment of substituting high drainage roadway with directional long drilling group to extract pressure-relief gas. *J. Cent. South Univ.* **2012**, *19*, 2591–2597. [CrossRef]

fire

Article

Numerical Computation-Based Analysis of Blasting Vibration Effects and Slope Stability in an Open-Pit Quarry

Botao Fu [1,2], Huaijun Ji [1,*], Jingjing Pei [1] and Jie Wei [1,2]

1 School of Engineering and Technology, China University of Geosciences (Beijing), Beijing 100083, China; 13126729896@163.com (B.F.); peijj@cugb.edu.cn (J.P.); wj2661573157@163.com (J.W.)
2 China Academy of Safety Science and Technology, Beijing 100012, China
* Correspondence: jhjun2015@cugb.edu.cn; Tel./Fax: +86-15701574758

Abstract: With the large-scale exploitation of mineral resources, the safety problems of open-pit slopes and dumps become increasingly prominent. In this work, the safety demonstration of the interaction between the eastside slope of an open-pit quarry and the dump was carried out. Firstly, the influence of blasting vibration on the stability of the eastside slope and the dump was analyzed, and the propagation law of blasting vibration wave and the attenuation law of velocity were obtained. Secondly, the finite element analysis of the seepage field of the slope was carried out, and the basic seepage parameters of the rock and soil were determined. Then, the limit equilibrium analysis method was used to quantitatively calculate the stability of the eastside slope of the stope under the load of the dump to assess landslide risk. The results show that no landslides would occur at the dump or affect the performance of the slope, and that the slope was able to reach the required safety reserve factor. Then, through the finite element analysis of the 3-3 profile, it was found that the deformation caused by the excavation unloading of the eastside slope was mainly concentrated in the lower part of the slope, and the displacement variation caused by the top area of the slope was about 0~10 mm. The research verified the stability of slope and dump, which provides technical support for slope project design and blast operation in open-pit quarries.

Keywords: numerical analysis; blast vibration; slope; dump; stability analysis; open-pit quarry

Citation: Fu, B.; Ji, H.; Pei, J.; Wei, J. Numerical Computation-Based Analysis of Blasting Vibration Effects and Slope Stability in an Open-Pit Quarry. *Fire* **2024**, 7, 420. https://doi.org/10.3390/fire7110420

Academic Editor: Haipeng Jiang

Received: 29 September 2024
Revised: 1 November 2024
Accepted: 2 November 2024
Published: 18 November 2024

1. Introduction

As an efficient and economical way of mining mineral resources, the open-pit quarry method has been widely used at home and abroad. However, with the continuous expansion of the mining scale and the increase in the mining depth, the stability problem of the slope of the open-pit quarry is becoming more and more prominent, and it has become one of the key factors restricting the safe production of mines. At the same time, as an important supporting facility of the open-pit quarry, the design, construction, and operation of earth-moving yards also have an important impact on the stability of slopes. Therefore, it is of great significance to analyze in depth the mutual influence between the slope of the open-pit quarry and the earth disposal site to ensure the safe production of mines.

Many scholars have carried out a lot of research on the stability analysis of quarry slopes. Yuan et al. [1] used a surrogate-assisted normal search particle swarm optimization algorithm for efficient slope reliability analysis. Lu et al. [2] proposed an integrated approach combining satellite interferometric synthetic aperture radar (InSAR), coupled numerical analyses, and field monitoring to more accurately assess the risk of landslides, taking into account the combined effects of mining and rainfall infiltration when evaluating the stability of slopes in open-pit quarries. Cai et al. [3] combined displacement retrospective analysis with strength reduction techniques to characterize slope heterogeneity and stability analysis using excavation-induced displacements.

With the rapid development of computers in recent years, the numerical analysis method is more and more widely used in analyzing slope stability. At present, the commonly used numerical analysis methods mainly include the finite element method, finite difference method, discrete element method, the rapid development and use of finite element calculation software represented by ANSYS and ABAQUS, finite difference software represented by FLAC^{3D}, and discrete element calculation software represented by PFC^{3D}, which have improved the breadth, depth, and accuracy of slope analysis. Wu et al. [4] used the discrete element method to model the slope using PFC^{3D} software to analyze the factors affecting the instability of soil slopes in high-speed coal measure. Yu et al. [5] used FLAC^{3D} software to simulate slopes, analyzed the stability of the slopes under different conditions (natural, heavy rainfall, earthquake), and proposed a reinforcement design plan. Su et al. [6] used the FLAC^{3D} software to analyze the effect of vibration on the staged excavation of high slopes and proved that the support works can increase slope stability under this condition.

The limit equilibrium method is based on the principle of statics, treating the soil as a plastic material, calculating the slope safety coefficient from the static equilibrium state of the soil on the sliding surface to the limit state of imminent destruction, and then determining the stability of the slope [7]. The Bishop method [8,9], Janbu method [10], Morgenstern–Price method [11,12], Sarma method [13], and transfer coefficient method [14] are commonly used limit equilibrium methods. Behera [15] et al. analyzed the stability of the drainage field in the Talche Coalfield using the limit equilibrium method. The study showed that the main cause of landslides in the dump was rainfall infiltration, and they proposed a series of disposal options. In order to extend the applicability of the finite element limit equilibrium method (FELEM), Liu et al. [16] proposed a 2D and 3D slope stability FELEM using an elastic finite element stress field for solving various slope problems.

This paper takes the eastside slope and dump of an open-pit quarry as the research object, focusing on the following three aspects: (1) testing and analyzing the impact of quarry production blasting vibration on the stability of the slope and dump; (2) discussing the possibility of landslide disaster on the quarry slope under the condition that the dump continues to be extended; and (3) whether the induced slope and dump displacement area meets the safety requirements when excavating and unloading on the quarry slope.

2. Geotechnical Site Characterization

2.1. Overview of Eastside Slope of the Quarry

Through the collection of topographic data and investigation of the open-pit quarry, 20 steps have been formed in the eastside slope of the quarry, with the bottom elevation ranging from −130.0 m to −145.0 m, the top elevation ranging from +41.0 m to +61.0 m, and the height of the side slopes ranging from 172 m to 194.0 m. According to the data, the natural angle of repose of the slope of the nearby andesite and amphibolite is between 29° and 34°. Figure 1 shows the current situation of the eastside slope of the open-pit quarry.

MSR high-precision slope radar is used for real-time monitoring and early warning of the eastside slope of the open-pit quarry. The slope radar monitoring area is divided into the normal area and key monitoring area according to the absolute value of displacement change (0–40 m area of the eastside slope). For the normal area, the Level 1 warning and Level 2 warning are set; for the key monitoring area, the Level 3 warning is set, and the three warning thresholds are shown in Table 1.

According to the radar monitoring, the displacement value of the key monitoring area of the eastside of the slope ranges from 0.03 to 1.25 mm, and the velocity value ranges from 0.00 to 0.05 mm/h, all of which are within the normal range, and the eastside of the open-pit area remains stable without large landslides or wall caving phenomenon.

Figure 1. Eastside slope of the open-pit quarry.

Table 1. Tertiary warning threshold.

Warning Level	Critical Displacement Value	Critical Speed Value
Level 1 warning	50 mm	8 mm/h
Level 2 warning	30 mm	5 mm/h
Level 3 warning	20 mm	3 mm/h

2.2. Overview of the Dump

The first-phase dump is 1250 m long from east to west, and 370–570 m wide from north to south, with the highest dump pile elevation of +150.0 m and a total volume of 18.86 million m^3. Under the premise of fully considering the use of the existing discharge capacity of the mine, the first-phase dump was expanded to serve as the second-phase dump, and the expanded dump is 1570 m long from east to west and 750 m to 1056 m wide from north to south, with an area of about 1.21 km^2. The current dumping elevation is from +45.0 m to +180.0 m, the maximum height of dumping is 135.0 m, and the volume of dumping is about 90 million m^3.

The maximum elevation of the west side of the dump is +129.0 m, and the minimum elevation is +46.0 m; the maximum elevation of the north side of the dump is about +180.0 m, and the minimum elevation is about 145.0 m; the maximum elevation of the south side of the dump is about +129.0 m, and the minimum elevation is about +45.0 m; and the maximum elevation of the eastside of the dump is about +134.0 m, and the minimum elevation is about 84.0 m. The heights of the current dumping steps vary between 3.0 m and 44.0 m, and the angle of the slope steps is a natural rest angle (about 35°). The west side of the dump is adjacent to the slope of the quarry, the height of the slope ranges from 10.0 m to 72.0 m, and the overall slope angle ranges from 5° to 14°.

3. Blast Vibration Testing and Analysis

Blasting vibration is an important factor affecting the stability of open-pit slopes, and the main impact of blasting vibration on slope stability is manifested in two aspects. Firstly, blasting constantly acts on the rock body, generating shear stress and tensile stress on the rock body of the slope adjacent to the blasting area, resulting in a reduction in the strength of the rock body, which affects the stability of the slope. Secondly, blasting vibrations generate acceleration and cause inertial forces that would increase the downward pressure on the slope and the stress on the potential sliding surface of the slope. This may reduce the safety of the slope and lead to landslides. Therefore, when analyzing and studying the stability of slopes, the effect of blast vibration on slope stability must be considered. Many scholars have conducted a lot of work on the effect of blasting vibration on slope stability. Wu et al. [17] calculated the effect of blasting vibration parameters on slope stability coefficients by the analysis of variance and proposed a prediction model for slope safety coefficients under different vibration parameters to better assess and control the effect of blasting vibration on slopes. Kesimal et al. [18] assessed the effect of blast-induced acceleration on slope stability in a limestone quarry. However, the process of vibration waves generated by blasting in the rock mass is quite complicated. Due to the wide variety of waveforms and propagation media, it is difficult to describe them with theoretical models. Therefore, on-site measurements are usually used in practice to analyze the blasting vibration parameters and the attenuation laws to study their effects on slope stability.

3.1. Blast Vibration Testing Principles

Shock waves are generated after the detonation of explosives, producing strong tensile and compressive stress in the rock mass, causing damage to the surrounding rock mass, to achieve the purpose of breaking the rock mass, while a large amount of residual energy is converted into seismic waves, which propagate out to the surrounding area in the form of waves, affecting the strength and stability of the rock mass around the center of the blasting. In practice, the method is usually measured, and then the measured results are converted into numerical form and then determine the degree of influence of blasting vibration on the stability of slopes.

3.2. Blast Vibration Testing Methods

Five NUBOX-8016 intelligent blasting vibration recorders are used for this measurement. Before each on-site blasting seismic wave vibration observation, the range of the automatic recorder, sampling frequency, and other parameters are set in advance, the sensor is connected with the automatic recorder, the blasting vibration recorder is opened 30 to 60 minutes in advance of the power switch, the automatic recorder is confirmed to be working properly and is then withdrawn from the quarry, the blasting is completed, then the quarry is entered to retrieve the blasting vibration recorder. The communication port of the blasting vibration recorder is connected to read out the blasting seismic wave measured time curve waveform for data processing and graphic editing.

3.3. Blast Vibration Testing Results

After the blasting vibration test data collection work is carried out, three tests are conducted along two measuring lines in the direction of the dump. Each time, 2~5 measuring points are monitored, and finally, 12 monitoring waveforms are acquired, including 10 valid waveforms. In this blasting vibration test, although the number of measurements is limited, the arrangement of each monitoring point is carefully designed with a view to obtaining the most representative results from the limited data. In addition, highly accurate monitoring equipment is used, and data are collected and analyzed in strict accordance with industry standards. The blasting vibration test data are shown in Table 2, and the velocity waveforms of the blasting vibration measurement points are shown in Figures 2 and 3.

Table 2. Blast vibration test data sheet.

Number	Monitoring Point	Monitoring Point Location	Blast Distance (m)	Maximum Dose (kg)
#1 monitoring	CZ1-1	Slope −78 m platform	279.6	360
	CZ1-2	Slope −78 m platform	176.8	
#2 monitoring	CZ2-1	Slope +6.0 m platform	549.3	440
	CZ2-2	Slope +6.0 m platform	484.5	
	CZ2-3	Dump +98 m platform	950.8	
	CZ2-4	Dump +129 m platform	1141.6	

Table 2. *Cont.*

Number	Monitoring Point	Monitoring Point Location	Blast Distance (m)	Maximum Dose (kg)
#3 monitoring	CZ3-1	Slope +6.0 m platform	332.9	360
	CZ3-2	Slope +6.0 m platform	305.8	
	CZ3-3	Dump +98 m platform	748.1	
	CZ3-4	Dump +129 m platform	938.4	

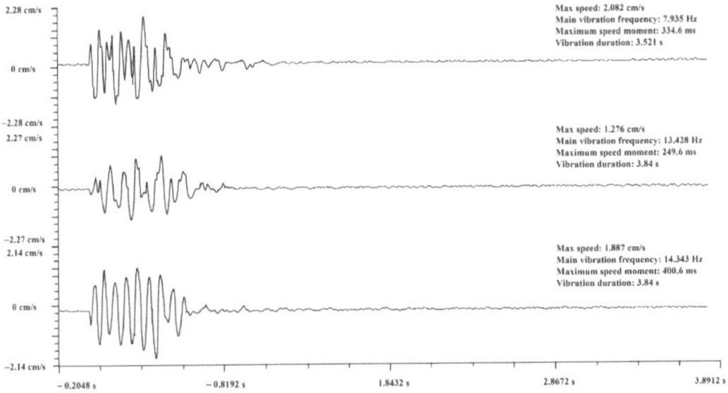

(**a**) Monitoring point CZ1-1

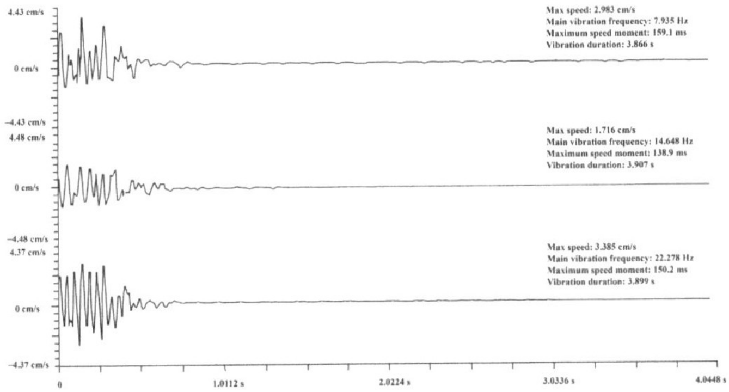

(**b**) Monitoring point CZ1-2

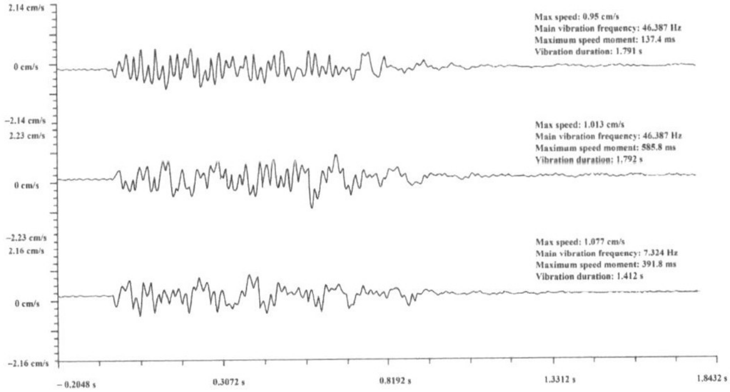

(**c**) Monitoring point CZ2-1

Figure 2. *Cont.*

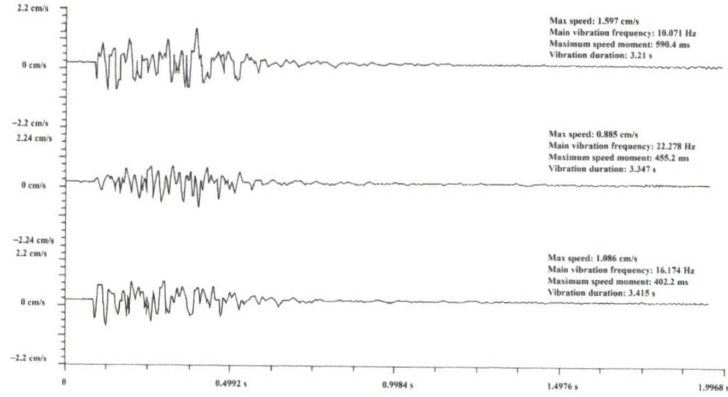

(**d**) Monitoring point CZ2-2

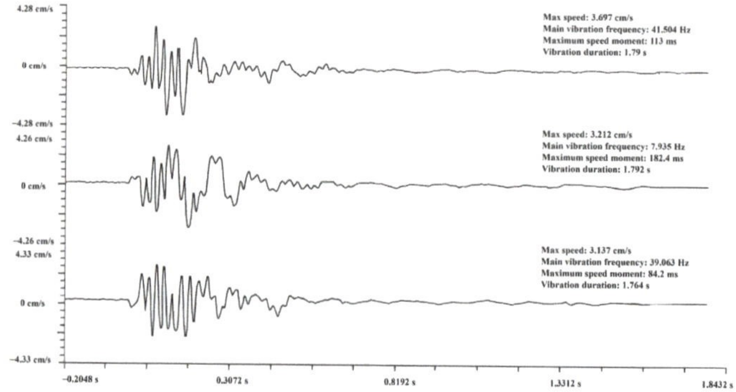

(**e**) Monitoring point CZ3-1

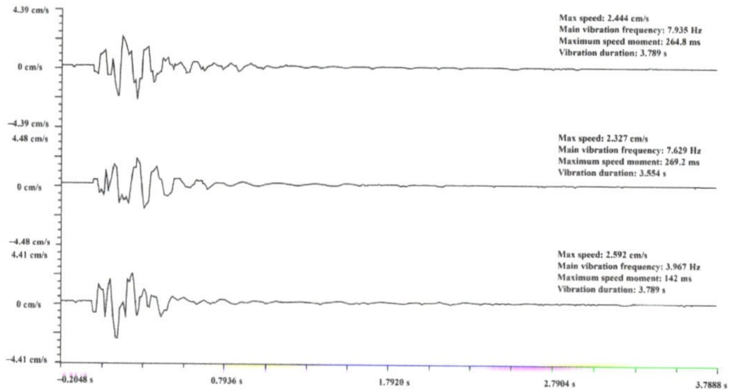

(**f**) Monitoring point CZ3-2

Figure 2. Waveforms of slope monitoring points.

3.4. Blast Vibration Propagation Analysis

Blasting seismic wave propagation formed by the performance of explosives, charge, charge structure, detonation mode, coupling conditions, filling conditions, bursting center distance, media properties, shape, etc., is analyzed. From the theoretical study of the impact of blasting vibration, media model generalization, and boundary valuation, simulation is challenging. In practice, many scholars are guided by theory, and put forward a variety of empirical formulas after field testing to obtain the empirical formula suitable for each site condition to describe the effect of blasting on the medium.

The main indicators obtained from the blasting vibration test are the time range waveform and the main eigenvalues of the blasting vibration, and the propagation and attenuation law of the blasting seismic wave in the medium is analyzed from these test results to analyze the effect of blasting vibration on the slope.

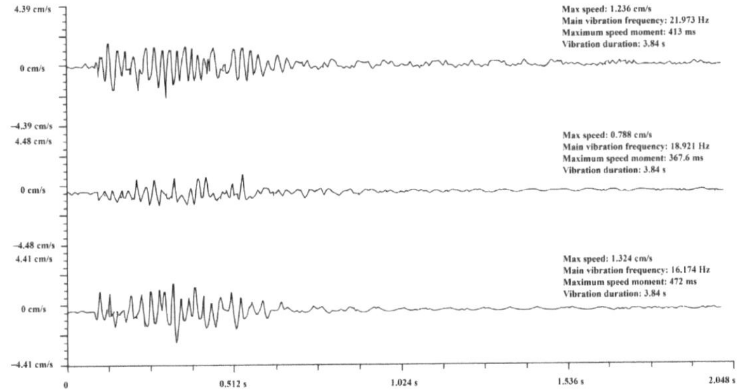

(**a**) Monitoring point CZ2-3

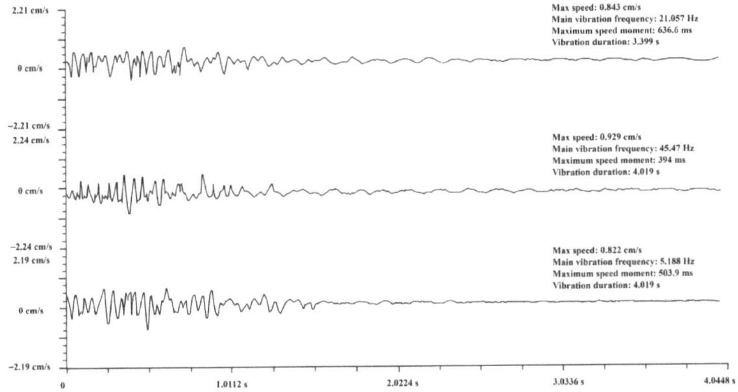

(**b**) Monitoring point CZ2-4

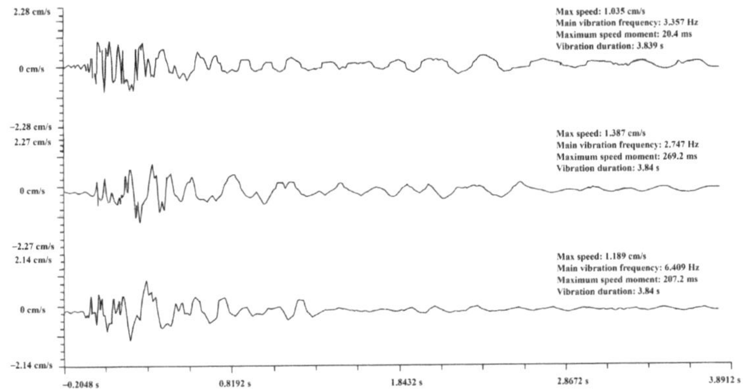

(**c**) Monitoring point CZ3-3

Figure 3. *Cont.*

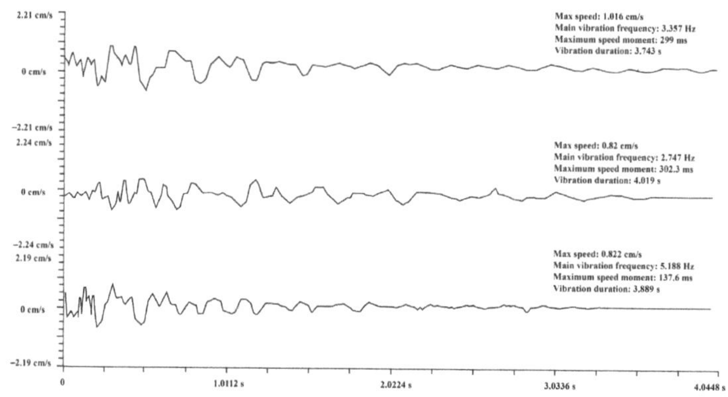

(**d**) Monitoring point CZ3-4

Figure 3. Waveforms of dump monitoring points.

According to blasting safety regulations and domestic and international research results, the blasting vibration propagation and attenuation laws in China generally adopt the empirical formula summarized by the Soviet scholar Sadovsky for regression calculation:

$$v = K\left(\frac{\sqrt[3]{Q}}{R}\right)^{\alpha} \quad (1)$$

where v is the peak vibration velocity of the mass (cm/s); Q is the maximum segment charge (kg); R is the distance from the measurement point to the source of the explosion (m); K and α are the impact factor and attenuation factor, respectively.

According to the blasting vibration monitoring data using the Sadowski formula regression analysis, the blasting vibration attenuation formula of the production blasting on the eastside of the quarry slopes and the dump is shown in Table 3.

Table 3. Blast vibration velocity regression equation.

Location	Regression Equation	Values of K and α	Correlation Coefficient r
Slope	$v = 87.88 \cdot \left(\frac{\sqrt[3]{Q}}{R}\right)^{0.93}$	$K = 87.88$ $\alpha = 0.93$	0.74
Dump	$v = 150.96 \cdot \left(\frac{\sqrt[3]{Q}}{R}\right)^{0.98}$	$K = 150.96$ $\alpha = 0.98$	0.95

3.5. Blast Vibration Safety Impact Analysis

Due to the different structures of the buildings to be protected, the permissible safe blasting vibration velocities and safe detonation dosage are different. Vibrations induced by rock blasting can cause damage to neighboring structures. The intensity of vibrations is influenced by factors such as explosive dosage, distance from the blast center, and geological properties, some of which are relatively stable in the short term, while others, such as dosage and distance from the blast center, change frequently. Therefore, an accurate assessment of these factors is necessary to implement effective controlled blasting techniques and to mitigate the effects on surrounding structures.

According to the measured data, and combined with the regulations, most of the main vibration frequencies are between 10 and 50 Hz. Concerning the blasting safety regulations, the permissible vibration speed of the permanent rock slopes of the selected quarry is 12.0 cm/s, the permissible vibration speed of the industrial buildings is 4.5 cm/s, and the permissible vibration speed of the slopes of the dump is 6.0 cm/s.

During the blasting vibration monitoring period, the measuring point of the dump is located at a greater distance from the quarry blasting, and the blasting vibration velocity values of the production blasting in the dump range from 0.5 cm/s to 2.0 cm/s, which are lower than the safety permissible value of 6.0 cm/s. The blasting vibration velocity values of the blasting in the side slope guide range from 1.5 cm/s to 5 cm/s, which are lower than the safety permissible value of 12.0 cm/s.

Following the on-site testing of blasting vibration for mine production, this study collects vibration velocity data in the direction of the eastside of the slope and the dump. Considering that critical industrial facilities such as belt conveyor systems and perimeter rock integrated utilization systems are arranged in the adjacent areas of the eastside slope and the dump, it is recommended that controlled blasting techniques such as pre-cracking blasting be used in production blasting to mitigate the potential impacts of blasting vibration on the slope and facilities.

4. Analysis of Slope Seepage Field

4.1. Finite Element Analysis of Seepage Field

The triangular element is often used for the simulation of 2D finite element seepage due to its simplicity and wide applicability [19]. A stable seepage region D is discritized into a finite number of triangular elements, as shown in Figure 4, and any one of the elements e, whose three vertices, i.e., the nodes, are numbered in the reverse clock direction as i, j, and m. Assuming that the linear interpolating function of the element is $\overline{H}(x,y)$, which is a planar equation of the space:

$$\overline{H}(x,y) = \alpha_1 + \alpha_2 x + \alpha_3 y \tag{2}$$

$$\begin{cases} \overline{H}(x_i,y_i) = \overline{H_i} = \alpha_1 + \alpha_2 x_i + \alpha_3 y_i = H_i \\ \overline{H}(x_j,y_j) = \overline{H_j} = \alpha_1 + \alpha_2 x_j + \alpha_3 y_j = H_j \\ \overline{H}(x_m,y_m) = \overline{H_m} = \alpha_1 + \alpha_2 x_m + \alpha_3 y_m = H_m \end{cases} \tag{3}$$

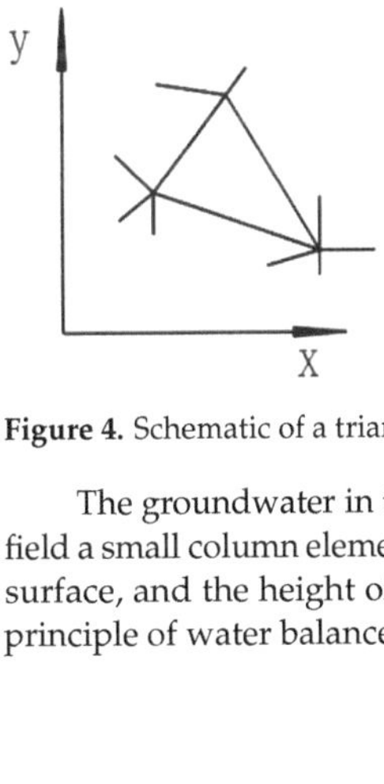

Figure 4. Schematic of a triangular element.

The groundwater in the mining area is mainly submerged. In the submerged seepage field a small column element is taken, as shown in Figure 5; the top surface is the submerged surface, and the height of the column is the thickness of the aquifer h. According to the principle of water balance, it can be obtained:

$$\frac{\partial v_x}{\partial x} + \frac{\partial v_y}{\partial y} = \mu \frac{\partial h}{\partial t} \tag{4}$$

where v_x, v_y are the seepage velocities in the x and y directions, respectively; μ is the degree of water supply; and t is the time.

Stability calculations are generally concerned with the most hazardous case, so it is sufficient to calculate only the steady flow, and the steady flow expression of Equation (4) is given as:

$$\frac{\partial v_x}{\partial x} + \frac{\partial v_y}{\partial y} = 0 \tag{5}$$

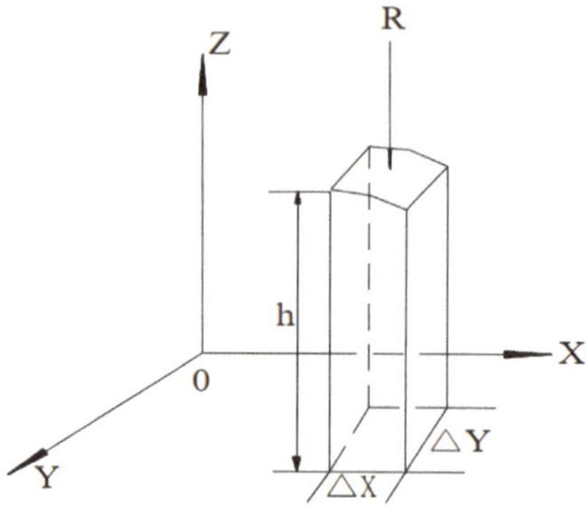

Figure 5. Small columnar elements in diving seepage fields.

According to Darcy's law: $v_x = K_x\frac{\partial H}{\partial x}$, $v_y = K_y\frac{\partial H}{\partial y}$, Equation (5) can be written as:

$$\frac{\partial}{\partial x}(K_x\frac{\partial H}{\partial x}) + \frac{\partial}{\partial y}(K_y\frac{\partial H}{\partial y}) = 0 \tag{6}$$

The groundwater level in the region is considered to be relatively stable over a certain period; therefore, the mathematical model of groundwater flow through the aquifer in a two-dimensional profile in the calculation area is as follows:

$$\begin{cases} \frac{\partial}{\partial x}(K_x\frac{\partial H}{\partial x}) + \frac{\partial}{\partial y}(K_y\frac{\partial H}{\partial y}) = 0 & \text{within } \Omega \\ H = H_b(x,y) & \text{within } \Gamma_1 \\ K_x\frac{\partial H}{\partial x}\cos(n,x) + K_y\frac{\partial H}{\partial y}\cos(n,y) = q & \text{within } \Gamma_2 \end{cases} \tag{7}$$

where Ω is the calculation area; Γ_1 is the first class of boundary; Γ_2 is the second class of boundary; H_b is the water head distribution on the first class boundary; q is the normal single-breadth flow on the second class of boundary, where inflow is positive and outflow is negative; and n is the outer normal direction on the second class of boundary.

The variational form of Equation (7) is:

$$E = \iint_{\Omega}\frac{1}{2}[K_x(\frac{\partial H}{\partial x})^2 + K_y(\frac{\partial H}{\partial y})^2]dxdy - \int_{\Gamma_2} q_b H ds \tag{8}$$

From the previous equation, the region Ω is discritized into M triangular cells with MN nodes, and, within each element, the basis function takes the form of Equation (9):

$$\widetilde{H}(x,y) \approx N_i H_i + N_j H_j + N_m H_m \tag{9}$$

Replacing the total generalized function with the sum of the individual element generalized function E^e,

$$E = \sum_{e=1}^{M} E^e \tag{10}$$

and

$$E^e = \iint_{\Omega^e}\frac{1}{2}\left[K^e x\left(\frac{\partial \widetilde{H}}{\partial x}\right)^2 + K^e y\left(\frac{\partial \widetilde{H}}{\partial y}\right)^2\right]dxdy - \int_{\Gamma_2^e} q_b H ds \tag{11}$$

Thus, the following equation can be obtained:

$$\begin{bmatrix} \frac{\partial E^e}{\partial H_i} \\ \frac{\partial E^e}{\partial H_j} \\ \frac{\partial E^e}{\partial H_m} \end{bmatrix} = \frac{K^e}{4A} \begin{bmatrix} b_i b_i + c_i c_i & b_i b_j + c_i c_j & b_i b_m + c_i c_m \\ b_j b_i + c_j c_i & b_j b_j + c_j c_j & b_j b_m + c_j c_m \\ b_m b_i + c_m c_i & b_m b_j + c_m c_j & b_m b_m + c_m c_m \end{bmatrix} \times \begin{Bmatrix} H_i \\ H_j \\ H_m \end{Bmatrix} - \begin{Bmatrix} F_i \\ F_j \\ F_m \end{Bmatrix} \tag{12}$$

or simply as:

$$\left\{ \frac{\partial E^e}{\partial H} \right\} = [G]^e \{H\}^e - \{F\}^e \tag{13}$$

where:

$$F_L = \int_{\Gamma_2^e} q_b N_L ds (L = i, j, m) \tag{14}$$

When the side e of the triangular cell coincides with the binary boundary, or when two nodes of the cell lie on a second-class boundary, the vector $\{F\}^e$ represents the water flux at the boundary nodes. When e is an internal cell, the vector $\{F\}^e$ is a zero vector.

The coefficient matrix $\{G\}^e$ is called the element infiltration matrix and its matrix elements are:

$$G_{i,j} = \frac{K^e}{4A} (b_K b_L + c_K c_L) \ (K = i, j, m, L = i, j, m) \tag{15}$$

The above element infiltration matrices are assembled to form the total infiltration matrix:

$$\sum_{e=1}^{M} [G]^e \{H\}^e = \sum_{e=1}^{M} \{F\}^e \tag{16}$$

or simply as:

$$[G]\{H\} = \{F\} \tag{17}$$

This results in a system of linear equations of order MN, and solving the system of equations yields the heads of the MN nodes.

4.2. Finite Element Calculation Theory

The basic idea of the finite element method to solve the problem is to discretize the slope body into a finite number of elements, i.e., the continuous structure expressed as the slope body can be replaced by the discretized structure, and the stability of the whole slope can be analyzed by the stress of these elements and the deformation of the elements [20]. The basis of finite element numerical calculation is the elastic–plastic theory. First of all, it is necessary to find out the ontological relationship of the slope geotechnical body, that is, the stress–strain relationship of the slope geotechnical body material. It is not only required to obtain the force balance of the element body but also to consider the deformation coordination conditions of the element body. The stress–strain relationship of the geotechnical body is not a one-to-one correspondence, but is nonlinear, so the numerical calculation of slope problems becomes very complex. Numerical analysis and calculation with the development of computer technology and software technology can carry out complex and fine calculations; the accuracy of the analysis and calculation mainly depends on the analysis of the slope geotechnical body provided by the parameters and the accuracy of the structure.

4.3. Finite Element Analysis Model

The properties of rock bodies can be expressed in terms of elasticity, plasticity, viscosity, or combinations of them, such as elastic–plastic, viscoelastic, viscoelastic–plastic, and so on. The corresponding model of the rock body ontological relationship also develops from a linear elastic ontological relationship to a nonlinear elastic model, nonlinear elastic–plastic model, and other consitutive equations.

With the gradual deepening of the mining, the original stress is gradually released, and the slope rock body is subjected to high stress, showing an elastic–plastic nature, so the material adopts the elastic–plastic intrinsic model. The equation of the nonlinear elastic–plastic model can be expressed by Equation (18):

$$[K]^e\{\delta\}^e = \{\sigma\}^e \tag{18}$$

where $[K]^e$ is the element's stiffness matrix, $\{\delta\}^e$ is the node displacement, and $\{\sigma\}^e$ is the load vector.

4.4. Generalized Geological Model and Parameters for Analyzing Profiles

This demonstration analysis takes into account the geological conditions, investigation data, the relative position of the slope and the dump, the height of the slope, and other factors, and provides a total of 1-1, 2-2, 3-3, 4-4, 5-5 five profiles for selection and analysis. The typical profile layout is shown in Figure 6.

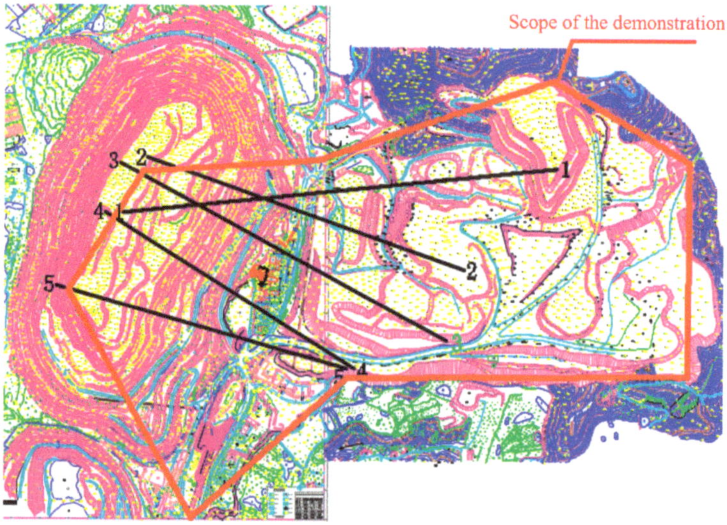

Figure 6. Layout of typical profiles.

This analysis focuses on the seepage field analysis of the representative slope profile consisting of the eastside of the open-pit quarry and the dump in the current and end state, which can be regarded as a stable seepage problem due to the slow extension of the mining depth. In the demonstration process, 3-3 is selected as the representative profile for slope seepage field analysis. The 3-3 profile slope generalized geological model is shown in Figure 7.

The upstream boundary, according to the previous slope engineering investigation borehole measured groundwater level value and local water level observation data, comprehensively determine the slope upstream boundary head value. Regarding the downstream boundary, the bottom of the quarry is the boundary of the fixed water head, and the water level is the elevation of the bottom of the quarry.

The infiltration parameters of each geotechnical layer are determined according to the data from previous investigation tests, and the specific parameters are shown in Table 4.

Table 4. Permeability coefficients of slope geotechnical bodies.

Number	Lithology Type	Penetration Parameter (cm/s)	Permeation Level
1	Drainage materials	2.9×10^{-3}	Medium permeable

Table 4. *Cont.*

Number	Lithology Type	Penetration Parameter (cm/s)	Permeation Level
2	Quaternary chalky clay	9.1×10^{-7}	Slightly permeable
3	Amphibolite	5.41×10^{-5}	Weakly permeable
4	Magnetite	4.32×10^{-5}	Weakly permeable

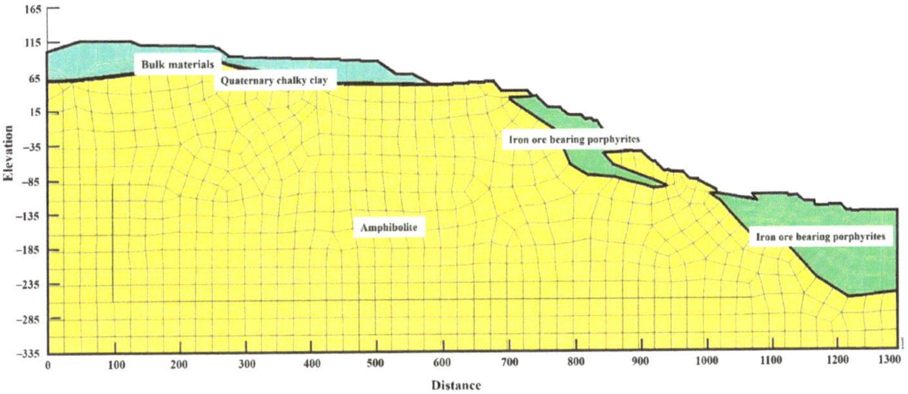

(**a**) The 3-3 profile generalized model of the current slope.

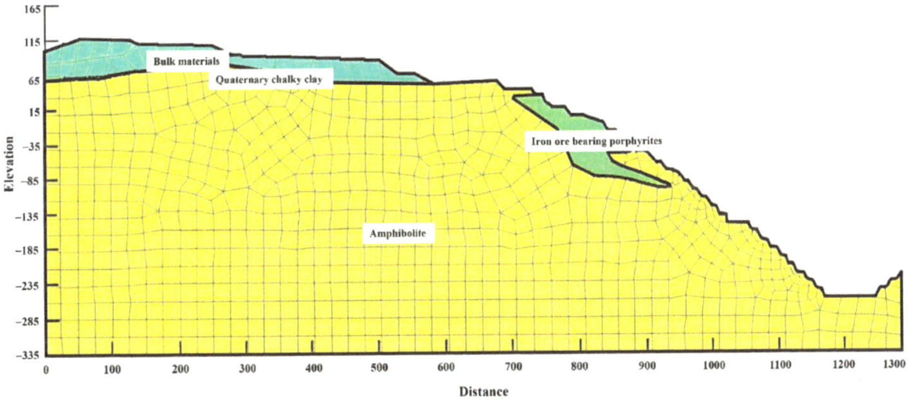

(**b**) The 3-3 profile generalized model of the end slope.

Figure 7. The 3-3 generalized geological model of profile slopes.

4.5. Calculation Results of Slope Seepage Field

Based on the generalized geological model, boundary conditions, and parameters, the seepage field analysis of the 3-3 profile of the current and end state of the slope is carried out by the numerical analysis method of finite element, and the groundwater infiltration line morphology of the 3-3 profile is derived as shown in Figure 8.

According to the analysis of the results of the calculation of the seepage field of representative profile 3-3 of the slope, it can be seen that the distribution of groundwater infiltration line is affected by the distribution of lithology of the slope, and the distribution mode is more consistent with the actual situation, which can be used as the basis for stability calculation.

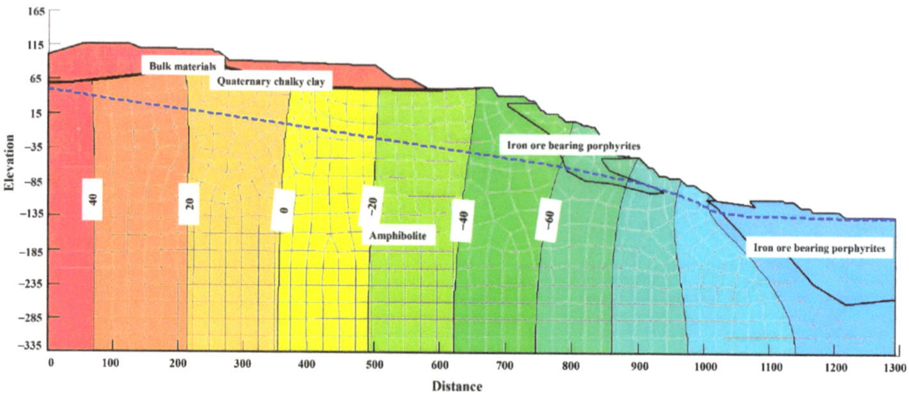

(**a**) Seepage results for the 3-3 profile of the current slope.

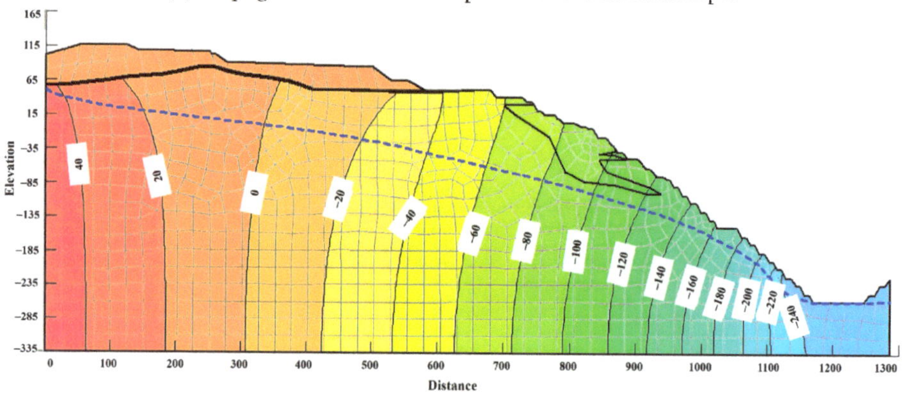

(**b**) Seepage results for the 3-3 profile of the end slope.

Figure 8. Seepage results for the 3-3 profile of the current and end slope.

5. Analysis of the Impact of the Dump on the Quarry Slopes

5.1. Analysis of Slope Failure Modes

The impact of the dump on the quarry slope mainly consists of the following two aspects: firstly, the geological disasters such as mudslides, landslides, and rolled rocks occurring on the adjacent side of the dump and the quarry slope, i.e., on the west side of the dump, have an impact on the quarry slope and the normal production; and secondly, whether or not an overall landslide occurs on the quarry slope under the effect of the dump's loading.

According to the site investigation, the eastside of the quarry has localized bulk structure slope damage, which is mainly in the form of arc-shaped collapse damage. Another kind of damage is wedge damage, which is formed in a block structure slope. Through the analysis of the actual geological data of the quarry, it can be predicted that the potential damage modes on the eastside slope of the open-pit quarry are as follows: (1) arc-shaped damage, which mainly occurs in the bulk structure type and the fracture structure type slope; (2) composite planar damage, which mainly occurs in block structure type slopes; (3) wedge-shaped damage, which mainly occurs in block structure type slopes, with both sides of the wedge formed by two sets of cross joints cutting the slope.

5.2. Stability Analysis of Slopes

To further analyze the overall stability of the quarry slopes under the loading of the dump, this demonstration adopts the limit balance method to analyze the current situation of the eastside of the open-pit quarry and the overall stability of the end slopes under the loading of the dump through the collection and analysis of the preliminary data and geological investigation.

The limit equilibrium method is one of the basic means of slope stability analysis. It is the slope stability problem assumed rigid body equilibrium problem to study, so there are the following basic assumptions:

(1) The composition of the rock and soil body of the landslide body as a rigid body, with the principle of theoretical mechanics to analyze the equilibrium state of its conditions must be met;
(2) Assuming that the sliding block along the sliding surface or sliding block between the surface of the misalignment of the body is in a state of limiting equilibrium, i.e., the forces acting on the sliding surface satisfy the Coulomb yield criterion;
(3) The sliding surface is simplified to be circular, plane, or folded according to the possible damage patterns. For rocky slopes, the dominant surface and its combined structure constitute the rocky slope separator and the sliding boundary, that is, the dominant structural surface controls the stability of the slope.

5.3. Safety Level of Slope Works

According to the technical specification for slope projects in non-coal surface mines, the design safety coefficients of overall slopes under different load combinations are shown in Table 5.

Table 5. Design safety factors for overall slopes under different load combinations.

Safety Level of Slope	Safety Factor for Slope Engineering Design		
	Load Combination I	Load Combination II	Load Combination III
I	1.25~1.20	1.23~1.18	1.20~1.15
II	1.20~1.15	1.18~1.13	1.15~1.10
III	1.15~1.10	1.13~1.08	1.10~1.05

Additions: load combination I—self-weight + groundwater; load combination II—self-weight + groundwater + blasting vibration force; load combination III—self-weight + groundwater + seismic force.

The safety level of the slope work on the eastside of the open-pit quarry is class I. According to the specification, the permissible safety coefficient in the case of load combination I is 1.20, in the case of load combination II is 1.18, and in the case of load combination III, the permissible safety coefficient is 1.15.

5.4. Slope Stability Calculation and Analysis

In this analysis, based on the generalization of the engineering geology of the slopes, the limit equilibrium analysis method is used to quantitatively calculate the stability of the current and end slopes of the eastside of the open-pit quarry under loads of the dump, and the stability analysis of the slopes of the 3-3 profile is carried out by using the Sarma method and the Morgenstern–Price method as shown in Figures 9 and 10, the results of which are shown in Tables 6 and 7.

Table 6. The 3-3 overall slope stability calculation results for the current profile.

Profile	Method	Load Combination	Stability Factor	Specification	Evaluation of Results
3-3 current profile	M-P method	I	2.038	1.20	Fulfilment
		II	1.962	1.18	Fulfilment
		III	1.834	1.15	Fulfilment
	Sarma method	I	2.037	1.20	Fulfilment
		II	1.960	1.18	Fulfilment
		III	1.833	1.15	Fulfilment

Table 7. The 3-3 overall slope stability calculation results for the end profile.

Profile	Method	Load Combination	Stability Factor	Specification	Evaluation of Results
3-3 end profile	M-P method	I	1.315	1.20	Fulfilment
		II	1.301	1.18	Fulfilment
		III	1.238	1.15	Fulfilment
	Sarma method	I	1.315	1.20	Fulfilment
		II	1.297	1.18	Fulfilment
		III	1.239	1.15	Fulfilment

Based on the generalization of the engineering geology of the slopes, this analysis uses the limit equilibrium analysis method to quantitatively calculate the stability of the current and the end slopes of the eastside of the open-pit quarry under loads of the dump, and the results show that the safety coefficients of the overall slope stability meet the requirements of the code, even there is a certain amount of safety reserve.

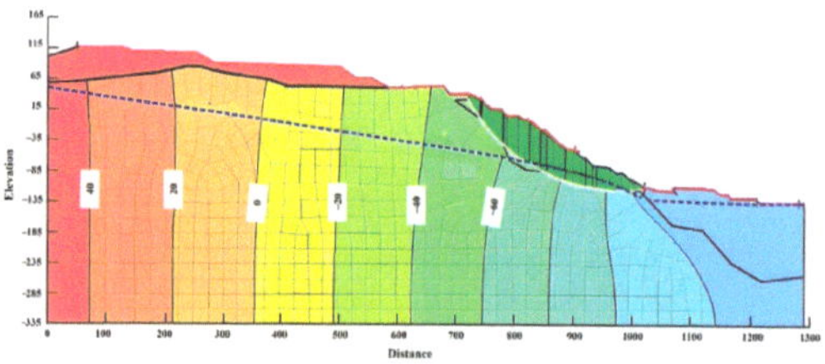

(**a**) Load combination I M-P method

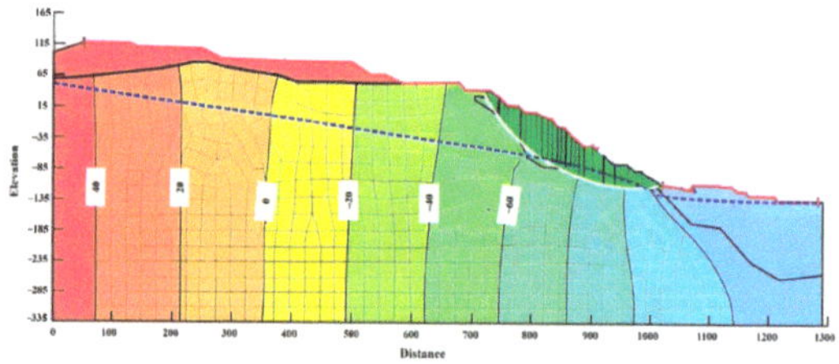

(**b**) Load combination I Sarma method

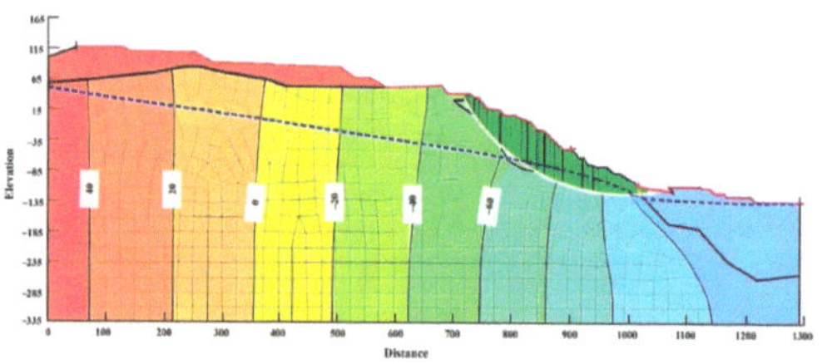

(**c**) Load combination II M-P method

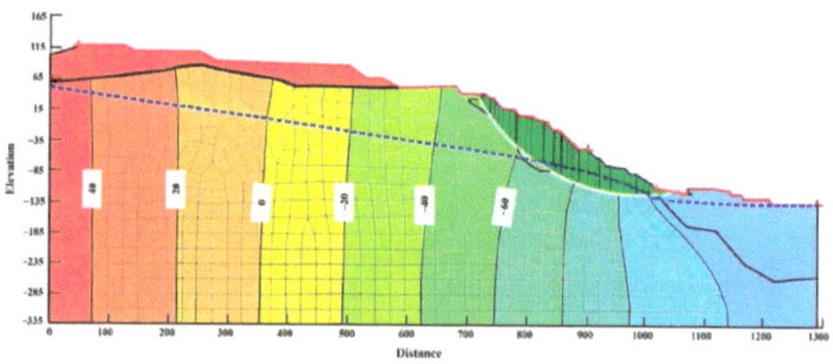

(**d**) Load combination II Sarma method

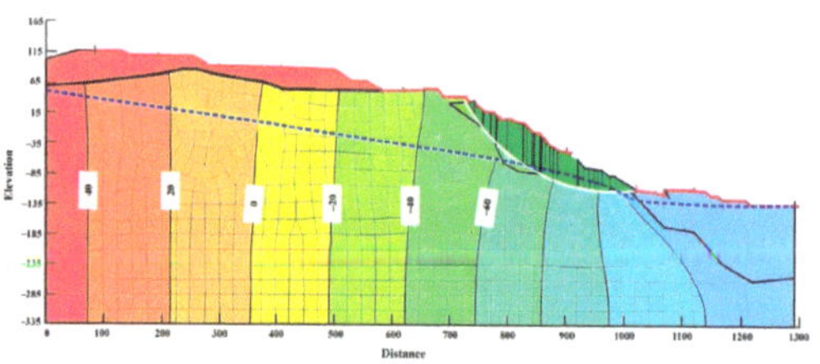

(**e**) Load combination III M-P method

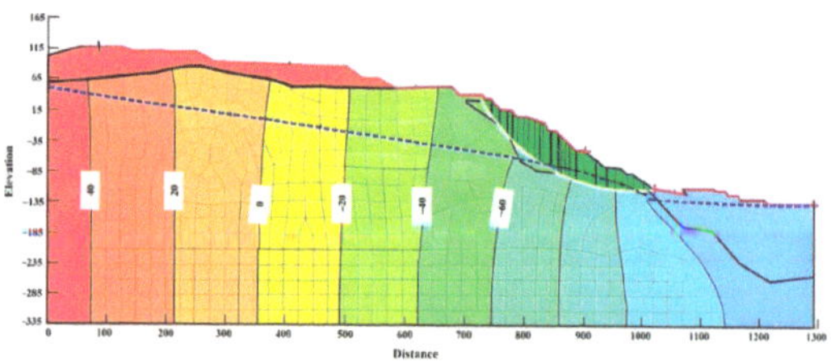

(**f**) Load combination III Sarma method

Figure 9. Stability results of the current slope in the 3-3 profile.

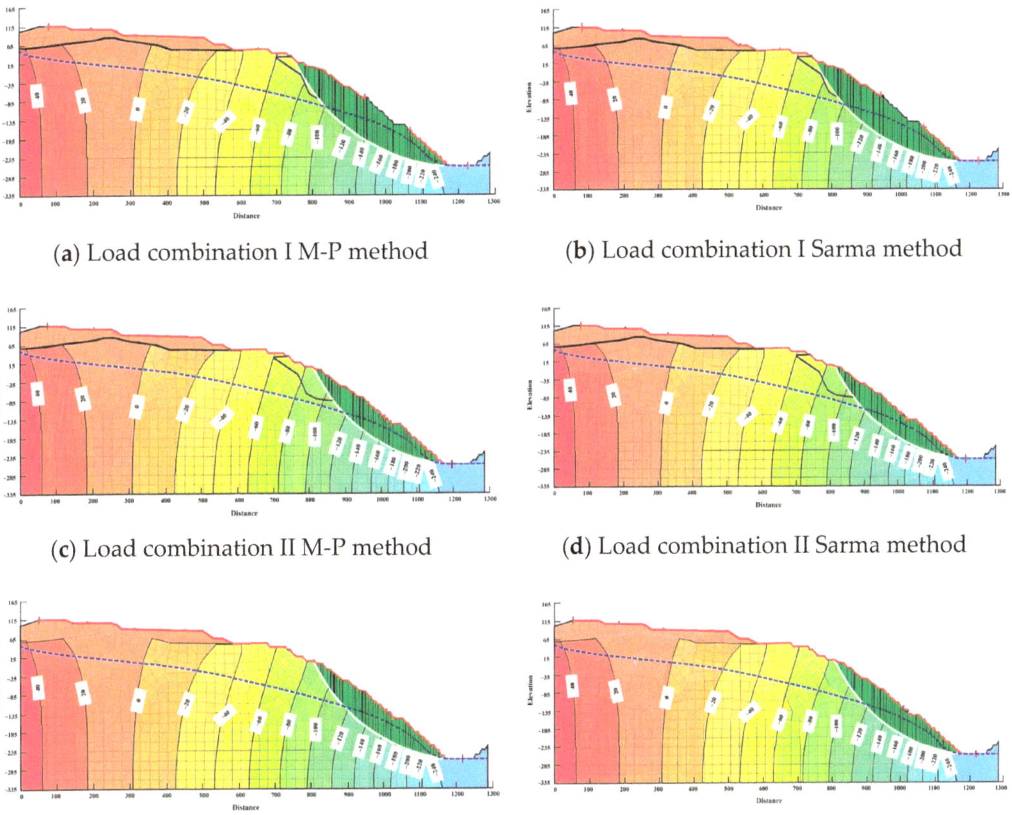

(a) Load combination I M-P method

(b) Load combination I Sarma method

(c) Load combination II M-P method

(d) Load combination II Sarma method

(e) Load combination III M-P method

(f) Load combination III Sarma method

Figure 10. Stability results of the end slope in the 3-3 profile.

6. Analysis of the Impact of Slope Excavation Unloading on the Dump

6.1. Yield Criteria

The most widely used yield criterion for finite element elastic–plastic stress–strain analyses is the Mohr–Coulomb yield criterion and its strength parameters are determined directly from tests [21]. The Mohr–Coulomb criterion is a yield preparation for isotopically hardened/softened materials. Assuming that the material yield criterion is as follows when the shear stress in a plane of the material is large enough to reach a particular value:

$$\tau = c + \sigma \tan \varphi \tag{19}$$

where τ is the shear stress; σ is the positive stress; c is the cohesive force of the rock; φ is the angle of internal friction of the rock. The criterion has been widely used in geotechnics due to the easy determination of the values of its parameters c, φ values, as shown in Table 8.

Table 8. Deformation parameters of rock mass for finite element analysis.

Lithology	Natural Heavy γ (kN/m^3)	Cohesive Force c (kPa)	Angle of Internal Friction φ (°)	Deformation Modulus (Gpa)	Poisson's Ratio ρ
Moderately weathered amphibolite	25.9	223	36.30	3.14	0.16
Magnetite	28.5	177	34.30	3.14	0.18
Silty clay	19.75	27.05	16.95	0.005	0.3
Bulk materials	20.5	18.0	29.5	0.017	0.21

6.2. Boundary Conditions and Parameters of the Computational Model

The boundary conditions of the computational model are mainly displacement boundaries. The displacement change of the slope is mainly on the surface of the slope and within a certain depth, so horizontal constraints are imposed on both sides of the model the bottom is fixed-supported, and the rest is a displacement-free boundary. The stress state of the model mainly considers the stress–strain and displacement condition of the slope under the action of self-weight after the excavation and disturbance of the slope.

The deformation parameters of the geotechnical body in this FEA calculation are specified in Table 8.

6.3. Displacement Field Analysis During Slope Excavation

As there are industrial facilities such as a peripheral rock comprehensive utilization processing system and peripheral rock crushing system between the quarry slope and the dump, to further investigate the impact of the quarry slope excavation disturbance on the above industrial implementation and the dump, this demonstration and analysis select the 3-3 profile to carry out the numerical simulation analysis of the slope excavation process and to analyze the displacement field after the excavation unloading. For the displacement of the slope excavation of the 3-3 profile in both the horizontal and the vertical direction, see Figure 11.

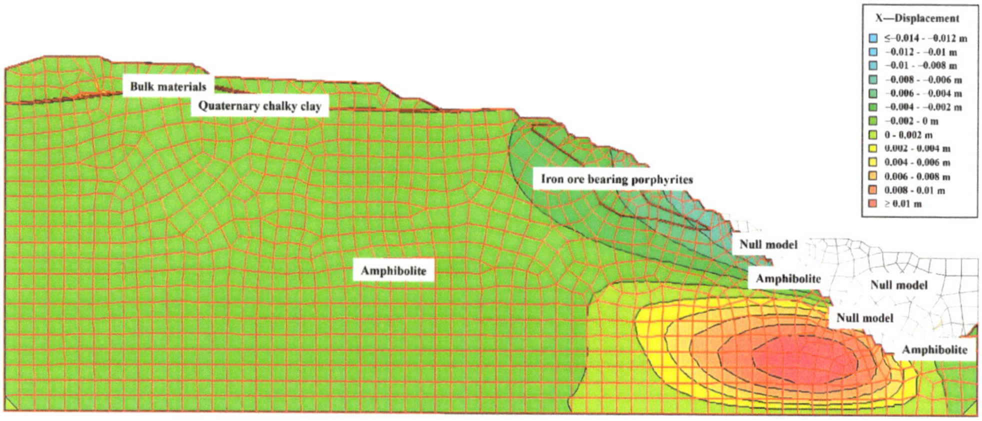

(**a**) Horizontal displacement diagram for 3-3 profile slope excavation.

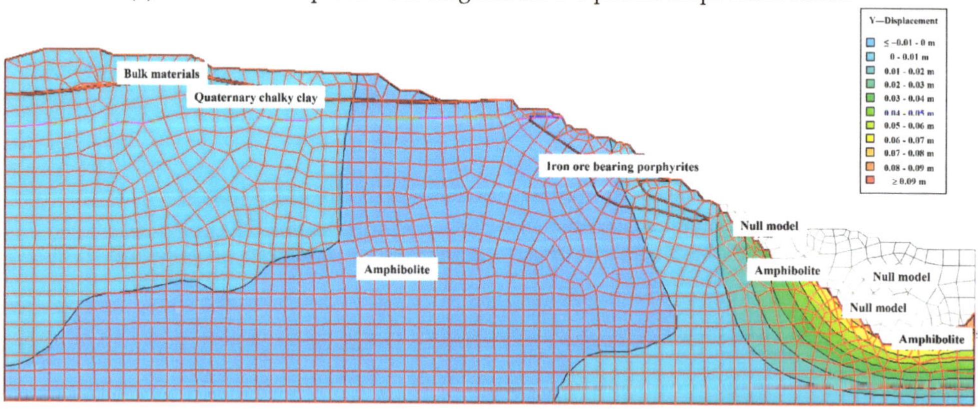

(**b**) Vertical displacement diagram for 3-3 profile slope excavation.

Figure 11. Displacement diagram for 3-3 profile slope excavation.

As can be seen from Figure 11, the deformation caused by excavation and unloading of the slope is mainly concentrated in the lower part of the slope, while the change of horizontal displacement in the top area of the slope is about 2–6 mm, and the change of vertical displacement is about 0–10 mm, and the change of displacement decreases with the increase in the horizontal distance from the excavation area, and the excavation disturbance does not have much influence on the industrial facilities and the dump.

7. Conclusions

Slope safety has always been a common but important issue for mine production management, where adverse geological phenomena may pose a threat to the safety of mine personnel and property. Based on the collection of previous investigation and design data, on-site geological investigation, and blasting vibration test analysis, and according to the technical specification for slope projects in non-coal surface mines, this demonstration analysis qualitatively analyzes and quantitatively calculates the mutual influence between the eastside slope of a certain quarry and the dump, and arrives at the following conclusions:

(1) The propagation velocity and attenuation laws of blasting vibration waves were obtained through on-site blasting vibration tests. Considering that the eastside slope of the quarry is adjacent to critical industrial facilities, pre-cracking blasting technology is recommended to reduce the impact of blasting vibration to improve the safety and efficiency of blasting in quarries.

(2) Under the load of the dump, the safety coefficients of both the current and the end slopes of the eastside of the quarry meet the requirements of the technical specifications for slope projects in non-coal open-pit quarries, and there is a certain amount of safety reserve. The location of the slip surface is mainly within the slope of the quarry, which is not involved in the boundary of the dump.

(3) The deformation caused by the excavation and unloading of the eastside of the quarry slope is mainly concentrated in the lower part of the slope, and the displacement change caused by the industrial site and the dump at the top of the slope is about 0–10 mm, which is lower than the alarm threshold of the slope radar, and the impact on the safety of the surrounding facilities is within the controllable range.

Author Contributions: Conceptualization, B.F. and H.J.; methodology, B.F.; software, J.W.; validation, J.P., B.F. and H.J.; formal analysis, B.F.; investigation, B.F.; resources, J.P.; data curation, B.F. and J.W.; writing—original draft preparation, B.F.; writing—review and editing, H.J.; visualization, H.J. and J.W.; supervision, H.J.; project administration, H.J.; funding acquisition, H.J. All authors have read and agreed to the published version of the manuscript.

Funding: Project of Special Funds for Basic Research Operating Expenses of China Academy of Work Safety (grant number 2024JBKY17).

Institutional Review Board Statement: Not applicable.

Informed Consent Statement: Not applicable.

Data Availability Statement: No data was used for the research described in the article.

Acknowledgments: We gratefully acknowledge the financial support provided by the special foundation for basic re-search operating expense of China Academy of Work Safety.

Conflicts of Interest: The authors declare that they have no known competing financial interests or personal relationships that could have appeared to influence the work reported in this paper.

References

1. Yuan, Y.; Hu, C.; Li, L.; Xu, J.; Hou, X. Efficient Slope Reliability Analysis Using a Surrogate-Assisted Normal Search Particle Swarm Optimization Algorithm. *J. Comput. Des. Eng.* **2023**, *11*, 173–194. [CrossRef]
2. Lu, Y.; Jin, C.; Wang, Q.; Li, G.; Han, T. Deformation and Failure Characteristic of Open-Pit Slope Subjected to Combined Effects of Mining Blasting and Rainfall Infiltration. *Eng. Geol.* **2024**, *331*, 107437. [CrossRef]

3. Cai, J.; Jim Yeh, T.; Yan, E.; Tang, R. Exploiting Excavation-Induced Displacement for Slope Heterogeneity Characterization and Stability Analysis. *Comput. Geothech.* **2024**, *170*, 106292. [CrossRef]
4. Wu, C.; Zhang, B.; Zhang, H. Stability Analysis of Coal-Measure Soil Slope Based on Discrete Element Method. *EES Sci.* **2021**, *719*, 042059. [CrossRef]
5. Yu, R.; Huang, Z.; Yang, Z. Surface-Mining Slope Reinforcement Design Based on GeoStudio and FLAC3D. *Min. Sci.* **2022**, *29*, 141–163.
6. Su, P.; Qiu, P.; Liu, B.; Chen, W.; Su, S. Stability Prediction and Optimal Angle of High Slope in Open-Pit Mine Based on Two-Dimension Limit Equilibrium Method and Three-Dimension Numerical Simulation. *Phys. Chem. Earth* **2022**, *127*, 103151. [CrossRef]
7. Liu, S.; Shao, L.; Li, H. Slope Stability Analysis Using the Limit Equilibrium Method and Two Finite Element Methods. *Comput. Geotech.* **2015**, *63*, 291–298. [CrossRef]
8. Salmasi, F.; Pradhan, B.; Nourani, B. Prediction of the Sliding Type and Critical Factor of Safety in Homogeneous Finite Slopes. *Appl. Water Sci.* **2019**, *9*, 158. [CrossRef]
9. Ji, J.; Zhang, W.; Zhang, F.; Gao, Y.; Lü, Q. Reliability Analysis on Permanent Displacement of Earth Slopes Using the Simplified Bishop Method. *Comput. Geotech.* **2020**, *117*, 103286. [CrossRef]
10. Azizi, M.; Marwanza, I.; Ghifari, M.; Trisakti, U. Low-Wall Slope Stability Analysis Using Planar Surface and 3D Simplified Janbu Method. *Technol. Rep. Kansai Univ.* **2020**, *62*, 569–580.
11. Ouyang, W.; Liu, S.; Yang, Y. An Improved Morgenstern-Price Method Using Gaussian Quadrature. *Comput. Geotech.* **2022**, *148*, 104754. [CrossRef]
12. Chen, Z.; Xie, C.; Xiong, G.; Shen, J.; Yang, B. Using the Morgenstern–Price Method and Cloud Theory to Invert the Shear Strength Index of Tailings Dams and Reveal the Coupling Deformation and Failure Law under Extreme Rainfall. *Sustainability* **2023**, *15*, 6106. [CrossRef]
13. Chen, M.; Lu, W.; Yan, P.; Zhou, C. New Method for Dynamic Stability Analysis of Rock Slope under Blasting Vibration Based on Equivalent Acceleration and Sarma Method. *Can. Geotech. J.* **2014**, *51*, 441–448. [CrossRef]
14. Svensson, L.; Ferm, M. Mass Transfer Coefficient and Equilibrium Concentration as Key Factors in a New Approach to Estimate Ammonia Emission from Livestock Manure. *J. Agric. Eng. Res.* **1993**, *56*, 1–11. [CrossRef]
15. Behera, P.; Sarkar, K.; Singh, A.; Verma, A.; Singh, T. Dump Slope Stability Analysis—A Case Study. *J. Geol. Soc. India* **2016**, *88*, 725–735. [CrossRef]
16. Liu, S.; Su, Z.; Li, M.; Shao, L. Slope Stability Analysis Using Elastic Finite Element Stress Fields. *Eng. Geol.* **2020**, *273*, 105673. [CrossRef]
17. Wu, T.; Zhou, C.; Jiang, N.; Xia, Y.; Zhang, Y. Stability Analysis for High-Steep Slope Subjected to Repeated Blasting Vibration. *Arab. J. Geosci.* **2020**, *13*, 828. [CrossRef]
18. Kesimal, A.; Ercikdi, B.; Cihangir, F. Environmental Impacts of Blast-Induced Acceleration on Slope Instability at a Limestone Quarry. *Environ. Geol.* **2008**, *54*, 381–389. [CrossRef]
19. Rebiai, C. Finite Element Analysis of 2-D Structures by New Strain Based Triangular Element. *J. Mech.* **2019**, *35*, 305–313. [CrossRef]
20. Lewis, H.; Papadimitriou, C. Elements of the Theory of Computation. *Prentice-Hall* **1998**, *29*, 62–78. [CrossRef]
21. Wang, H.; Zhao, W.; Sun, D.; Guo, B. Mohr-Coulomb Yield Criterion in Rock Plastic Mechanics. *Chin. J. Geophys.* **2012**, *55*, 733–741.

MDPI AG
Grosspeteranlage 5
4052 Basel
Switzerland
Tel.: +41 61 683 77 34

Fire Editorial Office
E-mail: fire@mdpi.com
www.mdpi.com/journal/fire

www.ingramcontent.com/pod-product-compliance
Lightning Source LLC
Chambersburg PA
CBHW041358210926
51850CB00013B/3117
* 9 7 8 3 7 2 5 8 3 5 5 7 7 *